D0820566

WHEN THE HEAVENS WENT ON SALE

ALSO BY ASHLEE VANCE

Elon Musk:

Tesla, SpaceX, and the Quest for a Fantastic Future

Geek Silicon Valley:

The Inside Guide to Palo Alto, Stanford, Menlo Park,

Mountain View, Santa Clara, Sunnyvale, San Jose, San Francisco

Vance, Ashlee,
When the heavens went on sale
: the misfits and geniuses rac
[2023]
33305257555445
ca 08/16/23

WHEN THE HEAVENS WENT ON SALE

THE MISFITS AND GENIUSES RACING TO PUT SPACE WITHIN REACH

ASHLEE VANCE

ecco
An Imprint of HarperCollinsPublishers

WHEN THE HEAVENS WENT ON SALE. Copyright © 2023 by Ashlee Vance. All rights reserved. Printed in the United States of America. No part of this book may be used or reproduced in any manner whatsoever without written permission except in the case of brief quotations embodied in critical articles and reviews. For information, address HarperCollins Publishers, 195 Broadway, New York, NY 10007.

HarperCollins books may be purchased for educational, business, or sales promotional use. For information, please email the Special Markets Department at SPsales@harpercollins.com.

Ecco® and HarperCollins® are trademarks of HarperCollins Publishers.

FIRST EDITION

Designed by Alison Bloomer

Decorative image on title page and part openers © alexkoral / Shutterstock

Library of Congress Cataloging-in-Publication Data has been applied for.

ISBN 978-0-06-299887-3

23 24 25 26 27 LBC 5 4 3 2 1

FOR MELINDA

I'm sorry you had to see that.

DEAR READER

This book covers about five years' worth of reporting across four continents and hundreds of hours spent with the subjects. The main characters in this book were generous enough to let me observe their worlds during the good times and the bad times and without any restrictions on my reporting. On many occasions, the subjects also took the risk of letting me peer into their personal lives, and for this, I'm grateful. It helped me understand their personalities, motivations, and perspectives in ways that reporters often struggle to achieve.

Every quotation in the book comes as a result of this firsthand reporting unless otherwise noted. As you will see, I have let the characters speak in their own voices and at length throughout the book. It was important to me to let them tell their own stories and for you to hear how they talk and think in their own words. On occasion, the wording of some of these quotations was edited either for brevity or clarity. In no case did the desire for readability supersede the need for accuracy. During this journey, I have also written stories on some of the figures mentioned in the book for *Bloomberg Businessweek* magazine and have in rare instances borrowed my own lines from these pieces for no better reason than that I liked a few turns of phrase.

I've made every effort to verify things that were told to me and to check and recheck the facts presented here. Any updates will be made to future editions of this book and noted on my website—ashleevance.com—where you can also contact me for feedback.

I hope you enjoy reading this as much as I enjoyed living it.

CONTENTS

AD ASTRA

THE MADDEST MAX

WHEN THE HEAVENS WENT ON SALE

PROLOGUE

A SHARED HALLUCINATION

Oh Earth, look up

Look up beyond the century's horizons, where the light of the millennium to come already stains the skies with colours strange and new

Look up: we have repealed the laws of gravity, torn off the ceiling of the world that was so very low

The skies are yours, new beaches made of cirrus-cloud, new valleys made of strato-cumulus

Lift up your heads! You were not made to gaze at gutters, mud and puddles all your lives, but have not dared to raise your sights in case the thing you longed for was not there

Look up and see it now, the shape that's haunted human dreams and legends since we first peered from the jungles long ago and wondered what might dwell upon those blue and distant hills, upon those mountains there . . .

Oh Earth, look up

—ALAN MOORE, *MIRACLEMAN*

> From an economic point of view, the navigation of interplanetary space must be effected to ensure the continuance of the race; and if we feel that evolution has, through the ages, reached its highest point in man, the continuance of life and progress must be the highest end and aim of humanity, and its cessation the greatest possible calamity.
>
> —ROBERT GODDARD, ROCKET PIONEER, 1913

So much of the excitement had given way to crushing anxiety and despair.

It was September 28, 2008, and a group of about fifteen SpaceX employees found themselves on a tiny tropical island preparing to fly the company's white Falcon 1 rocket into orbit. For many members of the group, the moment came as the culmination of six years of grueling work and should have represented a shot at achieving pure bliss. The problem was that they had been in such situations before, and things had not gone well. Three previous rockets had been launched from this small plot of jungle-covered land in the middle of nowhere and had either blown up shortly after launch or broken apart during flight. The wounds of those past failures had pushed many of the SpaceX engineers and technicians into a state of serious self-doubt. Maybe they were not as bright and creative as they had told themselves. Maybe Elon Musk, the SpaceX founder and CEO, had made a terrible, costly mistake by believing in them. Maybe they were minutes away from needing to look for new jobs.

The conditions for this type of operation were comically less than ideal from the start. SpaceX had set up its rocket-launching facility on Kwajalein Atoll, a collection of a hundred islands hanging out together in the middle of the Pacific Ocean with Hawaii and Australia as their theoretical neighbors. The island specs barely edge out of the water, and they all come with the thick humidity, inescapable sunshine, and salty spray welcome during a tropical holiday but abhorred while doing manual labor and toiling with machinery.

A couple members of the SpaceX team paid their first visit to Kwajalein in 2003, hoping to find a place where they could go about their wild rocketry experiments without too much interference. The location made some rational sense. The US military had run operations out of Kwajalein for decades, particularly around its radar and missile defense systems. To facilitate such missions, the army had built enough infrastructure on Kwajalein to support the daily lives of a thousand people and to conduct the complex weapons systems tests. Best of all, the locals were used to things being blown up and were perhaps more accepting of a group of unproven twentysomethings and a dot-com millionaire appearing on the scene with a big metal tube full of liquid explosives and crossed fingers.

The day-to-day reality for Team SpaceX, however, ended up being more *Gilligan's Island* than well-oiled military outpost, the main reason being that all the good stuff—the equipment, housing, stores, eateries, and bars—was located on Kwajalein Island, the largest of the islands. Meanwhile, SpaceX had been banished to Omelek Island, an eight-acre lump of land with an infrastructure that consisted of a couple boat docks, a helicopter landing pad, four storage sheds, and about a hundred palm trees. It was there that SpaceX would receive rocket parts being sent from its operations in California and Texas and assemble them before testing and eventually launching an entire rocket.

In 2005, the effort to transform Omelek into something more functional began in earnest. SpaceX employees poured a large concrete slab to serve as the rocket launchpad. They put up a tent to create a shaded place to work on the rocket and house their tools. Some 1960s-era portable trailers were reconfigured into living quarters and offices. Plumbing was available on a do-it-yourself basis, and meals consisted of prepackaged sandwiches or whatever they could haul out of the ocean.

Despite the difficult conditions, the SpaceX crew worked at an astonishing speed, particularly for the aerospace industry, which tends to measure delays not in weeks or months but in years. The once barren Omelek began to fill with large cylindrical tanks for storing the liquid oxygen and kerosene propellants needed to fuel the rocket and the helium used to pressurize various mechanical systems. Gas generators appeared almost like angels because they meant air-conditioning inside the trailers, which meant that you didn't sweat for a few minutes and tensions didn't run quite as high when things went wrong. A few industrious people installed real toilets and showers. And by early September 2005, SpaceX had constructed a tower of metal scaffolding that would hold an upright rocket in place and support it before launch.

About once a month, a cargo ship arrived with large containers full of gear. In late September, one of the ships delivered the first stage, or main body, of Falcon 1. By the end of October, the rocket had been assembled, moved out to the pad, and lifted to the vertical position. Most of the engineers were, well, engineers and not too bothered with the symbolism of the moment. Rather clearly, though, Falcon 1 looked like a religious totem—a bizarre aluminum obelisk poking out of a clearing in the jungle and making obvious its intentions to head up, up, up, as far as it could go.

It's at this point in every new rocket program where things keep going wrong for an extended period of time. The craft itself has already been designed and manufactured. The engines, usually the trickiest parts, have been tested somewhere else and fired again and again until it seems assured that they'll work when the big moment arrives. Many lines of software code have been written, debugged, and fine-tuned. The mass of wiring inside the rocket has also been finely tailored. The optimistic, intellectual hope is that you combine all these things together and they function in harmony. But the Rocket Gods never allow this to happen.

To get a rocket from fully assembled to ripping through the atmosphere, it must pass hundreds of tests on the ground. Quite often the tests are undermined by a relatively minor component. A $50 valve malfunctions and needs to be swapped out, which means opening a hatch on the rocket body and digging in to find the faulty piece of metal with sweat dripping from your brow. Or perhaps moisture oozed its way into a battery pack and that also now needs replacing.

Sometimes tests go awry or don't happen at all because of logistics. Huge volumes of liquid oxygen, for example, must be pumped into a rocket's fuel chamber again and again as the machine is prepared for launch. The trick is that LOX, as aerospace people call it, must be kept incredibly cold to remain liquid and begins boiling off instantly when it moves from a chilled, purpose-built tank into the rocket's fuel chamber, where the surrounding atmosphere heats it up. Quite often, the rocket will be filled with LOX and then people will fiddle around trying to fix one odd thing after another ahead of a test only to find out that by the time they're finally ready to proceed, too much LOX has boiled away to run the test. Only now you realize that you've gone through the same sequence of events about five times during the day, the LOX storage containers are empty, you're on a minuscule island out in the Pacific Ocean, and no one in two thousand miles cares whether you get more liquid oxygen before nightfall, nor do they have any means of delivering some to you quickly.

To an outsider, this slog part of the rocket-building process can appear absurd. For most intents and purposes, the thing is completed and ready to fly. Surely, there cannot be months' worth of small- to medium-sized issues that crop up one after another. But, yes, there are. The big joke is that the really difficult "rocket science" part of the problem, the physics, was figured out ages ago. The stuff holding the rocket back now is grunt work. What you need at this point are relentless, problem-solving mechanics, not PhDs.

From October 2005 until March 2006, the SpaceX team dealt with this exact scenario. Each day, the crew made their way to the rocket and battled with it from sunrise to well after sunset. The days were exhausting and often deflating, but the promise of a launch kept everyone going. SpaceX had been founded in 2002, and Musk—being Musk—had right away set the totally unrealistic deadline of launching the company's first rocket within a year. Still, even at four years in, the SpaceX team had been running at a historic pace for a new rocket program. The team fed off the energy. They fed off Musk's hyperbolic demands and his limitless support. They fed off the idea that they would prove the bureaucracy of old aerospace a relic and begin carving out a new path for the industry.

The Falcon 1 was by no means the most impressive rocket ever built. Far from it. Still, the machine had its charms. It stood 68 feet high and had a diameter of 5.5 feet. It had enough oomph to carry more than a thousand pounds of cargo into orbit and could do so for about $7 million per launch. The most notable thing about it was the price. Typically, rockets used to carry satellites into orbit cost $80 million to $300 million per launch. They're made up of parts supplied by hundreds of contractors all trying to extract maximum profits for their specialized hardware. SpaceX had flipped the whole equation by trying to build something pretty useful out of the cheapest parts available and to make as much of the rocket as possible by itself.

On March 24, the moment to test out the whole thesis finally arrived. Some people joined Musk in the mission control center based on Kwajalein Island, while the others were on call to respond to any mechanical issues on Omelek. The launch procedures began early in the morning as people went through their checklists and primed the rocket for its big moment, and at 10:30 a.m., the Falcon 1 lifted off. Its fiery rumble shook the temporary structures on Omelek for a few seconds before the rocket began its assault on gravity and headed skyward. For the SpaceX employees who were so emotionally invested in the Falcon 1, time began to stretch out. Every few seconds felt like minutes as eyes raced up and down the rocket's body to try and conduct visual checks on its health.

Even a casual observer,* though, could soon tell that something had gone wrong with the rocket. After taking off, its body began to rotate and wobble, which is a terrible sign when you're in the straight-and-true business. And then, thirty seconds into flight,

---

* About five thousand people watched the launch online.

the rocket's engine shut off. Instead of continuing up, the rocket paused for a brief moment and then proceeded to fall back toward the ground. At that point it was basically a bomb with Omelek as its prime target. The mass of metal and fuel smacked into a patch of reef two hundred yards from the launchpad and exploded. The rocket's payload, a small satellite made by the US Air Force, shot into the air and then ripped through the roof of a tool shed. Thousands of rocket pieces sprayed across Omelek, while others flew into the surrounding ocean.

The SpaceX employees were not thrilled with the result, but it was not unexpected. It's rare for a newly developed rocket to succeed on its maiden flight. The most humbling part of the experience came from the rocket's demolition on Omelek. If a rocket is going to blow up, it's best that the explosion occur at considerable height and over the ocean. No one on Team SpaceX wanted to suffer the indignity of going back to the island and picking up by hand all the reminders of their shortcomings as engineers.*

In the days that followed, people analyzed the data gathered from the brief flight and performed forensics on the rocket scraps. They soon discovered that an aluminum nut used to keep a fuel pipe in place had corroded as a result of being exposed to months of Kwajalein's hot, salty air. The part, which cost all of $5, had cracked and allowed kerosene to leak out, sparking a fire in the engine. In an ironic twist, SpaceX decided to fix the problem in future rockets by using even cheaper stainless-steel nuts.

It would take SpaceX another year to build a new rocket, go through all the tests, and try to launch again in March 2007. The second rocket performed much better and flew for more than seven minutes

* The army commander on Kwajalein was intrigued by the explosion and summoned Tim Buzza, the SpaceX vice president of launch and test, for a meeting. "I got a call from the top guy on the island, telling me to get down to his house right away," Buzza said. "I'm thinking, 'Oh, God, I'm in deep trouble.' This is some army colonel who has just been in Iraq. I ride my bike down to his house, and I can see him on the porch with two beers. I sat down, and he says, 'Well, it was a bad day, but you guys will recover. I'd like to talk more about what I saw on the video.'" And far from reprimanding Buzza, the commander wanted to discuss "how explosive that rocket is." He asked Buzza, "Is there something we can do with that in the army?"

before fuel began sloshing around inside it in an unexpected way and failed to feed the engine with enough propellant. Once again, the rocket tumbled back toward Earth, but this time it had the decency to burn up in the atmosphere. Almost eighteen months passed before SpaceX took a crack at its third launch, in August 2008. That rocket performed great all the way until the moment when the upper body of the rocket tried to separate from the larger, lower body but got stuck and triggered a failure before reaching orbit. "Falcon 1 blows it again," wrote one reporter covering the launch.

By that point, Team SpaceX was beyond frazzled. Life on Kwajalein had long transitioned from amusingly exotic to torturous. Late-night drinking sessions at the Snake Pit bar were no longer spent reviewing the work of the day or delving into aerospace nerdery. Instead, people were plotting their mutinous escapes. After putting down a couple Red Bull and vodkas, one engineer, for example, concluded that he might get kicked off the island if he went running across the airport runway nude. When everyone at the table agreed that that was probably true, the engineer found all the motivation he needed. He sprinted off and executed the plan. Unfortunately for him, the military personnel had seen worse, and it was back to Omelek the next day.

Publicly, Musk, NASA officials, and other people in the US government said all the right things. New rockets are supposed to blow up. SpaceX had been able to identify all the problems and fix them. The natural course of aerospace events was taking place. Privately, however, there were major concerns. Musk, for one, was burning through his personal fortune at an alarming rate and didn't enjoy what had become an annual tradition of meeting with the press to explain why SpaceX couldn't get to orbit. The government officials were also starting to wonder if, say, the guy in the SpaceX mission control with the orange mohawk held up by egg whites was representative of a corporate culture that was not so much amusingly eccentric as deeply dysfunctional. The voluminous amounts of beer and booze dotting the grounds of Omelek seemed to lend additional credence to this line of thinking.

"The third flight was the bottom of the bottom," said Tim Buzza, one of the key SpaceX figures behind the Falcon 1 and the efforts on Omelek. "Elon was running out of money and time. There was a lot of soul-searching going on, and it felt disastrous. That was the first moment when many of us thought that maybe it was the end. And then Elon had a teleconference call with the whole company. He said, 'I am going to borrow some money. We have one rocket left, and we have to launch it in eight weeks.'"

Pure dread: that's the feeling that comes with being told that you have to condense a process that had been taking a year into a couple months and that everything—the company, your career, the very idea of private space flight—hinges on your executing this rush job with precision. But Team SpaceX dug in and decided to make one last push.

The most immediate problem brought about by the obscene deadline was getting the fourth Falcon 1 rocket from SpaceX's headquarters in California to Omelek quickly. In the past, the rocket had arrived via the cargo ship that made its monthly appearance. But this time the rocket would need to be flown to the island, and that would require a very large plane, namely a C-17 military transport. Somehow, Buzza and other members of the team managed to locate a C-17 and some pilots and get the rocket loaded onto the plane shortly thereafter. That was the good news.

The bad news was that the pilots were ex-military flyers who liked to push aircraft to their limits for fun. Rather than easing the plane down onto the runway, the pilots brought the C-17 in as though it were a fighter jet. The rapid increase in air pressure caused the thin metal rocket body to begin collapsing in on itself. Horrified, a couple of SpaceX engineers lurched for some nearby tools and began opening up vents on the rocket to try and equalize the pressure between the inside of the machine and the conditions in the plane. Their quick actions worked to prevent any further damage, but the rocket arrived in anything but ideal shape.

The mood of Team SpaceX dipped even lower after this debacle. Some members of the island crew thought it borderline impossible to bring the battered rocket back to working condition before the launch. One poor soul had to call Musk and tell him what had happened. Per usual, Musk's response was to figure out a solution and keep going, going, going.

From the start of August 2008 until September 28, the SpaceX engineers and technicians gave their all to the cursed craft. Day by day, the Falcon 1's body was tended to and repaired, which allowed for the monotonous series of preflight tests to commence. A particularly large coconut crab three feet in length named "Elon" by the SpaceXers paid an occasional visit to the work area to watch the action, and that seemed like a good omen.

And so on the twenty-eighth, everyone took their places one more time. By now, the SpaceX team had experience on their side, although of all the rockets they had tried to send into orbit, this one might have been the least likely to succeed because of the mad sprint to get it onto the pad. Nonetheless, at 11:15 a.m., the Falcon 1's engine ignited, and the rocket surged into the blue sky and then up into space. Inside of mission control, everyone remained dead silent for most of the journey except for the occasional "Fuck, yeah" when the rocket did what it was supposed to at a critical juncture. Then finally—at long last—it became clear that the rocket had performed perfectly and reached orbit. Once in space, its pointed top opened up like a clamshell and deposited not a satellite but rather a dumb hunk of metal—because there were no longer customers willing to risk a real payload on one of SpaceX's machines.

In those first moments when it became clear that the launch had worked, the SpaceX team members on Omelek exchanged high fives but were muted in their celebration. They had to return to the pad to shut down fuel delivery systems and other machines. Meanwhile, the other SpaceX employees on Kwajalein Island hopped in boats and made their way to Omelek. Once the safety work was finished and the whole team had gathered together, someone started screaming "ORBIIIITTTTTTTTTT!!!!!!! ORBIIIITTTTTTTTTT!!!!!!! ORBIIITTTTTTTTTT!!!!!!!" Then everyone else started screaming, and the chant "Orbit, orbit, orbit!" took over the group like a primal war cry. The afternoon celebrations on Omelek shifted into evening and late-night celebrations on Kwajalein Island. Every so often, the chanting would begin again with drunken engineers funneling six years of struggle into a spectacular shared release of emotion. Rocket rapture.

THIS BOOK IS NOT ABOUT SpaceX, which might make you wonder why I just spent all those words on the company and its rocket. I would, however, argue that you need to know about the Falcon 1 and all that went into it because that machine set the action that takes place in this book in motion—and quite likely changed the course of human history.

In the most practical of terms, the Falcon 1 established SpaceX as the first private company to build a low-cost rocket that could reach orbit. It was an engineering milestone and an accomplishment many people in the aerospace industry had dreamed about for decades.

In more symbolic terms, the SpaceX engineers shattered the natural order of things. While not at all clear back in 2008, that first launch into orbit would emerge as an inciting incident. Like Roger Bannister besting the four-minute mile, SpaceX made people recalibrate their sense of limitation when it came to getting to space. The imaginations and passions of engineers and dreamers all around the world expanded. A turning point had occurred, and a space frenzy had been ignited.

The story of space since the United States and the Soviet Union had first started racing to the moon had largely centered on the actions of a handful of governments. It took the might of the United States, China, or the European Union to fund a rocket program. Those entities had turned space into a rare and precious commodity. The handful of wealthy individuals who had tried in years past to make their own rockets and alter the balance of power had failed. Without doubt, SpaceX received encouragement and financial backing from NASA and the US military. But it was Musk who came out of nowhere with $100 million of his own money and willed SpaceX into existence. He proved that a driven individual aided by a company full of bright, hardworking people could match, and maybe one day best, entire countries.

More broadly, SpaceX had rejected many of the "truths" held evident by the old, government-backed aerospace industry. It demonstrated that a novel approach to rocketry could work. Rockets didn't need to be made out of pricey "space-grade" equipment that had been certified as worthy for the job by specialist contractors. Consumer electronics had improved so much that off-the-shelf products were now often good enough to survive the rigors of space travel. Major advances in software and powerful computers also meant that engineers could now accomplish far more than they had in the past. Strip away layers of bureaucracy dating back to the 1960s and staid thinking, and you ended up in a place where the construction of rockets could be modernized and made more efficient. New things were possible.

Much of the existing aerospace community rejected those revelations. They still viewed SpaceX as an oddity, a minor-league player. The Falcon 1 could carry a thousand pounds of cargo into orbit, whereas the old guard's giant rockets could carry many tons. If SpaceX wanted to get more serious and make something bigger, it would be in for a world of hurt. The development costs would drain Musk's bank account. The engineers would fail to translate their skills and modern methods to more advanced machines. At best, SpaceX would end up bloated and expensive just like all the old-line companies. At worst, it would fall flat on its face trying, and they figured that scenario the most likely.

Hindsight has made it clear that traditional aerospace underestimated Musk and his SpaceX engineers in humbling or humiliating fashion—take your pick. A bit more than a dozen years after that Falcon 1 launch, SpaceX has gone on to build three more rocket families, each one bigger than the last. Its workhorse, the Falcon 9, now dominates the commercial launch industry, flying satellites into orbit week after week. The company has perfected reusable rocket technology, allowing it to bring rocket bodies back to Earth and fly them again, while its rivals continue to dump their once-used hardware into the ocean. SpaceX has also started a satellite business and manufactures and flies more of them than any company in history. In 2020, while covid-19 made the world stand still, SpaceX sent six astronauts to the International Space Station, restoring the United States' ability to put humans into space for the first time since the space shuttle was retired in 2011. Meanwhile, in south Texas, SpaceX is busy building Starship, a vehicle meant to fulfill Musk's ultimate ambition of forming a human colony on Mars.

The traditional aerospace players opted not to alter their businesses in dramatic ways as a result of SpaceX's presence. Their inaction, however, did not stop the impact of the Falcon 1 from rippling well beyond Musk's empire and changing humanity's relationship with space. Engineers, entrepreneurs, and investors saw what SpaceX had accomplished and began to have their own wild visions of what they might achieve. They, too, could ride the wave of ever-progressing electronics, computers, and software and create their own space companies. People all over the world—for better or for worse—began to see themselves as the next Elon Musk.

"The big guys controlled everything," said Fred Kennedy, a former colonel in the US Air Force and onetime director of the Defense Department's Space Development Agency. "I used to despair that if you didn't throw in with the big contractors, you were done. Then Elon showed that you could break through. He showed you could do something different. I think that caught everyone's imagination."

In the popular press, the increase in private space activity has tended to focus on Musk and his peers, such as Jeff Bezos, Richard Branson, and the late Paul Allen of Microsoft. Those men have all funded ventures varying from rocket companies to space planes. The fascination largely revolves around billionaires who are hoping to fire up space tourism businesses or, like Musk, setting off to colonize the moon or Mars.

What the public has paid less attention to is the frenetic amount of activity taking place among hundreds of other companies scattered all around the world who are building new types of rockets and satellites. These companies are locked in a race that feels more immediate and tangible than humans taking laps around the moon or doing their laundry on Mars. They are trying to build an economy in low Earth orbit, the stretch of space from 100 to 1,200 miles above our planet's surface, that would, in essence, be the next playing field in humankind's technological evolution.

From the 1960s to 2020, the number of satellites put in space had increased at a slow and steady pace, resulting in about 2,500 machines orbiting the earth. Most of them were sent up to perform jobs for military entities, communications companies, and scientists. Before being launched, each satellite was treated like an exquisite technological marvel. It took many years to design and build and often ended up as big as a van or a small school bus. The prevailing traditions in aerospace dictated that no expense be spared with those satellites because they were meant to perform their jobs for ten to twenty years and had to survive the harsh conditions of hurtling through space all that time. As a result, a single satellite could cost $1 billion or more.

From 2020 to 2022, something astonishing happened: the number of satellites doubled to 5,000. Over the next decade, that figure is set to increase to somewhere between 50,000 and 100,000 satellites, depending on whose business plans you believe. (It's perhaps best to take a moment and let those figures rattle around in your brain a bit.) A handful of companies and countries, including SpaceX and Amazon, want to put up tens of thousands of satellites to create space-based internet systems. The satellites will beam high-speed internet service to the 3.5 billion people who can't be reached by fiber-optic cables today. What's more, they will blanket the globe with an ever-present internet heartbeat that will let drones, cars, planes, and all manner of computing devices and sensors send and receive data no matter where they are.

Beyond the space internet, there are already hundreds of satellites orbiting the earth taking pictures and video of everything that happens below on a near-hourly basis. Unlike existing spy satellites that feed their images to governments, these new imaging satellites are owned by young start-up companies that allow almost anyone to purchase the pictures they take. Organizations have started collecting and analyzing the images by the tens of thousands to gain both political and commercial insights. They're assessing things ranging from North Korea's military activities and the amount of oil production in China to how many people are shopping at Walmart during back-to-school season and the rate at which the Amazonian rain forest is being cut down. The satellites, aided by artificial intelligence software, can peer down and monitor the sum total of human activity. They are, in effect, a real-time accounting system of the earth.

Much of the reason this is happening is because the satellites are smaller and cheaper than ever before—beneficiaries, once again, of the electronics and computing improvements that we've all witnessed in other facets of life and business. Instead of $1 billion each, the new satellites cost anywhere from $100,000 to a few million dollars. They range in size from something as big as a pack of cards to a shoebox to, say, a refrigerator. They're often designed to work as a group, or what the industry calls "a constellation," and to last for only three or four years in space before they deorbit and burn up as they reenter the earth's atmosphere. Their low cost means that the satellite companies can afford to send up new hardware all the time, replacing old gear with state-of-the-art technology instead of trying to eke out greater performance from something that's a decade or two old. It also means that more companies than ever can afford to try to do something in space, be it communications, imaging, science, or any manner of other applications. As a result, there are now hundreds of satellite start-ups at the ready, hoping to invade low Earth orbit with their wild ideas.

In a typical year, about a hundred rockets make their way into orbit to drop off payloads. China, Russia, and the United States account for about three-quarters of these launches, with Europe, India, and Japan making up the rest. But things are no longer typical when it comes to space, and there are simply not enough rockets to meet the demand of all the companies and governments looking to fly those tens of thousands of satellites into orbit.

It's for this reason that about a hundred rocket start-ups have appeared over the past few years, hoping to become what amounts to the FedEx of space. These young rocket companies tend to have radical ideas. They do not want to build large rockets that cost $60 million to $300 million per launch. And they do not want to launch once a month, as is the standard pace for traditional rocket makers. Instead, they want to build smaller rockets that cost anywhere from $1 million to $15 million per launch and can be fired off on a weekly, if not daily, basis. (The most radical idea here is a space catapult that could fling rockets into orbit at a cost of $250,000 per launch and do so eight times a day. Some very smart people even believe that this is a legitimate approach.)

The Falcon 1 was once meant to serve this FedEx-like function. But not long after that first successful launch in 2008, Musk decided to stop making small rockets and to focus his resources and energy on bigger machines. Back in 2008, that strategy made a lot of sense. There were not very many of the small satellites around at the time, and SpaceX needed to make real money to survive by launching big satellites for governments and communications companies. Beyond that, Musk's long-term plan has been to put people in space and then get them and thousands of tons of material to Mars. Neither task can be accomplished on a small rocket.

The rocket start-ups have rushed into the void left by the Falcon 1's demise with the theory that the time for cheap rockets that can be fired off at will has come. The company doing the best to prove this theory right is Rocket Lab, which was started by Peter Beck in Auckland, New Zealand. Beck does not date famous actors, nor does he have an electric car company or make wild pronouncements on Twitter. Yet his story is as remarkable and as unlikely as Musk's. He's a self-taught rocket scientist who never went to college yet somehow managed to build a rocket company in New Zealand, which had no aerospace industry on which to lean. Rocket Lab began launching its all-black, fifty-six-foot Electron rocket in 2017 and by 2020 was flying all the time, joining SpaceX as the only other private company putting satellites into orbit for paying customers on a regular basis.

Myriad other small-rocket companies want to join in the frenzy. Most of them are absurdly understaffed and underfunded and run by rocket hobbyists aspiring to something more magnificent. About ten of them, though, are legitimate and have actual shots at getting into the rocket game. A number of them are based in the United States, while others are in Australia, Europe, and Asia. Musk and then Beck unlocked the idea that any individual who is clever enough and relentless enough can build a rocket just about anywhere.

Without question, the companies manufacturing smaller rockets suffer from a major problem: they simply cannot take much cargo into space. If you pack a $60 million SpaceX rocket with hundreds or thousands of small satellites, it will get them into space all at once and at a lower cost per pound than a cheaper, smaller rocket. (Think a single eighteen-wheeler versus dozens of minivans.) What the smaller rocket makers are banking on is that many, many companies and governments will send a lot more stuff to space a lot more often if they know that there is always a cheap ride available. Instead of needing to apply to get onto SpaceX's launch manifest eighteen months in advance, they can just hop onto Rocket Lab's website and book a flight that can leave in a couple weeks. Once people know they can count on a system like this, the economy of low Earth orbit starts to change dramatically. The underlying infrastructure transforms from everyone fighting over access to a couple of railway systems to something much closer to mass transit.

Back in 2008, hardly any investment money flowed into private space endeavors. Musk and Bezos, with his start-up Blue Origin, were the main private rocket players, and very few satellite start-ups existed. Over the past decade, however, tens of billions of dollars have poured into the private space industry. The obvious transition has been from that of governments to billionaires and then to venture capitalists. Trying out an idea in space no longer requires congressional approval or some wild-eyed dreamer willing to risk his personal fortunes; it just requires a couple of people in a room agreeing that they're willing to spend someone else's money on a huge risk.

The future that all these space buffs have already started building is one in which many rockets blast off every day. Those rockets will be carrying thousands of satellites that will be placed not all that far above our heads. The satellites will change the way communications work on Earth by, for one, making the internet an inescapable presence with all the good and bad that entails. The satellites will also watch and analyze the earth in previously unfathomable ways. The data centers that have reshaped life on our planet will be transported into orbit. We are, in effect, building a computing shell around the planet.

Though this process has been emerging over the decades, the pace at which it has been actualized in recent years is breathtaking, inspiring, and unnerving. And the cast of characters behind the latest space movement often do not resemble their bureaucratic, measured predecessors. At the rocket start-ups, for example, you're more likely to find a welder who used to work on oil derricks or a Formula 1 race car engine builder than you are someone with an astrophysics PhD from MIT. These people are building rockets designed to ferry cargo into orbit, yes, but seen through another lens, they're building the equivalent of privatized intercontinental ballistic missiles, and their talents are now available to the highest bidder. It's the Wild West of aerospace engineering. Meanwhile, out in satellite land, we've already seen at least one company rush to sneak its devices onto a rocket and send them into orbit without obtaining any of the customary regulatory approvals. Better to get your stuff into space first and ask for forgiveness later if you want to seize your slice of low Earth orbit.

The rhetoric surrounding space has changed quickly, too. Nations used to spend billions upon billions of dollars to show off the abilities of their scientists and ensure the security of their citizens. Space activities were interlinked with nationalism and patriotism. When the billionaires such as Musk and Bezos came along, they pitched access to space as a noble, necessary quest that would fulfill humankind's destiny. They espoused the ideas that we are explorers by nature and we generate optimism among all people by pushing our intelligence and technology to its limits and journeying toward the unknown, if for no other reason than to make sure our species survives and thrives. Of course, these same motivations flow through the new work taking place in space, but so do much baser things. Silicon Valley's unending pursuit of wealth, control, and power has turned straight up. To put a fine point on things, space is now open for business. The heavens—like everything else—have gone on sale.

FOR THE PAST FEW YEARS, I've had a front-row seat in which to observe this peculiar moment in our shared history unfold. A journey that started off by following Musk and SpaceX has carried me to California, Texas, Alaska, New Zealand, Ukraine, India, England, Svalbard, and French Guiana and put me in rooms reporters are not usually allowed to inhabit. There have been late nights spent in grimy warehouses with engineers trying to ignite their rocket engines for the first time all the way up to glorious rocket launches from South American jungles. There have been private jets, communes, gun-toting bodyguards, hallucinogens, a troop of male strippers, a rotting whale carcass in a bathtub, espionage investigations and federal raids, space hippies, and multimillionaires guzzling booze to dull the pain as their fortunes disappear.

In this book, I've tried to place you in the heart of the action as people the world over become obsessed by a grand new quest. The story follows four companies—Planet Labs, Rocket Lab, Astra, and Firefly Aerospace—on their missions to build new types of satellites and rockets. The companies, their leaders, and their engineers find themselves in an uncharted world not unlike the early days of the personal computer or the consumer internet. They can sense that something fantastic is within their grasp and that they have a chance to play a part in history.

Much of the story within is inspirational. Planet, for example, has changed space technology and the low Earth orbit economy in as dramatic a fashion as SpaceX. Meanwhile, there are people such as Brigadier General Pete Worden, who arrived on the scene well before Elon Musk and worked in the background to set this revolution into motion. There are idealists and do-gooders and very smart people doing exceptional things. Some of the characters go on the Hero's Journey and overcome tremendous odds. I will warn you, however, that not all ends well for our main characters. There are healthy doses of comedy and tragedy along the way. The stories that follow try to capture the spectacular madness of it all.

And it is madness. For as much as I've argued that space is becoming a real business, it remains unique in terms of the pursuits people choose to make money. Space carries with it centuries of mythology and fantasy. The Falcon 1 on Kwajalein resembled a totem because it was a totem—a Promethean tube full of fire that spoke to the core of human striving. Even the most cynical welder who claims to be working until 2:00 a.m. just to collect a paycheck revels in the idea of one day telling his friends that he helped put something into the great void hanging over us all. The chief engineers, the CEOs, the wealthy investors see themselves as being adventurers. They're taking ridiculous risks in a bid to overcome every obstacle set before them, to overcome physics itself and to prove that even Earth cannot limit their wills. At a visceral level, they want to conquer something. On a more abstract level, something about space allows humans to perceive themselves as being part of a timeless story and casting their lot in with the infinite.

I've come, then, to view this current incarnation of the space industry as being powered by a type of shared hallucination. If you ask people during quiet moments if all the rockets and satellites make sense or if their businesses will one day turn a profit, they will sometimes confess that no one really knows if any of this shit will work. Still, billions upon billions of dollars keep flowing in, and the new ventures become ever stranger. Idealism, passion, invention, ego, greed: the usual suspects are all there, inciting action. But so, too, is the underlying bond that propels the great illusion forward—that you not ask too many questions, think about consequences too long, or let reality interfere with your hopes and dreams. After all, this is space. It's best just to say, "Fuck it! Let's do this thing because we kind of have to."

THE GREAT COMPUTER IN THE SKY

CHAPTER ONE

WHEN DOVES FLY

Robbie Schingler went to India to make history.

It was February 2017, and he'd landed in Chennai, the chaotic city of 7 million people situated on the country's eastern coast. On the cusp of forty, Schingler could have passed for your typical tourist. He was of average build and wore jeans and a short-sleeved collared shirt, topped by a pair of sunglasses resting on his head of brown hair. Upon his arrival, Schingler checked into a nice hotel and attempted to stave off his jet lag and adjust to local conditions by walking around to take in a few sights. The heat, humidity, and sensory overload of Chennai are powerful, though: just a few steps away from the hotel's gates were throngs of people going about their day, tuk-tuks racing by, and colors, smells, and sounds coming in relentless waves. So after the walk, Schingler succumbed to his jet lag and napped.

To nap on February 13 struck me as impressive. Schingler had cofounded a company called Planet Labs, which built satellites. In two days' time, eighty-eight of the shoebox-sized machines were meant to fly into orbit aboard an Indian rocket known as the Polar Satellite Launch Vehicle, or PSLV. Alongside Planet's satellites would be sixteen other satellites from universities, start-ups, and research groups. No rocket had ever flown as many as 104 satellites to space before, and the Indian press was treating the record-setting event as a major source of national pride.

Though setting records is nice, Planet's very survival as a company was at stake. Founded in 2010, it had been trying to revolutionize both the satellite industry and our understanding of the earth. In the most basic terms, the satellites Planet built were cameras that orbit around us and snap constant pictures of what is happening below. Much larger, more expensive versions of these image-taking satellites had existed for decades. But there were not many of them, and the places they could look were limited. What's more, the images they produced usually went first to governments or the military and then to a small number of corporations that could afford to purchase them.

Planet's big idea was to make many smaller, cheaper satellites and arrange them in what the company called a constellation. By having hundreds of satellites orbit the earth in a particular pattern, Planet could snap pictures of *every* spot on the ground *every* day. Such a technological achievement would carry massive implications. Photos of the activities taking place here on Earth would no longer be rare and traded among the few. Instead, Planet would create a constant record of everything happening on the ground and offer the photos through an online service that anyone could use. Be it photos of troops amassing in Crimea, cargo ships sailing across the ocean, buildings going up in Shenzhen, or even North Korea testing missiles, Planet would have pictures of these day-to-day machinations available for a modest fee and instant download.

This sounds like the stuff of espionage and intelligence gathering, and certainly such a constellation would benefit such activities. But Schingler and his cofounders, Will Marshall and Chris Boshuizen (pronounced BOSH-hausen), were a combination of space nerds and space hippies. They envisioned their satellites as being a force for good. People could use the images produced by the machines to monitor rain forests, measure methane and carbon dioxide levels in the atmosphere, and track the movements of refugees in war-torn regions. To the extent that the satellites would be used for intelligence matters, it would hopefully be to provide an objective truth about something such as a weapons test or an environmental disaster, preventing a government from trying to hide the incident or spinning it in a disingenuous way. It was with all of that in mind that Planet decided to call its satellites Doves.

Leading up to the 2017 rocket launch, Planet had already put dozens of its satellites into orbit to test out the basic ideas behind its thesis and improve its underlying technology. This launch, though, would complete the constellation and make it possible to see everything all the time. If Planet's machines worked as billed, they would set several major milestones. A start-up would emerge as the operator of the most satellites orbiting Earth, allowing Planet to join SpaceX as the next New Space maverick of consequence. The company would also confirm that small, cheap satellites working together could equal or best the large, expensive machines that had forever dominated the industry. And space would be democratized in a fashion previously thought unimaginable. Anyone with a computer could examine the earth in fine detail and analyze the sum total of human activity.

When the next day rolled around, naps were no longer an option. A government-provided SUV picked Schingler up at his hotel in the morning and began the almost three-hour drive north toward the Satish Dhawan Space Centre.

Among spacefaring nations, India ranks high. The country has immense engineering talent combined with low labor costs. This makes the PSLV, its workhorse rocket, a reliable and affordable choice for both homegrown satellites and those made by India's numerous allies, including the United States. Each year, roughly three to five PSLV rockets ferry cargo into orbit with the missions managed by a government-backed entity, the Indian Space Research Organisation (ISRO). The achievements of ISRO are so lauded at home that a picture of Mangalyaan, the first Asian spacecraft to orbit Mars, can be found on the 2,000-rupee banknote.

India has a few rocket launch sites to pick from, but the Satish Dhawan Space Centre is perhaps its most exotic. The center opened in 1971 on the island of Sriharikota in the Bay of Bengal. From above, Sriharikota looks like a snake that's in the middle of digesting a goat. It has thin stretches at the top and bottom of its seventeen-mile-long coast and a swollen center that's five miles across. To reach the launch complex from Chennai, you drive along a highway that functions on the "give no fucks" principle where pigs, cows, semitrucks, motorcycles, buses, and women carrying plastic water buckets on their heads all vie for the right of way on the road. Eventually, you turn off the highway and onto an outlet road that leads to a causeway surrounded by marshlands, salt ponds, and mud pocked by opportunistic vegetation.

Every rocket launch site I've ever been to triggers the same sense of confusion. Your brain clicks into rocket mode and expects to be fed with images of slick, futuristic objects. After all, you're witnessing the home turf of one of the pinnacles of humankind's scientific and engineering achievements. The launch complexes, though, tend toward rough and rugged rather than fit and finished. This is mostly because space agencies place their launchpads in remote spots near coastlines, where wayward projectiles are less likely to kill people or cause major damage. It's also because many of the sites were built during the go-go days of the Space Race and have not benefited from much modernization in the intervening decades.

True to form, the Satish Dhawan Space Centre that greeted Schingler looked more defunct disco than sci-fi fantasy. He pulled up to a security gate where a couple of police officers asked for identification. The guards then demanded that everyone in the car get out and remove their electronic items such as laptops and cell phones and wrote the serial number of each device by hand into a ledger. A mango tree provided shade during the lengthy process, while a couple of white cows lumbered around the property as they pleased. After that check, Schingler was sent to a nearby second office to receive some credentials. There light bulbs hung from the ceiling via bundles of exposed wires, and yellowing posters of rockets and scientists were placed randomly on walls. A pair of barefoot clerks rose from their desks, gathered Schingler's documentation, and returned a while later with his entry permit.

After Schingler dropped his bags off in a dormitory-style room, a couple of ISRO's top officials came by to take him on the rest of his adventure. Since he was paying many millions of dollars for a rocket launch, he got the grand tour, which included a visit right up to the rocket and a look at the mission control center. At each stop, ISRO had cleared patches out of the dense tropical forest to make room for its buildings. The sounds of monkeys scrambling among the trees could be heard along the entire journey, and on occasion, the government car had to stop and wait for a cow or two to cross the road.

In the evening before the launch, there was not much to do but wait. A couple of Planet employees had flown to India to observe the launch from outside the complex, as they were not allowed onto the space center grounds. They managed to acquire eighty-eight statues of Ganesh in honor of the satellites and phoned Schingler to tell him about their purchase. Schingler felt sure that the idols would bring good luck.

On the morning of the launch, Schingler doubled down on boosting Planet's karma. He awoke before sunrise, ate breakfast at a cafeteria, and then went to a temple near the dormitory. He meditated and prayed. Planet had been particularly unlucky during previous launches, when its satellites had been destroyed after first an Antares rocket and then a SpaceX rocket had exploded. In a weird way, the explosions validated Planet's approach to satellite making. Since it produced lots of small, cheap devices, Planet could afford to lose them now and again. Its predecessors, which had often spent a decade building a single $500 million machine, could not say the same. Nonetheless, losing eighty-eight satellites in one go would be horrendous and would greatly impede Planet's desire to race forward faster.

Having made his peace with the Space Gods, Schingler now put his faith in India and its superb engineers. He traveled with the ISRO officials to the mission control center, which looked just like anything you've seen of NASA on TV: a few rows of desks with their computers and screens, and people in lab coats either sitting down and thinking hard or walking around taking care of various tasks. Schingler settled into a theater area just behind all of it, separated from the mission control by glass. Many Indian dignitaries were in the audience as well, and I camped out next to Schingler as the first foreign journalist ever let so far into the Satish Dhawan Space Centre.*

The rhythm of a rocket launch is that of prolonged tension marked by a sudden flurry of excitement. On edge and full of nerves, Schingler watched as the engineers went through their last checks over the course of about ninety minutes. He could do little more than fidget and make small talk as tens of millions of dollars of satellites hung 140 feet in the air at the top of the PSLV rocket. About thirty minutes before the launch, however, time started to become fuzzy. You were still watching people fiddle around like before, but the minutes seemed to dissolve in chunks. Five gone all of a sudden. And then seven gone. And then, my God, this is really going to happen, isn't it?

It was while that very thought was going through Schingler's mind that someone opened a couple of large doors at the side of the viewing theater and began ushering everyone outside. Dozens of people gathered on a semicircular patio with miles of forest in front of them. A sound system behind them pumped out the mission control chatter. Thirty seconds. Fifteen seconds. And right into the final ten-second countdown. It took a few agonizing seconds more to see anything, and then there it was, the rocket rising up from the trees and heading into the clouds at 9:28 a.m. People applauded and let out a few hoots and hollers. Those accustomed to seeing rocket launches quickly turned around and went back inside. Schingler lingered a few moments longer and hugged a coworker as a permagrin stretched across his face. "I am happy," he said. "Let's go see it."

* Or so I was told.

By "Let's go see it," Schingler meant that we needed to return to the mission control to find out if the rocket would do what it was supposed to. The machine had pulled off the tough work of fighting through the earth's gravity but still had plenty of mission ahead. It would need to fly to the correct orbit and eject all of those satellites safely in the right places. That meant more time waiting around in a heightened, existential state while hoping for the best.

After about a half hour, word came back that the satellites had been placed safely in orbit three hundred miles above the earth. Covered in white paint, the Doves had tumbled out of the rocket's cargo chamber one after another and created what looked like a string of pearls traveling across a black backdrop. Employees at Planet's headquarters in San Francisco began communicating with the satellites through a network of antennas the company had placed at ground stations all around the world. The first step was to check on the Doves and make sure they were alive and functioning correctly.

No organization had ever put up anywhere close to eighty-eight satellites at the same time. The going rate was one or two and on rare occasions four or five. As such, Planet had needed to invent many of the methods for locating, controlling, and commanding its flock of Doves as they whirled around the earth at incredible speeds.

For this mission, it had picked three "Canaries" to receive the first major round of commands. As the health checks were sent, the three satellites were told to turn on their magnetorquers, small devices that create a magnetic field around the satellite. The purpose of the exercise was to stop the satellites from tumbling end over end by using the man-made magnetic field to interact with the earth's magnetic field and create torque that would right the satellites into a stable position. The magnetorquer and a reaction wheel were then combined to point each satellite toward the sun as they unfurled solar panels on either side of their bodies. The Doves got their wings. After that, a number of onboard sensors worked together to calibrate the satellites' positioning systems and cameras by looking for things like constellations and the moon.

Those processes identified a couple of errors, which Planet

engineers fixed by writing new software and sending it up to the machines. Commands then went out to a larger batch of satellites and then another, until they were all configured for action.

By the time the mission was completed, Schingler would no longer be in India. It would take a couple months for the Doves to slowly spread out and end up evenly spaced from one another as they formed a picture-snapping ring around the earth. Rather astonishingly, they moved around in space not by thrusters but by a technique called differential drag. The solar panels worked sort of like sails and pushed against the faint trace of atmosphere in space. When in a vertical position, the panels produced five times as much resistance as when horizontal. Using differential drag to control a bunch of satellites in orbit like this had been a mostly theoretical concept until the big brains at Planet proved it worked.

Before all that happened, however, Schingler found time in India to celebrate the more immediate accomplishments. He did interviews with local TV stations and reporters while the ISRO officials held a press conference. Then all the dignitaries shared lunch. After that, Schingler gathered his belongings and hopped back into the SUV for the ride to Chennai.

During our journey back, Schingler asked the driver to stop by a roadside shop so that we could buy a few Kingfisher beers to celebrate. A short while later, after he'd toasted the launch, we cruised into an intersection along with a half-dozen other vehicles all trying to turn this way or that at the same time. It became obvious that we were going to crash into one of the cars, but neither driver chose to address the situation, and we eased into a slow-motion wreck. The drivers hopped out, looked at both vehicles, and decided to press on with their lives. Schingler kept right on smiling through the entire event, refusing to let the vagaries of Earth get into the way of the miracle of mathematics and physics he'd just witnessed.

Over the next couple of days, the hippie part of Schingler's space hippie personality came out in full force. He'd forgotten to book a hotel for the night after the launch and realized it only late into an evening of drinking. With the success of the launch behind him, he had become well and truly a multimillionaire, but he ended up crashing in the room of one of his employees. The next day he took a trip to the coast to splash around in the water and visit some ancient temples. I went home after that, while Schingler continued on, visiting the "utopian community" of Auroville. He couldn't find a bed there, either, and spent the night sleeping in a shed on a concrete floor while curled up next to an old ice cream machine.

The Indian press had made a big deal of the launch, and a few other reporters around the world took note of both the record number of satellites put into orbit and Planet's aspirations. Few people outside the hard-core space community, though, really understood the import of what had just transpired. Not since SpaceX had launched the Falcon 1 rocket had a private space company enjoyed such a monumental moment.

From its founding in 2010 to this launch in 2017, Planet had managed to place hundreds of satellites in space. Some of them had run their course, tumbled back toward Earth, and burned up in the atmosphere. But around 150 of them were now doing their intended work and constantly photographing the blue spinning ball below them as if it were a movie star at a never-ending film premiere. A start-up comprised of a couple hundred young employees had charged into space and gobbled up a large swathe of its most valuable territory. After the launch from India, Planet's machines accounted for almost 10 percent of all the functioning satellites orbiting the globe. That had become possible as a result of the idealism and brashness of its founders and a total rethinking of how to approach the design and construction of satellites.

Because he's Elon Musk and because rockets are cool, SpaceX consumed most of the public's attention when it came to new things happening way up above and the notion that the economics of space might be changing. People closer to the space industry, though, were just as enthused by Planet's accomplishments. It appeared that the rules of engagement were changing—fast—regarding both how we get to space and what we can do upon arriving in orbit. When added together, SpaceX and Planet cemented the belief of those who wanted to believe that private industry could push governments out of the way and come to dominate activity in space. The notion that a new economy was being formed in low Earth orbit felt more real than ever. From 2017 on, billions upon billions of investment dollars began to flow toward space start-ups, with each new company envisioning itself as the next SpaceX or the next Planet.

The questions a curious onlooker might have asked around that time were: How did Planet come into existence? How did a guy who sleeps on shed floors and his two equally unusual chums end up building a system capable of recording the earth's every undulation?

As I would discover, the answers to those questions would not actually start with Schingler or his cofounders. The revolution in space that seemed to come out of nowhere had been decades in the making. It had been shepherded along by a genius general who had a tremendous talent for pissing everyone off. He was one of those figures that few people had ever heard of but who played the role of supreme puppet master—and managed to usher spectacular things into existence.

CHAPTER TWO

SPACE FORCE

A *New York Times* front-page story ran on February 19, 2002, with the headline "A Nation Challenged: Hearts and Minds; Pentagon Readies Efforts to Sway Sentiment Abroad." The story revealed that the US Defense Department had created something called the Office of Strategic Influence. The goal of the office, according to the article, would be to try and shape the global opinion of the United States' military actions following the 9/11 terrorist attacks. In other words, the United States hoped to use propaganda to make the War on Terror more appealing, particularly in Islamic countries, by placing media stories that would carry a pro-US spin without leaving traces back to the Department of Defense.

Though very vague on the details, the story also suggested that the Office of Strategic Influence would spend millions of dollars on more nefarious programs, using the internet, advertising, and covert operations to spread misinformation. Right away, people questioned the legality of such an enterprise, while foreign journalists were none too thrilled to learn that they might end up as unwitting participants in a big-budget psychological operations campaign. Defense Secretary Donald Rumsfeld and others denied that the new office would be up to anything dodgy and tried to present the program as an analytical approach to winning over hearts and minds. It wouldn't be just propaganda; it would be high-tech, measured propaganda that would get the best bang for the taxpayer buck.

Nonetheless, the public disclosure of the program immediately undermined its cloak-and-dagger nature and stirred up too much political trouble. Just a week after the *New York Times* story appeared, the Office of Strategic Influence met its demise. "The office has clearly been so damaged that it's . . . pretty clear to me that it could not function effectively," Rumsfeld said at the time. "So it's being closed down."

That was not an ideal turn of events for Air Force Brigadier General Simon P. Worden, who had been in charge of the Office of Strategic Influence. But then Worden, whose friends call him "Pete," had become accustomed to uncomfortable situations during his thirty years in the air force. An astrophysicist, he had bounced around from doing weapons research and conducting black ops missions to conducting purer pursuits closer to his educational background, such as studying the nature of the universe. At each stop, Worden had built up a reputation as a very bright, very unconventional thinker who had the audacious tendency of trying to make bureaucratic institutions less bureaucratic. His personality resulted in a career pattern in which he would be appreciated up until the point when he rubbed a high-ranking bureaucrat the wrong way for too long and would then be shunted to a new outpost.

On this occasion, the government decided that Worden's next stop would be the Space and Missile Systems Center in Los Angeles, which focuses on the application of military technology in space. Worden took over a team of fifty people who were tasked with thinking up wild new ideas that might push space weaponry forward in unexpected ways. The major goal was to write interesting papers and hope that they would one day catch the eye of someone important in the military. "You would do a study and brief a bunch of senior folks, and they'd say, 'That's very nice,'" Worden said. "Often they would just put the study in a closet or something, and then maybe six months later or five years later a new challenge would pop up, and someone would remember that one of the studies could help and pull it out."

Worden did not mind crafting studies and saw the value in these exercises, but he preferred taking action. He had long been thinking that the dramatic improvements in electronics and computing were opening up new possibilities not just in satellites but also in rocketry. His thesis was that if someone could build a small, capable satellite and put it onto a small, capable rocket, there could be a major breakthrough in what the military called "responsive space." In a nod to what would eventually become formalized as the space force, Worden wanted the ability to deploy space assets with the same speed and precision as other tools in the United States' military arsenal.

"If you have a sudden crisis in, say, Botswana, just to pick a random place," Worden said, "the problem is that you don't have satellites optimized for Botswana. If you know that you're going to deploy the army and the air force to somewhere like that in a few days, it would be game-changing to launch a satellite at the same time for support."

The military, though, had a self-defeating mechanism built in when it came to moving fast and cheap in space. Going back to the 1960s, the ethos of both NASA and the military had been that every rocket and every satellite had to work and they would pay whatever it cost to ensure that happened. When something did go wrong, people were blamed, new codes and regulations were written, and more procedures were put into place to guarantee that the same mistake would never occur again. As Fred Kennedy, the former air force space whiz, put it, "A zero-defects culture had built up over forty years. The only way to fix it was to rip everything down and start over."

The Defense Advanced Research Projects Agency (DARPA), the research-and-development arm of the Defense Department, had grown increasingly frustrated by the way the old guard operated. It's DARPA's job to think ten, twenty, thirty years ahead and develop sci-fi-level military technology. Its leaders wanted to experiment with all kinds of mad scientist ideas in space but could hardly ever manage to hitch a ride on a rocket because the military contractors like Boeing and Lockheed Martin were slow and launched so infrequently. "We had people at DARPA saying 'Let's go buy fifty small rocket boosters and launch one a day and thumb our nose at this idiotic community that can't get its act together,'" Kennedy said.

As Worden made the rounds in his new post, he soon encountered the like-minded folks at DARPA, and they began to brainstorm together. One intriguing development that caught their eye was a very rich guy named Elon Musk, who had started a company called SpaceX that intended to launch as many small, cheap rockets as it could.

Soon enough, Musk materialized in Worden's office, and the two men hit it off. "He said he would have this rocket called Falcon 1 ready in a couple of years," Worden recalled. "He really just wanted to know if we would use it." As perhaps the highest-ranking space tech nerd in the military, Worden had encountered every "out there" pitch and wacko inventor imaginable, ranging from guys building ray guns in their garage to folks convinced that their flying saucer would be the next great military vehicle. But Worden identified Musk as legit and something of a kindred spirit. They both hoped that humans would one day colonize Mars and expand even farther into the universe, and they enjoyed batting theories back and forth on how to accomplish such things. "Elon was a visionary, and there were a lot of visionaries around at the time," Worden said. "But there was something about him where I thought, 'This is not a boloney artist. This is the real thing.' The other thing that was different about him was that he really understood the rocket stuff and how they worked."

At Worden's urging, DARPA decided to give SpaceX a contract to launch a small satellite on its behalf*—a gesture that both lent some prestige to the start-up and let DARPA keep an eye on its work.

* The one that ended up crashing through the shed roof.

Over the next couple of years, the Defense Department tasked Worden with monitoring SpaceX's operations on Kwajalein. Now and again, Worden would make the long trek from California to Hawaii and then to Kwajalein Island and Omelek and report back on what he observed. Much of SpaceX's approach to rocket building appealed to Worden. He liked that the company had kept its team lean. He liked the energy of the employees and their ingenuity under difficult conditions. He was less impressed, however, with what he perceived as a general lack of rigor in their operations. The SpaceX crew did not seem to document any of their procedures. They had not set up a steady chain of supplies but were reliant on infrequent cargo ships and on Musk's private jet for emergency deliveries of crucial parts. Even for an unconventional military man who enjoys a good chat over a good amount of scotch, Worden found the amount of drinking taking place around the launchpad concerning.

"I'm watching these kids with tennis shoes fiddling with the rocket and crawling all over it," Worden said. "I went over to the little sleeping trailers and opened up a closet door, and there were a couple cases of beer. I love beer, but not when you're trying to launch rockets. People were telling jokes over the mission control communications feed. It reminded me of a bunch of Silicon Valley kids doing software. That is fine because if the software doesn't work when you compile it, you start again, and it doesn't cost you anything. But with a rocket it's millions of dollars and takes six months. The devil is in the details, but so is salvation."

Worden's worries were conveyed to the Defense Department and to Musk, who did not care for the criticisms. "Elon said, 'You're an astronomer. You don't build rockets,'" Worden recalled. "I said, 'I'm not criticizing your propulsion technology or designs. But I've spent billions of dollars on this stuff as an air force officer and observed a few things about operations. I predict you will fail.'"

After wrecking one rocket and then another and then another, SpaceX began to implement some of the things Worden and others had suggested. The young version of SpaceX had no intention of embracing all of Old Space's ways and baggage. That would defeat the whole point. But the company's engineers and mission control crew knew that the operations could be improved, and not long after a dash of professionalism was added, the Falcon 1 made it to orbit.

Even before the Falcon 1 flew into space, Worden had seen enough to know that a revolution had begun. He'd spent years banging on the military complex to think different and to piggyback on the glorious, constant improvements of consumer technology. Now it was obvious that the people who had designed our modern computers and software were intent on moving into aerospace and showing up the bureaucrats. Yes, the Silicon Valley types could perhaps be overconfident and cavalier at times, but they came with ambition, original thinking, and loads of money. Worden, then in his late fifties, thought that, like Musk, he could be an agent of change and play a major role in the movement as something of a go-between linking two worlds: Old Space and New Space. All he needed to do was find a place where he could combine his deep knowledge of space and the inner workings of the government with Silicon Valley's speed and drive. As luck would have it, a job opened up at just such an idyllic location.

CHAPTER THREE

WELCOME, LORD VADER

New arrivals in Silicon Valley are usually disappointed. You leave San Francisco and head south, hoping to discover a tech wonderland. Bright lights. Antigravity trains that travel at nine hundred miles per hour. Gleaming, futuristic skyscrapers and awe-inspiring monuments that celebrate the immense wealth generated in the area over the past sixty years. But no. It's strip malls, low-slung office buildings, and neighborhoods full of dated houses dotting dated roads. The most sci-fi touch may be the abundance of self-driving Teslas inching forward through freeway traffic.

Beyond not celebrating its present, Silicon Valley does not celebrate its past. It has no time for history. When a tech giant falls from its position of dominance, a new company simply takes over its buildings and begins the cycle anew with no homage paid to the predecessor. The first transistor factory*—the birthplace of the whole damned tech revolution—was turned into a fruit and vegetable shop years ago, and then that was ripped down to make way for yet another office building.

---

* Shockley Semiconductor Laboratory at 391 San Antonio Road in Mountain View. If you're in the neighborhood, go to Chef Chu's.

To the extent that Silicon Valley does have an iconic temple to its past and present, it's NASA's Ames Research Center, located in Mountain View. Anyone driving past it on Highway 101, Silicon Valley's main artery, will notice the unusual two-thousand-acre complex. It begins at the water's edge, where a series of enormous hangars and a long runway emerge from the flat marshlands and salt ponds of San Francisco Bay. Dozens of office buildings spread across the grounds, but they are far from the main attractions. The largest wind tunnel ever built—1,400 feet long and 180 feet high—dominates the site with its giant trapezoidal form looking like the nerd equivalent of a pyramid at Giza. There are multistory flight simulators, quantum computing centers, and all manner of giant metallic orbs used to store fluids and gases for experiments. Viewed from above in its entirety, Ames appears to have a split personality where a stodgy government compound meets a Bond villain's evil lair.

The history of Ames dates back to the late 1930s, when Charles Lindbergh advised the US government to set up a new national aeronautics research center on the West Coast. At the time, the south Bay Area was concentrated far more on agriculture than on technology. It was known as the Valley of Heart's Delight for all of its fruit and nut orchards. "There were strawberry fields for miles and miles and miles," said Jack Boyd, an aeronautical engineer, who arrived at Ames in 1947 at a starting salary of $2,644 per year.

As the United States exited World War II and then made its way into the Space Race with the Soviets, Ames started booming. The center had put an emphasis on building wind tunnels, and they became key to research on subsonic and supersonic planes. Next the Ames engineers came up with a series of technological advances that made the Apollo, Mercury, and Gemini missions possible. With semiconductors still in their infancy, Ames represented the pinnacle of the Bay Area's high-tech efforts and had no problem attracting talent. "The average age of the employees was about twenty-nine," Boyd said. "And the people here were among the most brilliant in the world. It was exciting. You felt like you could almost do anything." The glory days for Ames continued through the 1980s.

Ironically, however, the South Bay tech industry that Ames helped nurture eventually started to drain the center of its talent. Graduates of Stanford and the University of California, Berkeley, saw working at semiconductor, PC, and software start-ups as more exciting, and certainly more lucrative, than heading to NASA. The shift was exacerbated by the Space Race slowing down during the waning days of the Cold War. What's more, Ames had been outmaneuvered politically by other NASA centers, namely the Jet Propulsion Laboratory (JPL) in Southern California. As JPL's budget increased alongside its share of the interesting new work, Ames went in the opposite direction. By 2006, as Musk and his SpaceX team were getting ready to fire off the first Falcon 1, things had gotten so bad that NASA began to consider whether it should shut the once glorious Ames down.

The NASA director at the time was Michael Griffin, who had known Worden for decades. Both men had experience working on the Strategic Defense Initiative (SDI) during the Ronald Reagan years. Dubbed "Star Wars," SDI proposed to fill space with a variety of futuristic weapons designed to take down enemy missiles before they could reach the United States. Like Worden, Griffin had a fondness for doing weird stuff in space and an appreciation of the Silicon Valley start-up ethos. Griffin began to think that perhaps Worden could shake up Ames and bring some New Space, tech-friendly vibes to the place, so he tapped him as Ames director in May 2006. "Mike said, 'I want you to come and work for me, but you can't keep making those statements about NASA out loud,'" Worden said.

"Those statements" referred to Worden's long history of publicly bashing NASA. Among other things, he'd once given a talk about NASA entitled "On Self-Licking Ice Cream Cones." The major argument in the talk was that somewhere along the way NASA had lost its focus on advancing the United States' space capabilities in favor of becoming a bureaucratic jobs program. Powerful politicians, Worden said, had taken over NASA by lobbying for expensive efforts such as the development of the space shuttle and the Hubble Space Telescope to take place in their states and blocking rival attempts to do things cheaper and faster.

A self-licking ice cream cone,[*] then, serves no other purpose than to keep itself alive, and Worden found NASA to be perhaps the worst example of this self-perpetuating lifestyle choice. What's as enlightening as the rhetoric, though, was that even back in 1992, Worden had such a clear picture of where space technology should and would go. He'd been calling for investments in cheaper rockets and small, cheap satellites. He had also urged NASA to move away from its sole focus on costly long-term science missions and to consider putting up a series of rapid-fire experiments in space to measure things like climate change before it was too late. Needless to say, NASA ignored his advice.

When Worden took over Ames, he agreed to temper his public tongue-lashings of NASA. He did not, however, have any intention of operating the center in the same fashion as his peers. Worden had twenty-five years of pent-up frustration to unleash upon the space agency. Now he had a chance not only to tell the people there how to behave but also to show them. His performance would end up being legendary.

IT'S AT THIS PART OF the story that the unfolding of the great Ames shake-up is meant to begin. And we will get there shortly, I promise. But before you learn about what Worden did and why he did it, you need to learn about Worden and how such a man came to be. For as much as the Falcon 1 was the inciting incident to the rise of commercial space, Worden played the role of the prime instigator behind the scenes, always nudging and agitating against the status quo and opening up new opportunities for people who thought differently.

Worden was born in Michigan in 1949 and grew up in the suburbs of Detroit. His mother taught school, and his father worked as a commercial pilot. When he was fourteen, his mother died of cancer. Since his father was often away from home for long stretches of time and since Worden was an only child, he found himself quite alone. "I

[*] The internet appears to give Worden credit for originating the phrase, although Worden said he had learned it from his dad, who, in turn, said that the army air corps had used it during World War II.

didn't have any friends," he said. "I was probably overly demanding of attention. Part of that was probably being an only child, and part of it was just my personality."

Astronomy fascinated Worden right from the start. The first two books his mother purchased for him were entitled *Stars* and *Planets*, and he devoured them. He kept right on reading everything he could find on the subject and plowed through plenty of science fiction books, too. Near the height of the Apollo program, he enrolled at the University of Michigan and obtained a dual degree in physics and astronomy. He also signed up for the Air Force Reserve Officer Training Corps (AFROTC) because it let him dodge the school's gym requirement. "I hated athletic crap," he said. "And in the ROTC building, they had a pamphlet with a picture of the galaxy on the front of it that talked about the Air Force Office of Aerospace Research. That persuaded me to get involved and become a science officer."

After Michigan, Worden joined the air force and went to the University of Arizona to earn a doctorate in astronomy. There he turned into a hot recruit within various air force divisions. One job offer, at the National Reconnaissance Office, proved too good to pass up. "It was exciting because I went to Los Angeles for the interview, and it was like something out of *Get Smart*," he said. "You go in, and there's these multiple vault doors with codes, and then they turn these flashing lights on and there's a siren that announces a person without clearance has come in, and the colonel wouldn't tell you exactly what they were doing, but he said, 'Trust me. It's cool.'"

The NRO spies on things, and Worden began working on covert projects that he still can't reveal. What's clear, however, is that he did well and played a role in overseeing efforts with multimillion-dollar budgets. He also built valuable relationships throughout the military and with key players in Washington, DC.

Worden's quick rise through the ranks became obvious in 1983 when President Reagan unveiled the Star Wars program, or the Strategic Defense Initiative (SDI), via a speech from the Oval Office. The man tapped to lead Star Wars was Air Force General James

Abrahamson, who had been running the space shuttle program, and Worden became his special assistant.

Many, many people considered Star Wars a crazy idea. It depended on a wide array of technology that did not yet exist. Sophisticated radar systems on Earth would communicate with a network of satellites to spot and analyze Soviet missiles just as they were launched and then track their flight paths. Lasers on the ground were meant to shoot up to mirrors in space, which would reflect the lasers to other mirrors, which would then fry the Soviet missiles. If the lasers weren't enough, missiles orbiting in space would spring into action and do the trick. Neutral particle beams were involved, too, because they sounded intimidating and cool—that is, unless you were the Soviets or not a nuclear war enthusiast. Because then the Strategic Defense Initiative and its expansive catalogue of space mayhem sounded horrifying.

Unlike such worrywarts, Worden was a Star Wars fanatic. The arms race, as Worden described it, had turned into a chess match that favored the Soviets' strategic nature. They had "frozen the rules" through arms control agreements and were comfortable in their position. Americans, by contrast, he said, have the disposition of poker players, "who can do really well if you change the rules of the game." Star Wars would change the rules of engagement by creating a new battlefield with new weapons—in space. "That was our poker advantage," Worden said. "You know—lasers are wild this week."

Part of being a special assistant entailed defending Star Wars to its varied critics. Worden went up against politicians, scientists, and the Soviets in public and private forums. He excelled at the job and was promoted again. "I have an enthusiasm for controversy," Worden said. He also fully believed in the cause but mostly on pure space nerd grounds. "I viewed this as a major military space initiative," he said. "I felt then and still do that focusing the military on space will really advance our future in space. That was my major interest, more so than the missile defense aspects."

By 1991, Worden had been promoted to colonel and took on a new role within the SDI effort, this time as the head of technology with a $2-billion-per-year budget at his disposal. He opted to aim a large chunk of the money at advancing SDI while also pursuing his love of space. He concocted the Clementine mission, which would send a spacecraft into orbit around the moon for two months to test out a suite of SDI sensors and map the entire lunar surface. The United States had not funded a moon mission in twenty years, and Clementine proved to be a smashing success. The spacecraft, developed by NASA and the SDI team, cost about one-fifth of other similar probes because Worden had stressed the use of commercial software and hardware to build it. The machine not only produced a detailed map of the moon's surface from pole to pole but also provided some of the first evidence of water in lunar craters.

Along with Clementine, Worden helped fund the Delta Clipper, a reusable launch vehicle that could perform vertical takeoffs and landings. "I'm always amused that Elon and Bezos say they're the first ones to do the reusable stuff," Worden said. "Well, rubbish. We did it twenty-five years ago."

SDI would eventually be canceled in 1993. Its critics maintained that the technology to build such a complex system had never really been feasible, although Worden and others close to the project fully believed it could be done. Their experiments with reusable launch systems and other successful tests with missile interceptors proved as much, he said. But whether it would work or not, the mere existence of SDI arguably helped bring about the end of the Soviet Union by forcing Russia to spend money it did not have on combating a futuristic, ambiguous defense system. "It turned out that what we could afford was a lot more than what the Soviets could, which we did not know at the time," Worden said. "We did enough experiments to show that this thing was actually moving forward."*

---

* A number of generals had tried to convince Reagan to cancel Star Wars as a type of peace offering to the Soviet Union, arguing that it would aid arms negotiations around banning further nuclear weapons development. Worden and other SDI advocates, however, felt that the program provided much-needed leverage. To that end, Worden wrote a paper that outlined the technical feasibility of Star Wars and placed it in *National Review*, which Reagan always read cover to cover. Worden is convinced that the paper helped sway Reagan toward backing SDI during the Reykjavík summit in 1986 with Soviet premier Mikhail Gorbachev. "It was my major contribution to ending the Cold War," he said.

Once the Soviets realized that they were unable to stop the program, Worden believed, it was pretty clear that they would have to take a radically different path. "It opened up a lot of seams in the Soviet power structure and really led to the ultimate collapse," he said. "I know that a lot of the people who hated the Star Wars program doubted that it had much of an impact. But I think it was *the* key. And it fundamentally changed the game on a strategic level from a nuclear missile competition to a nonnuclear space competition."

According to Worden, SDI fostered the United States' advance beyond the exploration of space and toward the application of technology in space. Along the way, it shifted a large amount of space technology research away from NASA and to a new class of people and companies who were trying out novel ideas. "The SDI program was a huge impetus of cash and focus that enabled us to begin to do what we are doing today with new technologies and different approaches," he said. "It was sort of the start of a new space movement that was outside of the traditional aerospace companies. In that sense, it was probably one of the most phenomenally successful defense expenditures ever. We never had to build anything, and we changed the game."

In the years that followed SDI's demise, Worden had a dizzying number of jobs that came with truly wonderful job titles. He was a deputy for battlespace dominance; the director of analysis and engineering at the Space Warfare Center; the commander of the 50th Space Wing. At various points, he controlled fleets of satellites and managed thousands of people. He impressed many folks at the air force and in the highest echelons of the government. He infuriated others. NASA tried to have him fired at least three times.

After the terrorist attacks occurred in 2001, the government tapped Worden's special talents for the ill-fated Office of Strategic Influence. As he described it, "There are two kinds of generals or admirals in the Pentagon. The majority of them are bureaucrats, which is useful most of the time because most of the time we're at peace. But when you go to war, you want somebody that is going to change the game on the enemy and do something radical or different. The military knows this, so they keep a few people like this around. I was apparently one of these people. You know, 'In case of war, break this pane of glass for the crazy person.'"

To this day, Worden has maintained that the information warfare program may have been the United States' best attempt at fending off terrorist threats. "The terrorists have to be persuaded that they don't want to do terror," he said. "You aren't going to kill all of them."

But, he added, "People accused us of doing disinformation. That was rubbish, but it caused enough controversy. So I had the honor of being fired by the president. I got out of the air force as a brigadier general. Not too bad."

IN OCTOBER 2002, HOUSTON, TEXAS, played host to the annual International Astronautical Congress. The event dated back to the 1950s and brought together many of the leading space figures for the typical conference fare of speeches and seminars.

Fresh off being fired by President George W. Bush, Worden decided to attend the event and update his space contacts. He sat in on some sessions, made small talk, and then, when the evening rolled around, went to a bar with a friend. As he sipped a scotch, Worden could make out some of the loud chatter coming from a nearby table. It was full of space buffs who were part of the Space Generation Advisory Council, an organization of students and young people working in the aerospace industry who sought to "employ the creativity and vigor of youth in advancing humanity through the peaceful uses of space." In other words, space hippies.

True to form, those youthful space nerds were in a bar deep in the heart of Texas, decrying the very concept of space warfare. "I could hear a lot of statements about the evils of space weapons," said Worden. Thing is, Worden likes a good fight, so his friend went over to the group and asked if any of them would like to meet Brigadier General Pete Worden, former SDI executive, former deputy for battlespace dominance, bringer of orbital doom. "My friend said,

'Darth Vader himself is sitting over there if you would like to have a real chat,'" Worden said.

Among the young people at the event were Will Marshall, Chris Boshuizen, and Robbie Schingler, the future founders of Planet Labs, and George Whitesides, who would go on to run the space tourism outfit Virgin Galactic. Some of these men began to form a semicircle around Worden as he held court, and then Worden and Marshall really got into it. They were both space wonks and shared common ground around their desire to settle the moon and advance humankind deeper into the solar system. The space weapons, though, were an area where Worden and Marshall fiercely disagreed. Marshall made his youthful case for peace throughout the universe, and Worden perceived him as naive. "We both deployed the standard arguments, and it quickly became clear to me that Will had this view that keeping space pristine from space weapons would allow for a new utopia," Worden said. "And I had the crusty old military officer's view that there's usually a lot of evil in utopias." He tried to convince Marshall that military systems were not just weapons. They were a means of influence and control, and they needed to be viewed with a level of nuance.

The conversation was animated and contentious but ended with no hurt feelings. Quite the opposite. Worden had taken a liking to the idealistic youngsters, and they enjoyed his intellect and stories. Marshall and a few of the other students stayed in touch with Worden over the next couple of years. They would meet up in Washington, DC, California, or wherever a space event was taking place, dive right back into their space dialogues, and end up in a more informed place through their conversations. "Whether in business, politics, science, or whatever, the best thing to do is find somebody who is smart and disagrees with you," Worden said.

The initial meeting at the bar and those subsequent gatherings would prove fortuitous for just about everyone involved. By the time Worden took over Ames in 2006, most of the members of that band of space hippies had graduated from university and were looking to advance their careers with interesting space work. At the same time,

Worden had concocted a grand vision for how he wanted to revamp Ames. He thought some fresh blood would inject new life into the center and shake up its operations. One by one, he began reaching out to idealists such as Marshall, Boshuizen, and Schingler, coaxing them to head west.

It was no easy sell. NASA still had appeal and evoked positive feelings among the wider public, but it did not call to many young engineers. Like most of the government-backed space agencies around the world, NASA had a time capsule quality to it. Its people did things the same way they had for decades, and the status quo trumped all. Though NASA should have been the home of cutting-edge technology and radical thinking, it most certainly was not. It was slow and bureaucratic and behaved far more like a military contractor than a plucky home of scientists charging after the final frontier.

Worden, though, could be charming and persuasive, and he convinced the youngsters that they could become part of something bigger than working for yet another technology start-up. As time would prove, Worden's crew would indeed go on to form a new core of irreverence and innovation at the NASA center. In the years that followed, the group would deepen their friendships and accomplish more than Worden could have ever expected. They would move well beyond the confines of Ames to create a wave of pioneering satellite and rocket companies.

As he settled into the new job, Worden also began recruiting a handful of the freer thinkers from his air force days. Pete Klupar, who had worked on helicopters, planes, and satellites at the Air Force Research Laboratory, became the Ames director of engineering. Alan Weston, a weapons designer for the SDI, became a director of programs. Worden also managed to ferret out the few odd souls who had been at Ames for years and were still looking to do big things, having not been fully beaten down by bureaucracy and dysfunction. One such figure was Creon Levit, a researcher with a mess of unruly curly hair, who had already been at Ames for twenty-five years and then signed on as Worden's special assistant. "Pete had collected all these people young and old and convinced them that they could transform the place," Levit said.

Worden inherited a number of troubled programs and thousands of civil servants, many of whom had fallen into a state of complacency that the general found intolerable. He also inherited the ill will of higher-ups at Ames who were appalled that they'd been passed over for the director's job. Right away, Worden imposed his will on the situation. Levit recalled that he was very businesslike: "If an answer was not forthcoming on something like a deep question about a contract, he would demand a full report on the situation by the next week. He was very experienced at running an organization. That was obvious to me. He would start quiet and soft and then get angry. He would say, 'Goddammit, I am the center director. Don't tell me *No*. Tell me, *Yes, if*.'"

Worden was a general used to giving orders, but this was the civil service. "There was a lot of resentment and uncertainty around his authoritarian attitude and his frustration with the obstinate slowness and needless delays," said Levit. "And he had brought in all these new people and hired them on top of a lot of other people. Eventually, what sort of happened was that the place bifurcated. He was saying all this stuff about 'Ames is going to be a space flight center, we're going to do small satellites, we're going to revolutionize the way we do things at NASA, we're going to fix these busted programs and streamline things and allow people to get new things done.' So it kind of split into the people who loved Pete and the people who hated Pete."

Worden was fifty-seven when he arrived at Ames, and his look and mannerisms were well baked in after all those years in the military. He kept his graying hair short on the sides and longer on the top, parted at the right. Never big on exercise, he had developed a paunch on his average-sized frame. His most defining characteristics were his voice and his countenance. When he talked, the words seemed to originate from inside a well deep in his throat and to be pushed out with considerable effort, giving them a gravelly moderated tone. As for his face, well, he displayed a look of constant mild consternation. The whole package presented as gruff and grizzled—although if you got him talking about something or someone he liked, warmth and enthusiasm came out in generous volumes.

While it's true that a huge swath of Ames took an immediate dislike to Worden, many others of the 2,500 or so employees were pleased with his arrival. Even though it had hit a rough patch, Ames still had many top-class researchers who, like Worden, liked to go against the grain. "The knock on Ames was that it was packed with people who were not team players," Worden said. "Well, I think the idea of team players is just the most disgusting thing because it's a way of saying 'Don't do anything. Be a team player.' Ames was full of people who wanted to do things and were kind of troublemakers. There needs to be someplace within NASA that runs on the motto of, you know, 'Proceed until apprehended.' And that's what we did."

One of the first projects Worden hoped to tackle was an effort to send a robot to the moon. President George W. Bush had tasked NASA with retiring the space shuttle by 2010 and then coming up with a successor to the shuttle that could conduct crewed missions to the moon by 2020. Before humans stepped onto the lunar surface, robotic probes would perform scouting missions, and Worden wanted Ames to build some, if not all, of the probes. He proposed applying some of the low-cost development techniques being pioneered by SpaceX and others to build the cheapest robotic machines in NASA's history.

Right away, though, the bureaucracy of Old Space and government cronyism derailed Worden's plans. Richard Shelby, a powerful senator from Alabama, caught wind of Worden's ambitions. In addition to having NASA's Marshall Space Flight Center in his home state, Shelby had long been beholden to companies such as Lockheed Martin and Boeing. Since those companies have factories in Alabama, Shelby had fought to thwart just about every New Space–type project out of fear that his cronies would lose out on valuable contracts, and he did this despite their continued lethargy, greed, and incompetence.*

* For example, he was one of SpaceX's main detractors in Congress.

In the case of the lunar mission, Shelby called Griffin and demanded that he take Ames off the robotics project and direct the work to the usual suspects. "Shelby is the worst pork-barrel pig in Congress," Worden said. "The first thing he did was steal the project away from me. To my mind, that was one of the most disappointing and disgusting examples of bad government."

Just a few months into the job at Ames, Worden almost got fired—for the first time—as he publicly railed against Shelby and against NASA's capitulation. True to form, however, he did not accept the decision and began a secret program within Ames to prove just how wrong the bureaucrats were.

Instead of making a small robot that would poke or prod the moon's surface for a few moments, Worden greenlit the creation of an entire lunar lander. Historically, such a machine would have cost hundreds of millions or billions of dollars if done to NASA's standards. Worden, by contrast, wanted to prove that a lunar lander could be made for, say, $20 million and still do stunning work.

To lead the project, he tapped Alan Weston. That is to say, the legendary Alan Weston.

An Australian by blood, Weston had traveled the world with his parents before ending up at Oxford University to study engineering. While studying there in the 1970s, he became one of the main members of the Dangerous Sports Club. This group of drunken, often drug-addled adventurers performed a wide range of stunts varying from hurling themselves through the air with a trebuchet to crossing the English Channel while riding on a pink inflatable kangaroo held aloft by giant helium balloons.

In 1979, the club invented the modern form of bungee jumping after Weston ran some computer simulations that suggested a human would likely survive a leap from a bridge while tethered to a stretchy cord. He confirmed the theory correct by leaping off a bridge in England and then went to San Francisco and hurled himself off the Golden Gate Bridge.* To avoid being arrested, he slid

* His sisters tried to stop the proceedings by calling the police and telling them that Weston intended to commit suicide.

down an extra line tied around his chest and into a waiting boat, which sped him to shore and a getaway car.

That very same Alan Weston somehow ended up in the US Air Force as a weapons designer—and somehow ended up working on Star Wars, where he had the job of making a space-based projectile system that could fire in multiple directions and annihilate many Soviet missiles and decoy warheads at the same time.

That research, oddly enough, came to serve as the basis of the low-cost lunar lander. While in the air force, Weston and his team built a trash bin–sized weapon system that had its own thrusters for maneuvering in space. If all the destruction bits and pieces were stripped away, the resulting machine had much of the skeletal structure for something that would be needed to land on the moon and deposit a small rover that could travel over the lunar surface.

Weston recruited Will Marshall and several other people to the top secret project. They took over an old paint shop at Ames and turned it into their skunkworks lair.* The machine they were building was about four feet across and three feet high and had a trapezoidal shape. The plan was to equip it with a thruster on the bottom to handle the landing and thrusters on the sides to make small adjustments to its flight path. In the workshop, the engineers built a large enclosure of netting material so that they could perform tests on the device without its flying out of control into the walls or into mostly innocent bystanders. By 2008, great progress had been made with the Micro Lunar Lander, as it became known. The machine could take off, stay aloft in a violent hovering motion, and then land safely and smoothly on the ground.

The lander became a celebrity at Ames. People were amazed at how much the small team had accomplished in such a short time. Google cofounders Larry Page and Sergey Brin came by to see it, at Marshall's invitation, and so, too, did some astronauts, officials, and other old buddies of Worden. Weston, Marshall, and a couple of other lead engineers on the project wrote a paper about the machine, sug-

* They even named it Area 51.

gesting that the small, cheap lander could accomplish most of the research jobs NASA needed to do ahead of astronauts setting foot on the moon in 2020.

The Ames team had built the lander prototype for less than $3 million, including all of the materials and personnel salaries. They thought it would cost a bit more to perfect the machine and that it could be flown to the moon for a total of $40 million, including the cost of the rocket launch, as they hoped to hitch a ride on SpaceX's Falcon 1.* "Suddenly it was possible to do these sorts of missions at such a low cost," Marshall said. "It was really cool."

With the prototype working, Worden and Weston decided to reveal their secret project to NASA. Surely, they thought, the space agency would be thrilled to learn about this wonderful low-cost lander that could potentially revolutionize the United States' lunar research and set the country up for a huge success with the crewed missions to come.

NASA and dear old Senator Shelby, however, had quite the opposite reaction. The science teams at NASA dismissed the lander as incapable of being able to do real experiments. The exploration teams said that they should have been in charge of the effort. And Shelby was outraged to discover that Ames, and not the Marshall Space Flight Center in Huntsville, Alabama, had been doing lunar lander development. "Once Shelby found out about it, it was terminated for the second time," Worden said. "I was fit to be tied."

Thankfully, one of the people Weston showed the lander to was Alan Stern, the NASA executive in charge of science, who controlled a $4.4 billion budget. He proposed that the team change the vehicle from a lander to something that would orbit the moon and conduct experiments. Since it would only go around the moon and not actually touch the surface, Shelby figured that Marshall Space Flight Center's lander projects would be safe and eased up on his fury at Ames. And so the Ames machine got a new mission and a new name:

* That was a few months before the Falcon 1 made it into orbit in 2008, but the Ames team were hopeful that SpaceX would succeed.

Lunar Atmosphere and Dust Environment Explorer, or LADEE. Stern doubted that Ames could build such a thing for $40 million, so he doubled the budget.

The blessing was that the work could go on; the curse was that the $80 million check from Stern arrived with all of NASA's bureaucracy attached. "I'll never forget," Marshall said. "I gained twelve management layers overnight." A team of about a dozen people swelled as safety and management types showed up to support what had become an official NASA mission. Then the technical requests began to swell, too. At first, NASA wanted a machine that could orbit the moon while carrying an ultraviolet spectrometer. Then it also wanted the machine to carry a dust detector. Then it wanted a mass spectrometer, which weighed twenty-five pounds and would require Ames to make its machine much bigger, redo all of its calculations and tests, and purchase a ride on a larger rocket. "We said, 'No, we really can't take that,'" Marshall said. "And they said, 'Well, if you don't, we'll cancel the mission.' And we're like 'Oh, you fuckers.'"

The process continued in various forms for years until LADEE eventually made its way to the moon in 2013 at a cost of $280 million. For NASA it was still a bargain basement price, but it undercut much of what Worden and his crew had hoped to prove. "It's not that we didn't care about the science, because we did," Marshall said. "Still, we wanted to show that you could do a bunch of missions for forty million dollars each instead of needing a billion for every flight. The whole thing flew in the face of that technology demonstration idea that we started with."

Even with Senator Shelby on its side, Marshall Space Flight Center would watch in the years to come as all of its lunar missions were canceled. Ames, meanwhile, flew a second mission called Lunar Crater Observation and Sensing Satellite (LCROSS), in which an old rocket body was slammed into the lunar surface and then followed by a small spacecraft that analyzed the debris. That mission, which Will Marshall also worked on, confirmed not only that the moon had water but that it had far more water than scientists previously expected. In many ways, LCROSS rejuvenated interest in the moon and hopes of setting up lunar colonies.

AS THE YEARS TICKED BY, Worden refashioned NASA Ames into one of Silicon Valley's scientific hotspots. The center obviously focused on its near-term space missions and its tradition of helping with the design and testing of aircraft. It also, however, began to align more futuristic work with Worden's long-held fascination with the human exploration of deep space.

Worden set up a series of new research labs at Ames. He created a synthetic biology center that allowed NASA scientists to toy with DNA and see what it might take to send custom microbes and bacteria into space. In conjunction with Google, Ames also built a quantum computing center, hoping to foster breakthroughs in fields such as artificial intelligence. Ames, Worden thought, might one day be able to send smart, self-replicating biological machines to other worlds and have the organisms form their own settlements.

The computing partnership with Google represented another front in Worden's Ames revival, namely developing stronger ties with Silicon Valley. Google's headquarters in Mountain View are next door to Ames, and Worden managed to strike a deal in which the company's top executives would pay to land and house their jets on the Ames runway and to expand Google's campus onto Ames by leasing forty acres of land for $146 million over forty years. Ames also opened up some of its unused buildings to start-ups. For a relatively modest fee, young companies could acquire office space and, as needed, team up with NASA on projects. In a similar vein, Worden seized on opportunities to codevelop technology based on NASA inventions with the likes of SpaceX and other budding commercial space players. At one point, he even tried to bring the Tesla car production facility to Ames, but the NASA lawyers couldn't quite figure out how to match the sweetheart deals other locales were offering without committing a federal crime.

One of the key people Worden enlisted to butter up the Silicon Valley elite was Chris Kemp, who started working at Ames in 2006 and would later cofound a rocket company called Astra. Kemp hailed from Alabama and had been running computing and internet businesses since he was a teenager. Just before joining Ames, he'd spent six years as the founder and CEO of a travel website called Escapia that assisted people with renting out their vacation homes. An entrepreneur to the core, Kemp was also a space nerd and, through an odd set of twists and turns, had met Will Marshall years earlier and fallen in with the space hippie crew.

Kemp, then in his early twenties, had run into Worden while hanging out with Marshall and Robbie Schingler at a space conference in Los Angeles. Mike Griffin had just appointed Worden as the head of Ames, and Worden, in his most gruff-general way, was bemoaning the thought of trudging to LAX to grab a flight up north. "He's sitting at this table with us and says, 'Ugh, I don't want to get on a fucking plane. Does anyone have a car?'" Kemp said.

Kemp offered Worden a ride up to the Bay Area, caravanning with Marshall and Schingler along the California coast on Highway 1. For most of the ride, Worden lectured Kemp on his desire to settle the solar system. "He's going on and on about how we have a moral obligation to do this and how he's going to use Ames to chase after this goal," Kemp said. "This was fascinating to me. It was like a dream come true to be having a one-on-one conversation with this guy for hours. And near the end of the drive, he asks me what I'm doing. I said that I was running this travel company up in Seattle. He said, 'Well, that's bullshit. Why don't you come down to NASA and join me, and you can run an information technology department?'"

Although unconvinced by the offer, Kemp agreed to cancel his flight to Seattle and stick around at Ames for a few days. He remembered driving up to Ames for the first time to drop Worden off that night with armed security guards greeting the men at the entrance and lights flashing across the campus to scan for intruders. In the following days, Worden took Kemp to see a heat shield facility where plasma was raised to thousands of degrees and blasted into various materials, the largest supercomputer on the West Coast, and the multistory flight simulator. Marshall and Schingler joined them for parts of the tour, all three of them wondering why Worden was taking such care to show them around.

"We're all like 'What's the catch?'" recalled Kemp. "And the catch was just that he wanted to bring in people that would otherwise never have this opportunity and that we would be loyal to him as he took over the center. We didn't care about politics. We weren't looking for a GS-13 to GS-15 promotion.* We didn't even know what that was. Pete was just so inspirational. He told us to come in and do what we wanted."

Kemp accepted the job running a technology department at Ames and brought with him all the chaos Worden had hoped for.

As one of his first major actions, Kemp called an all-hands meeting for the department and informed the hundreds of tech staffers that they were underachieving. In fact, he thought the entire information technology department could be run better with about half as many bodies doing twice as much. When the employees began barking back at Kemp, he started playing a video of testimonials he'd gathered from Ames employees voicing their opinions of the center's technology operations. People complained about taking six months to receive a computer, only for it to take six months more to receive a monitor to go with it. The network was ancient. The software sucked. On and on it went.

"People got up and walked out of the room," Kemp said. "They're like 'Who the fuck are you to ask people how they feel about us?' And 'This isn't our fault.' I said, 'Yes, it is.' I'm about a third of the people in the room's age, and they see me as this insolent child leader who for some reason has been placed in charge of all this stuff. I didn't make a lot of friends."

As Kemp began firing people and cutting costs, he started to make enemies outside Ames as well. Ross Perot, the businessman and onetime presidential candidate, ran a technology services company that had huge contracts with NASA, and one afternoon, the

* These are designations related to the civil service pay scale.

billionaire summoned Kemp to his office in Plano, Texas, to yell at him in person. "He said, 'What's going on over there? You're spending half as much money. We milk $180 million out of you every year,'" Kemp said. "I replied, 'Yeah, not anymore, sir.'"

The more glamorous side of Kemp's job involved meeting influential Silicon Valley executives and offering them a glimpse into the shinier sides of NASA. He would, for example, take people like Google's CEO, Eric Schmidt, to Florida to watch space shuttle launches from the VIP suite at Cape Canaveral. Given that launches can be extended affairs, Kemp would have hours in which to chat with his target, trying to find some common ground where NASA and the technology companies could work together. Quite often, he managed to arrange deals in which the companies would pay NASA for its imagery or research or technology services. A couple of the tie-ups with Google came from exactly that type of shuttle schmoozing.

The trips also provided Worden with a venue to teach the then-naive Kemp valuable lessons about how the government works. Worden insisted that he and Kemp always drive their own rental car to the launch site rather than hopping into a chauffeured SUV alongside the tech execs. At dinners with rich executives, who had ordered $500 bottles of wine, Kemp made sure to grab the check first and pay the bill, always on his own dime. Worden had experienced too many congressional investigations to have accusations of bribes or even trivial exchanges of favors hang over him or his employees. "Pete also warned us that it may seem minor, but you never wanted to let your enemies get you on the details," Kemp said. Those practices would prove crucial to saving Worden, Marshall, and Kemp in the years to come when factions within NASA and the US government came after them.

Beyond hiring eccentric young engineers, Worden brought other flourishes to Ames that many people would find puzzling to discover on a NASA campus. Ames began an annual tradition of inviting the public to its campus for a space rave replete with science exhibits, art installations, electronica, dancing, and booze, and people showed up by the thousands to catch performances by the likes of Common and the Black Keys. "This was a NASA director saying 'We're all in. We're going to create a Woodstock type event but for space,'" said Alexander MacDonald, the chief economist at NASA, who worked at Ames at the time. In 2008, Ames also became the headquarters of Singularity University, an unconventional, borderline cultish school dedicated to celebrating and exploring the rapid advance of technology.

One of the more fortuitous turns of events that resulted from Singularity University's formation was the arrival at Ames of Chris Boshuizen, who would later go on to cofound Planet Labs. Boshuizen, like so many of Worden's twentysomething hires, had been part of the Space Generation Advisory Council, the collection of youthful space idealists. He had been present in Houston and heard of Worden's spat in the bar with Marshall; in later years, he morphed into an elder statesman and organizer of the group. As Worden looked to set up Singularity University at Ames, Marshall recalled Boshuizen's strong organizational skills and suggested that Worden recruit him.

Fresh off designing a space telescope for his PhD in physics at the University of Sydney, Boshuizen had never entertained the idea of working somewhere as bureaucratic as NASA—that is, until he heard Marshall and others discussing their jobs at Ames during Space Generation meetings. "Deep down, I just really didn't want to go to NASA," he said. "It took about a year of listening to Will for it really to sink in that it wasn't normal NASA or the normal way of doing things." When the Singularity University gig popped up, it certainly fit the not-normal NASA credo.

Boshuizen's arrival at Ames came with immediate signs that his life had taken a drastic turn. He'd just spent years toiling away as an academic, begging for funds to further his research. Suddenly he was palling around with people such as Larry Page and General Worden and in charge of building an entire university—of sorts—from scratch. Boshuizen dived into the project and in a matter of weeks had helped recruit people to the school, shape the curriculum, and figure out the logistics of where on Ames the classes would take place. One of the more daunting tasks he spearheaded was raising enough money to make Singularity University real. Worden suggested that $2.5 million should do the trick. Much to Boshuizen's shock, the money poured in during the first fundraising dinner, starting with a $500,000 contribution by Page.

Boshuizen had come to Silicon Valley in the hope of finding a slightly more vibrant version of NASA. As the months ticked by, however, he started to realize that something more dramatic was underway. All of the people whom Pete Worden—Darth Vader himself—had drawn to the Bay Area shared not only a love of space but also a romantic view of what life could be like here on Earth.

The technology industry, through its dot-com-era obsession with stock prices and acquiring users, had lost much of the counterculture luster that had driven it in the 1960s. Back then, people like Apple's future cofounders, Steve Jobs and Steve Wozniak, had been hacking the phone system to stick it to AT&T, while others were championing computers as the tools that would let "the people" take back power from the government and corporations. The new arrivals at Ames were, like everyone else, part of a much more commercial era of technology, but they also held on to many of the beliefs that had long made the Bay Area an epicenter of social revolutions. Weston, for example, engineered the world to keep up with his free spirit. Marshall was a borderline Communist, Kemp a part-time anarchist. And their friends such as Levit were chasing mind expansion by any means necessary.

"Pete got us all here and created something like a crusade," Boshuizen said. "There was that 1960s feeling that everyone had moved here from different parts of the country and the world to do big stuff. There was this weird movement, and it was really fun, man."

Part of the movement was tied to giving people more freedom to express their creativity and experiment. Another part, though, went deeper. Many of Worden's acolytes saw space as a canvas on which they could express their ideas about how society should operate and how humanity should evolve. Marshall would soon emerge as the frenetic, charismatic force that would carry what Worden started at Ames out into the real world and then into the heavens.

THOUGH WORDEN HAD THE BEST of intentions with the raves, Singularity University, and all the rest, he knew there would be repercussions for such projects. Many members of NASA's old guard considered holding a rave on the Ames campus to be abhorrent behavior. Some of these same people despised the clubby "cult of Pete" that had formed. There were complaints that Worden had given his young followers too much power, and some of the more patriotic Ames employees began to raise concerns about the number of foreigners such as Marshall and Boshuizen who had been brought onto sacred NASA grounds. Worden had a tendency to make the situation worse by flaunting his favoritism for certain people.

The Google deals also brought unwanted scrutiny upon Ames. Kemp was one of the key people who had helped Worden attract Google's plane fleet to the center and had finalized the partnerships that resulted in Google Moon and Google Mars. He and Worden were impressed with themselves for discovering a new revenue opportunity for NASA and thought such arrangements made perfect sense. NASA had tons of data but not always the means or will to organize it and make it useful. Google specialized in exactly those things. Why not let the public have the feeling of exploring the moon or traveling through a canyon on Mars in 3D?

But people at NASA criticized Kemp for playing favorites by doing the deal with just Google instead of a number of technology companies. When NASA finally allowed the projects to proceed, it wasn't even sure how to account for the money Google paid for the data. "I vividly remember going over to Google, meeting with Eric Schmidt, walking to their accounting department, and bringing a several-million-dollar check back to NASA," Kemp said. "They were like 'Wait, the money goes the other way. Do we give this to the

Treasury? What do we do with this thing?'" Members of the press were flummoxed as well. Surely big old bloated NASA was being taken advantage of by Google. What's more, taxpayers had funded the probes and telescopes that were taking all of these pictures. How could a public corporation then take those assets and turn them into consumer services, albeit free ones?

The coverage of Ames's arrangements with Google went far beyond the local and technology press. The late-night talk show host Jay Leno made a joke during his nightly monologue about using Google Moon to pull up directions to a Starbucks on the lunar surface. Sometimes the top executives at NASA appreciated coverage like that and Ames's sudden superstar status; sometimes they didn't. When Kemp struck the deal to turn Ames into Larry Page and Sergey Brin's private airport, for example, word of the arrangement leaked to the press before Kemp had notified the NASA higher-ups. "That story made the front page of the *New York Times*," Kemp said. "There was this picture of a giant 757. Pete got in my face, and I could taste his spit as he yelled, 'What the fuck have you done?' I told him that we'd made a good deal for taxpayers. I didn't get fired. He didn't get fired. And everybody came out in a better position."

In 2009, the criticisms of Worden and fractures within Ames reached a crisis point. Will Marshall had flown to Vienna for a space conference and returned to the San Francisco airport, at which point customs officers detained him. At first, Marshall thought it was all part of a random security check. The officers peppered him with questions and asked to look through his bags. After half an hour of prodding, however, Marshall began to realize that things were more serious, as he was led into a backroom for a full-on interrogation. "They started asking technical questions about the lunar missions we were working on," he said.

The officials demanded that Marshall hand over his laptop and give them his password. "On the one hand, NASA had told me never to give my password to anyone," he said. "On the other hand, there's a government official telling me to do it. I felt quite conflicted about the whole thing."

Al Weston, aka Worden's right-hand man for doing crazy shit, had come to the airport to pick up Marshall and wondered what was taking the voluble Brit so long to get out of the terminal. He kept phoning Marshall's cell phone, and finally, the officials allowed Marshall to take one of the calls. Weston, who was Marshall's boss at the time, directed him to turn over the password. "We never saw that laptop again," Marshall said. The questioning continued for another six hours until the officials finally let Marshall go.

It turned out that a faction within Ames had written a report accusing about two dozen people of being Chinese spies. Worden and Weston were on the list, and so, too, was Marshall.

Since the lunar lander had a Star Wars weapons propulsion system as its base, it fell under an export control policy called International Traffic in Arms Regulations (ITAR). ITAR is a fuzzy thing open to various levels of interpretation, but it's mostly meant to prevent anyone except for US citizens from looking at weapons systems or even seeing documents or photos related to them. The concern was that foreign nationals like Marshall were passing information about missile defense systems to nations such as China or at least being lax about how they guarded materials related to the lunar lander.

The report had come from some of the more patriotic members of the lunar lander team who found Marshall's loud, brash, idealistic ways appalling. Marshall would do things like hang a UN flag in the lunar lander workshop, while the team was meant to be building the machine for the glory of NASA and the US of A, not out of the goodness of their hearts for all nations. Through a superpatriot's lens, Marshall was either a careless space weirdo putting the nation's secrets at risk or, even worse, part of a suspect, sinister element that Worden had gone out of his way to bring to Ames.

The evidence against Marshall did not appear very damaging. He'd gone to China at one point to teach classes during summer sessions of the International Space University in Beijing. On his more recent flight to Vienna, he'd also taken the NASA laptop, which did contain sensitive information. Marshall had not realized that NASA personnel were meant to obtain permission to travel with their laptops.

Still, things escalated quickly and dramatically from that point. A group of Ames employees handed Congress a fifty-five-page report that, according to Worden, suggested the existence of a far-reaching conspiracy to destroy the US space program. Not only were Worden and his buddies allegedly involved in the conspiracy, but so, too, were President Barack Obama, Elon Musk, and Lori Garver, the deputy administrator of NASA, who had been a major advocate of SpaceX and private space exploration. Fueled by the document, the FBI kicked off an investigation that ended up taking four years and dragged Ames through the press as a bastion of spies.

For three months, the authorities banned Marshall from setting foot on the secure part of the Ames campus and using his work email. Marshall said, "I'll never forget asking my lawyer that I thought you were innocent until proven guilty. He said, 'Yeah, that is a common misconception.' I'm like 'What? I grew up believing this. If this is true, it's a big, fucking deal, and we should tell the whole fucking world.' He's like 'Yeah, that is a common misconception.' I thought to myself, 'Stop fucking saying that!'

"It's true that I didn't get the permission slip for the laptop and had broken that rule, but no one had ever even heard of these fucking permission slips. In any case, I had broken internal rules. I had not broken any law. And I wasn't trying to spy for China. I wasn't trying to spy for anyone, thank you very much."

It appeared that neither the FBI nor the US Attorney's Office was ever able to find evidence of wrongdoing. And although anything is possible, the idea that Worden, a brigadier general who spent much of life trying to protect the United States, had plotted to distribute Star Wars weapons plans to other countries was daft. "Fortunately, it finally managed to get some adult review, and a US attorney in San Francisco dismissed the whole thing," he said. "I found out later that I had been cleared early on, and even though they were still after a bunch of folks, the attorney eventually said, 'Look. There's nothing.'"

Worden was annoyed that Marshall had flown with the NASA laptop to another country. Marshall had received the same warnings that Worden provided to Kemp about following the rules so that your detractors couldn't use minor infractions to start larger controversies. Marshall, though, never cared much about paying attention to bureaucracy or niggling details. And just about everyone on Team Pete assumed that Worden's renegade actions at Ames were the true underlying cause of the investigation. Marshall had simply provided an avenue for people to vent their frustration in public.

"Like any change agent, pretty soon you trigger antibodies," Worden said. "There were probably a few dozen people at Ames who had decided I was the most evil person in the world and had to be stopped. A lot of them focused on Pete's Kids, of which Will was probably the most visible and not so attuned to regulations."

A final incident proved how silly the NASA and US government aversion to Worden had become. As a favor to an Ames employee who sidelined as a photographer, Worden and about a dozen other people dressed up like Vikings and staged a land assault in the marshy areas around the Ames campus. When the pictures of Worden, sword in hand, charging through mist to slay imagined marauders, appeared on the internet, they were enough to prompt Iowa senator Chuck Grassley to call for a federal investigation into the photo shoot. Grassley wanted to know how much federal time and taxpayer money had been wasted on the frivolous exercise. It turned out that everyone had volunteered their time and the shoot had been done on a weekend. The investigation cost more than $40,000.

Worden ended up running Ames from 2006 until 2015. During that span, he took a NASA center at death's door and turned it into the space agency's most famous—and occasionally most infamous—research facility. His actions helped Ames make use of Silicon Valley's most valuable assets: its people, technology, and wealth. They also stretched Ames into new, fruitful technological areas that would ensure the center played a prominent role in missions for decades to come. All the while, he was also a relentless advocate for reducing the cost of space missions and a champion of private space.

The sum total of Worden's often controversial positions ultimately led to the close of his time as Ames director. NASA had received too many complaints about Worden yelling at underperformers, and the union did not appreciate his trying to fire people who seemed unwilling to do their jobs. Some people thought he was still up to his old NASA-hating ways and that he leaked information to the press to make parts of the space agency look bad. Others were upset that he applauded the likes of SpaceX while criticizing NASA's decision to continue making its own absurdly expensive rockets. In the end, too many of the people on high who had protected Worden departed from NASA, which left him alone and politically vulnerable. Worden opted to retire and pursue a simpler life.

"One of my favorite quotes comes from Machiavelli, which is unfortunate because everybody hates him," Worden said. "It's paraphrased, but it says, 'The hardest thing to do is change the order of things because everybody that has something to gain will be a lukewarm supporter, and somebody that has something to lose is a vicious enemy.' You end up with a very difficult challenge. That's fundamentally why I think Elon and I are kindred spirits. He's probably done a lot more to change things, but we both know what it feels like to be a target."

Still, Worden had survived NASA's bureaucracy long enough not just to revamp an institution with built-in defense mechanisms against revamping but also to make an impact well beyond the center's walls. He had plucked smart, enthusiastic young people from all over the world, brought them together, given them a sense of purpose, and taught them how to overcome obstacles. He'd created a once-in-a-generation backdrop in which great ideas could flourish and the strongest of friendships could be formed. As we'll see, the youngsters he collected at Ames were fated to stick together, and they were destined to pick up where Musk had left off with the Falcon 1. Pete's Kids would go on to trigger the next major revolution in private space.

CHAPTER FOUR

THE RAINBOW MANSION

As Pete's Kids descended on Silicon Valley in 2006, they needed a place to live. Since several of them already knew one another, they began to entertain the idea of living together. Will Marshall, Robbie Schingler, and Jessy Kate Cowan-Sharp had been doing as much during a stint in Washington, DC, along with a few other friends. Why not, they figured, keep it going and bring their communal lifestyle to California?

Marshall took the lead on the project, poring over Craigslist ads in a bid to find a place near NASA Ames that could accommodate an unusually large number of housemates. He quickly stumbled on a listing for a very big house—a mansion, in fact—located at 21677 Rainbow Drive in the suburbs of Cupertino, aka Apple's hometown. The owners of the house had been involved in the tech business, but they'd moved elsewhere, and the property had sat idle for years in the wake of the dot-com bust. If an unusually large group of twentysomethings wanted to rent the place, so be it, so long as they were willing to put down the first month's rent, the last month's rent, and a security deposit for a grand total of $20,000.

At first blush, the lofty figure seemed ridiculous. Get ten or more people in the house, however, it started to match up well with the going rate for rent in the Bay Area. Cowan-Sharp had also been turning communal living into something of a science and refining her methods for finding roommates, divvying up rent, and splitting other costs. Best of all, Chris Kemp also wanted to live in the house and had already cashed out of a couple of start-ups. He had the money to cover the first payment and set things on their way.

The house sat on a hill overlooking Silicon Valley. It resembled a Mediterranean villa with its terra-cotta tiled roof and cream exterior. Anyone walking up to the front door would first notice a modest moat running through the front yard replete with a small drawbridge and koi cruising in the water underneath. On the inside, there was an enormous master bedroom that took up an entire wing of the house along with a few smaller bedrooms and living spaces. It seemed as though the owners had left everything in the house except their dishes. It had a piano, a bar with an acrylic countertop, some furniture, and a home theater with a projector. A pastel color theme ran throughout the house with lots of light pinks and light blues. Take everything in, and you ended up with a five-thousand-square-foot McMansion in a very nice tree-filled Silicon Valley suburb that would soon have an unusual set of rotating inhabitants.

Marshall served as the core around which the group assembled. He'd met Schingler and Cowan-Sharp when they were all college students taking part in the youth space meetings, and they'd become fast friends. (Cowan-Sharp would end up marrying Schingler in 2010 and taking his last name, so from here on out, they will be Robbie and Jessy Kate to avoid confusion.) Even though he grew up in England and went to school there, Marshall had also befriended Kemp years earlier during an internship in 1998 at the Marshall Space Flight Center in Alabama, where Kemp lived at the time. The two men had hit it off and would often go on hikes and field trips around Alabama, and Marshall had pulled Kemp into his space geek club in the years that followed. One of the other initial housemates was Kevin Parkin, another Brit who had started his undergraduate studies at University of Leicester alongside Marshall in 1996 and been recruited to Ames by Worden a decade later.

Robbie and Jessy Kate moved into the master bedroom. Marshall took over a Japanese-style tatami room with reconfigurable floors and walls. Kemp and Parkin went upstairs and ended up in what had formerly been kids' rooms that shared a bathroom. "It even had a whirlpool tub," Kemp said. "And it had a window that overlooked this Japanese garden behind the house." Not sure about the longevity of his Ames job, Kemp had left most of his possessions in an apartment in Seattle and was pleased to find that his room already had a bed and items such as closet organizers. "The room had built-in everything," Kemp said. "All I had to do was put my nerdy clothes away. It was great."*

Pete's Kids dubbed their new home "The Rainbow Mansion" because of its location on Rainbow Drive and set to work finding more people to help pay the bills. They decided to make use of a pair of bunk beds in one of the rooms and turn it into a youth hostel open to folks traveling through Silicon Valley or looking for a short-term place to stay. A couple other living areas were converted into bedrooms. The goal became to find ten or more people to live in the house at any given time.

The concept of a communal house was certainly not new, nor was the idea of communal living at all novel for the Bay Area. It would, however, turn out that the Rainbow Mansion revived the idea for the engineers and software developers flocking to Silicon Valley. In large part thanks to Jessy Kate, so-called hacker houses were about to become a thing. Some people were simply trying to deal with the Bay Area's ever-rising rents. Some wanted more of a sense of community. Others used the houses more as networking

* One friend described the Kemp of the Rainbow Mansion days as "Mr. Government Nerd." He'd come from the world of information technology and looked and dressed the part. Wire glasses, slightly pudgy, buzz cut. Very Microsoft. Still, he was said to always be scheming and planning to take over the world. He also built up a reputation among the group as a problem solver, the kind of person who could find his way out of any situation. When trapped in an elevator one time, for example, he punched out a panel in the ceiling, climbed up into the elevator shaft, pried open the second-floor door, and then started pulling his fellow passengers out to safety.

tools for their start-ups. By 2013, enough of those types of houses would appear to warrant a trend piece in the *New York Times* with the title "Bay Area Millennials Are Flocking to Communes—No Tie-Dye Required" in which the Rainbow Mansion and Jessy Kate made a prominent appearance.

The Rainbow Mansion of 2006, though, was different from the clones that followed. It had a hint of magic. At the heart of the house was a group of friends who would form bonds as tight as those of any family. They shared a love of space and also something more profound: they were united by idealism. Led by Marshall and the Schinglers, they genuinely believed they could change the world for the better and tried to infect anyone who passed through the Rainbow Mansion with the same spirit.

Outside the core group, an ever-changing cast of characters lived in the house. At any given moment, there might be a couple of people from NASA, someone from Apple or Google, and a couple more people working on their start-ups. Some of them lived in their own rooms, while others lived in the makeshift hostel that had been set up replete with bunk beds.

Many of the new Rainbow Mansion residents were drawn to the house by the unusual ads Marshall posted on Craigslist: "Seeking a driven, passionate, young female who wants to change the world" or "Roommate needed for an Intellectual Community." Instead of describing the house and its layout first, the ads jumped right into describing the people living there. They also included questions like "Imagine you came home on a Wednesday night to a group of 15 people having impromptu dinner in the library. How would you feel?" and "What are 2 things you'd like to impact or contribute to during your life?"

When new people came into the house, Marshall sometimes interrogated them about their work and life choices: What are you doing? Why are you doing that? Why are you doing it that way? What's the end purpose? Part of this stemmed from Marshall's insatiable curiosity. He simply wanted to learn from people and understand their way of thinking. Part of it, though, was a form of the Socratic method, with Marshall challenging people to contemplate how they spent their time. While he didn't intend for it to be a painful experience, he did manage to have at least one person crumble into the fetal position in the corner of a room after a grilling.

The spirit of the Rainbow Mansion aligned with what Pete's Kids were doing at NASA. At work, they were trying to shake up the space industry and put more power into the hands of individuals rather than governments and the military. At home, Jessy Kate, Robbie, and Marshall were proponents of new social structures. They viewed the house as an "intentional community" that was designed to spur discussions of big ideas that might alter the status quo and reorganize society for the better. The counterculture vibes of the 1960s had certainly been dulled by the engineers and investors who had descended upon the Bay Area, but the remnants of that era were alive to some extent within the house.

Marshall regarded communal living as more in keeping with humankind's tribal nature. "I love community living," he said. "It's a loving environment and one where there are lots of ideas flowing and where I could learn so much every week. My real question is why people want to be limited to the nuclear family unit or whatever you call it. It's a peculiar, very recent invention by humans and not a particularly smart one in my opinion."

People who lived in or just visited the Rainbow Mansion reveled in the communal energy. It felt like a permanent summer vacation. Almost every night, housemates and their guests gathered for family-style meals whipped up by whoever decided to take the lead that evening. Marshall had a special flair for transforming leftovers into new dishes and could somehow feed thirty people on a whim. At various times, the guests included heads of state, astronauts, scientists, billionaires, and inventors.

After the meals, people often collected in the mansion's large library, which was filled with books that ranged from philosophy to chemistry to home building and had a variety of colorful, eclectic paintings adorning its walls. Tea would be brewed. Scotch bottles would be opened. Discussions would break out on a wide array of topics, from the coming threats of artificial intelligence to the perils of space debris.

Somewhat more formally, the Rainbow Mansion functioned as a shared research-and-development lab for artistic projects and technology. It was common to walk into the house and find new art installations being hung up on the walls. The entryway, for example, once had a giant tetrahedron dangling from the ceiling that had been constructed from toilet paper and paper towel rolls. Many of the house members were part of the open-source software movement in which the underlying code of applications is shared so that people can use it and modify it as they wish. They hosted myriad hackathons, and coders would descend upon the mansion and take it over for an evening or an entire weekend.

Celestine Schnugg, a onetime Rainbow Mansion housemate (and now venture capitalist), used to sunbathe by the koi pond only to have her tanning sessions interrupted by hordes of engineers. "People would be carting in boxes and boxes of electrical equipment and lighting up the house," she said. "Hundreds of people would sprawl out and take it over for twenty-four hours. You either joined in or worked around them. There was something authentic about it all, with people following their passions and trying to make something useful. It didn't have to be software. It could be a personal project. People were so generous at the house in general about teaching each other and learning. I called it 'Nerd Mansion.'"

On the weekends, Marshall would inevitably think up some adventure and invite any willing participants to go along. "Will was always on these impatient missions where everyone was welcome," Schnugg said. "He'd be like 'We're going here! Are you with me?' And everyone would jump in some beater of a car and drive to NASA and chase goats and see Will's crazy fucking lunar lander ricocheting off some trampoline. And then we'd take another beater car up to San Francisco to go salsa dancing."

Those adventures with Marshall often benefited from what his friends described as an "unreality field." It seemed as though Marshall could generate good fortune from even dire situations and happy coincidences followed him wherever he went.

Not long after moving into the mansion, for instance, Marshall decided to go on a hike with his then girlfriend. The two-day trek required the couple to walk fifty miles from Cupertino, through the Santa Cruz Mountains and down to Waddell Beach. He packed a couple of bottles of water, a sleeping bag, and a bag of nuts for the journey. He had no notion of how they would get home from the beach or how they would survive if something went even slightly awry. No matter. After hiking for a day, sharing a sleeping bag at night, and hiking another day, the couple made it to their destination. "As me and my girlfriend were walking down the final hill, it really started to sink in that we had no cell reception, no way back, and no plan," Marshall said. "We almost broke up along the way because she was upset that I didn't bring any food. But then I saw some people kitesurfing down on the beach, and I said, 'I hope they're our friends.'"

The unreality field kicked into full effect at that point because, yes, one of the people kitesurfing was Don Montague, a pioneer in the sport, whom Marshall knew. What's more, Montague was giving lessons to Google founders Larry Page and Sergey Brin, who lived near enough to the Rainbow Mansion to offer Marshall and his girlfriend a ride home. As Marshall described it, "I'll never forget that on the way back, Sergey asked, 'So what exactly was your Plan A?' I was like 'Welllll, there wasn't much of one.' It ends up being us, Larry, Sergey, and a dog in their Toyota Prius. My girlfriend asks them what they do, and they told her that they work at Google. They never gave anything away.

"Then Sergey sent me a note afterwards. He'd analyzed our route on Google Maps and was wondering why we'd chosen certain paths. He thought we'd taken a crazy route. After that, we ended up hanging out at Burning Man, and we've been friends ever since."

As in any communal setup, the house had its tensions and interpersonal debates. Dishes were often stacked up by the sink, awaiting a volunteer to clean them and upsetting the more fastidious members of the group. People shared the food in the house, and you needed to slap your name on any precious items that were for personal consumption only. Still, some people ignored such territorial markers. Questions such as whether the front door should be locked or not would go up for debate during house meetings. Some people would argue in favor of security. Marshall, though, always came down on the side of the symbolism of the lock. Leaving the door unlocked signaled that the house was open to anyone.*

The communal life did not suit everyone—least of all an introvert like Parkin, Marshall's former university classmate. "It was exasperating; it was like being in a reality TV show without any cameras. One diplomatic crisis was when Chris and I both bought locks for our doors. That was deemed antisocial. I was told that it caused a disturbance in the force. They have a different sort of expectations, method of organization, and boundaries. I didn't really know what I was getting into. I'd been friends with Will for years, but you don't really know someone until you live with them."†

Parkin also found it distasteful when Marshall consumed his cereal and failed to replace it with a new box. "Will thinks the rules don't apply to him," Parkin said. "And, for the most part, they annoyingly don't."

To the extent that the Rainbow Mansion had an archnemesis, it was the closest neighbor, Rita. She owned an even bigger house at the top of the hill. No one knew Rita terribly well, but it was clear that she resented the weirdos next door. Ahead of every party at the mansion, the residents would vote on who should walk over to Rita's

---

* Even many years later, after the original Ames crew had departed, the door remained unlocked, and old residents would walk right in to meet the mansion's new inhabitants.

† Like the rest of the core Rainbow Mansion crew, Parkin had been recruited to Ames by Worden. After doing his undergraduate studies in physics at Leicester, Parkin earned a doctorate in aeronautics from Caltech and specialized in exotic propulsion systems for rockets. Upon arriving at Ames, he set out to create a more modernized version of a mission design center. That meant building the computing and software systems that engineers could use to come up with new spacecraft and simulate how they would work and how much they would cost. Parkin took over an old library at Ames to build the center, and after that, he designed a new type of rocket based on his university research.

and inform her about the impending noise. As one person put it, "Someone would have to go over to Rita's and manage her feelings."

The attempts at cajoling Rita usually failed, and she was known to call the police when things got out of hand. One September, on the occasion of International Talk like a Pirate Day, the housemates hoisted a skull-and-crossbones flag up a pole in the front yard. Rita phoned the cops and told them she felt threatened by the flag. She also did not appreciate the ritual of the UN flag ceremonies in which members of the Rainbow Mansion would march out to the flagpole and blow a conch shell before holding a small parade to celebrate the citizens of Earth.

While that could all sound ridiculous to some, the Rainbow Mansion housemates took their world-changing aspirations seriously. The housemates strove to codify their world-saving agendas. They held regular meetings in which they listed ways they could positively impact the earth and all its peoples. They set dates to accomplish their goals. They also tried to hold one another accountable for their actions. No one took it all more seriously than Marshall.

To meet his own exacting standards, Marshall developed a spreadsheet system for quantifying various aspects of his life, which his friends called the Marshall Matrix. The prime motivation for the spreadsheets came from Marshall's desire to analyze which of his actions would have the most impact on the world and the betterment of all people. He literally wrote an algorithm designed to result in a meaningful life. As he described it, "Basically, you list your goals. You list each project and say, 'How does this project help goal number one, how does it help goal number two?' Then 'What's your probability of success? How much money and time will it take?' You divide by that factor, then you multiply by a series of other factors.

"I tried to factor in things that you usually miss, like your abilities as compared to other people or, say, difference in participation. Like, is someone else going to do this anyway even if you don't? If so, you should reduce the weight of that project. There were other things like your level of interest or how much work something would be.

"What I tend to find, if people will do this, is that it's not just that one project is twice as likely to succeed or have a big impact on the goal. It often ends up being three, four, five, even six orders of magnitude more likely.

"We—myself and my community—have ambitious goals about trying to help the world. This sort of analysis system helps you to focus on those projects."

Marshall had similar analytical routines for almost every facet of his life. He built a Marshall Matrix for dating. He also tried to optimize his driving routes from the Rainbow Mansion to Ames by keeping track of his choices and journey times.

More controversially, Marshall spent about five years recording nearly every conversation he had. He kept a constantly running audio recorder in his shirt pocket and wore a large sticker on his shirt that informed people about the device's presence. He found it sad that moving discussions about space or philosophy with his friends would vanish into the ether and tried to fix the problem by taping and cataloguing his chats. Most of his friends chalked the experiment up to yet another odd thing that Marshall liked to do and had no problems with it. His detractors at Ames, however, pointed to the recorder as more evidence that he might be a spy, while other friends were unmoved by the audio life logging. "I did not approve of that," Parkin said. "I asked him to stop."

Marshall's peculiarities and the unconventional living taking place at the Rainbow Mansion may strike some as trivial or silly. In this case, though, they point to the mysterious underpinnings behind invention. It's unlikely that the idea for Planet Labs could have originated anywhere else than among a collection of do-gooders egging one another on. And it's unlikely that anyone other than Will Marshall could have ended up in charge of the company.

CHAPTER FIVE

PHONING HOME

Born in 1978, William Spencer Marshall grew up in the countryside of southeastern England, the middle child between a pair of sisters. The family lived what Marshall's sisters described as "an aspirational middle-class" existence in a modest home with a vegetable garden out back and room for a small menagerie of animals, including sheep, goats, and guinea pigs. Things became tougher in Marshall's teenage years when his parents went through a bitter, protracted divorce.

For much of his youth, Marshall was the skinny, nerdy, ginger-haired kid in class. He did well at math and science while showing less aptitude for the humanities. Part of his struggles stemmed from his poor handwriting. It was bad enough that one of his teachers thought he might be dyslexic; he also simply did not care for fiction or the vagaries of the social sciences and would often argue that studying them was a waste of time. From early on, he had firm theories about what mattered in life and could be combative against people and ideas that he deemed illogical. His sisters theorized that he had trouble with social cues and had to learn how to navigate them and how to deal with people over time.

Outside school, Marshall was a ball of energy. He liked to run through the countryside and always seemed to make his way high up in any tree worth climbing. His carefree spirit could spook those around him, as he seemed to fear absolutely nothing. "He's an adventurer and would do things like stand on the edge of a cliff and not even worry about it," said one of his sisters. "He is very much the epitome of his philosophy that the only risk in life is to take no risk."

Marshall's parents fostered a love of nature, which he embraced. He helped take care of the family animals, rode horses, and went on many a camping vacation. His father oversaw conservation programs to help protect gorillas in Rwanda, and Will became interested in that work. "I've long been concerned about protecting nature and have had a deep-seated sense of justice for the interests of the people and creatures that don't have a voice," he said.

Marshall had his precocious moments, particularly in math and physics, where he scored high on standardized tests and impressed with his coursework. A couple of teachers pulled his mother aside and informed her that Will seemed to think differently and behave less conventionally than the other children. But according to his family and longtime friends, he did not really stand out from his peers in a dramatic way. "None of the people in our friendship group would have said, 'Oh, there's a guy who is going to be a kingpin of Silicon Valley,'" said an old friend.

The most obvious signs of where Marshall might be heading can be found in his early interest in space. Young Will decorated his room with space posters and had all sorts of space and science magazines scattered across the floor. He would sometimes drag a mattress on top of the family's old Land Rover and lie alone for hours, looking up at the stars with a pair of binoculars. At sixteen, he spent months trying to save up enough money from his jobs working in a pub and a hardware store to buy a telescope. After figuring that it would take too long to secure the necessary funds, Marshall decided to build the telescope himself. He constructed every bit of the device by hand but could not afford the $1,600 lens needed to complete it.*

In an early unreality field moment, Marshall's school helped procure the lens by writing to a famous British astronomer named Patrick Moore to ask for help. An eccentric, Moore hosted the popular astronomy television show *The Sky at Night*. He not only provided Marshall with the lens but also turned up at a school awards ceremony where Marshall unveiled his creation.† After the visit, Marshall kept the relationship with Moore going, with the two men swapping letters back and forth and Moore advising Marshall on how to further his studies.

In 1996, Marshall left home to attend the University of Leicester to pursue a degree in physics. He picked the school because it offered an accelerated program that let him obtain a master's degree in four years. Leicester also had the best undergraduate space program of any school in Europe, and the students in the physics department had a chance to build real satellites and instruments for the European Space Agency.

The buildings the students lived in at Leicester were mansions once owned by wealthy textile families who had moved on and turned the properties over to the school. Marshall ended up in a large octagonal room and treated it like an engineering experiment. "He had rigged it so that everything he needed in life was at the end of a string pull," Parkin said. "All of these strings were running everywhere, and if he needed a shirt, he could pull a string and a shirt—a badly ironed shirt—would slide down. Same thing if he needed to open the door or turn off a light." Marshall also had a go at Parkin's dorm room so that a firecracker banged every time Parkin did something like opening a drawer or closing his curtains. "He cared," Parkin said. "He took the effort to rig my room to explode. How can you not find that endearing?"

Early in his university days, Marshall emerged as a leader among

---

* During that time, Marshall told his family and friends that he'd informally changed his name from William Spencer Marshall to William Space Marshall.

† Marshall's school still has the telescope on display.

his cohort and began blending science and politics. “Will was just in an entirely different sphere of activity in his own right,” said Parkin. “He had a political dimension to him and wanted to have a voice in policy.”

Marshall did things such as drafting a letter to the prime minister, urging the United Kingdom to become more involved with the International Space Station. He also wrote a letter to the United Nations Office for Outer Space Affairs and as a result received an invitation to a conference to help organize a group of young people to present their thoughts on what should happen in space in the years to come.* He also helped organize field trips to the United States and Russia and throughout Europe during which the students visited various space agencies. “Sometimes we couldn’t afford accommodations, so we had a compulsory ‘pulling policy,’” he said. “You had to try and find someone you could hook up with. Except we were physicists, and so we never succeeded at this at all. It was a very cool time of getting drunk on trains, sleeping on random floors, and visiting space things.”

Marshall’s organizational skills caught the attention of a committee that tracked young, inspirational Brits, and one day he received an invitation to join a couple hundred other such people for tea with Queen Elizabeth II. Marshall being Marshall, he prepared a letter to take to the tea that contained four pages of suggestions on how the United Nations could be improved, ranging from strengthening India’s position on the Security Council to making the organization more representative of all people. He did not plan to give the letter to the queen personally. He simply figured that someone important would be at the meeting who might appreciate his wisdom. Sure enough, Prime Minister Tony Blair was in attendance, and Marshall went up to him and pulled the letter from his jacket pocket. “I had a discussion with him, and he took the letter and said he would read it,” Marshall said. “I don’t think he ever did.”

That event captured much of Marshall’s developing personal-

* That work evolved into the creation of the Space Generation Advisory Council and to Marshall’s meeting many of the Rainbow Mansion members.

ity. Attributes such as celebrity and wealth did not faze him. Breaking from British stereotypes, he felt as comfortable approaching the queen or Tony Blair as he did anyone else and treated them like regular people rather than special objects. At the same time, he never failed to seize on an opportunity. He would enter a room and head right for the most important person in it to try to further one cause or another. Because he came off as intelligent and enthusiastic, people tended to listen to him and be drawn into whatever the young man was pitching at the time.

By 1999, Marshall had emerged as a significant force among young space enthusiasts and scientists. Along with his space road trips, he'd spent a couple of summers doing internships at the Jet Propulsion Laboratory in California and the Marshall Space Flight Center, where he'd met Chris Kemp. He'd also befriended scores of future space entrepreneurs, researchers, and academics, including some of his future Rainbow Mansion roommates, at a UN conference held in Vienna. The two-week-long event brought together students from around the globe to focus on the exploration and peaceful uses of outer space. At the end of the event, the students helped produce the Vienna Declaration on Space and Human Development, which emphasized that space should be a shared resource of all people and urged humans to exploit space in a responsible manner.

Marshall considered the Vienna event a seminal moment in his life. He'd found a group of kindred spirits who cared about space and wanted to find ways to use their passion and knowledge to better life here on Earth. The youngsters cracked open beers and stayed up late into the night, batting their idealistic ideas back and forth. "It was just the best thing," Marshall said. "It felt like we were doing something really important and that we were unstoppable. We decided that we wanted to find a way to live like this all the time, and so we decided to form a space kibbutz. We wanted to live in the same place and combine forces, and that is what eventually became the Rainbow Mansion."

In 2000, Marshall began pursuing a doctorate in physics at Oxford. He studied under Roger Penrose, a Nobel laureate who conducted groundbreaking physics research alongside Stephen Hawking. Marshall spent four years aiding Penrose with some of his concepts for experiments into the fundamental nature of the universe. All the while, he kept up his relationships with the young space crew, while also feeding off the political and societal discourse at Oxford. By the end of his studies, he concluded that he could not keep pace with the world's top theoretical physicists and would be better served by directing his talents toward accomplishing something tangible in space.

After obtaining his doctorate, Marshall spent a couple of years studying space policy, and then Pete Worden came calling. Worden and Marshall had hit it off since their first meeting in Houston at the International Astronautical Congress in October 2002. They had kept in touch through emails, phone chats, and gatherings in the communal houses. Worden had Marshall high atop his prized recruits list. Of all of those who would go on to be Pete's Kids, Marshall stood out as the one most likely to do big things.

With his wiry frame, glasses, and scruffy appearance, Marshall did not look like much. But his enthusiasm for science, space, and bucking conventions was infectious. People wanted to be around him because interesting things happened to him and those in his vicinity. Marshall had the combination of wits and an unusual breed of nerd charm that made him perfect for Silicon Valley. He could hang with the engineers, and he could hang with the billionaires. He was destined to be at the center of the action.

At Ames, Marshall bounced around among different projects. He spent a lot of time building the lunar lander alongside Al Weston and others. He helped build the spacecraft that would orbit the moon and then later the spacecraft that would collide with the moon and discover water. At each step of the way, he felt drawn to Worden's overarching call to try to modernize spacecraft and make trips to space much cheaper.

In 2009, Marshall fell into an unlikely opportunity to push the cheap-space idea to its limits. A group of students from the International Space University, a nonprofit that offers programs in space education, visited Ames. Marshall and Chris Boshuizen became two of the leads in charge of showing the students around Ames and finding something for them to do. Initially, the two men had hoped to let the students see the lunar lander and perhaps work on its development. Ames officials, however, quickly shot down the idea since many of the students were foreigners and would be working with the old military technology at the lander's core. That was when someone suggested that Marshall and Boshuizen provide the students with a box of parts and have them assemble a satellite from scratch. "I thought it was the dumbest suggestion anyone had ever given me," Boshuizen said. "It was almost insulting. Like, 'We're going to work on a fake satellite instead of doing something real.' But at the time it seemed like the best we could do."

Marshall, Boshuizen, and the students ended up building satellites out of Legos. They took the Lego Mindstorms NXT kits, which come with a variety of sensors and robotics systems, and combined the bits and pieces with gyroscopes, magnetometers, and a camera. Boshuizen also borrowed some of the LADEE software tools and tweaked them to run on the main Lego computing unit. For $900, the group built a prototype satellite about the size of a lunch box that had the Lego computer at its front and metal scaffolding around it to hold the other parts. To demonstrate the machine's abilities, the students hung it from a string attached to the ceiling, sent it tasks, and watched as the gyroscopes spun and adjusted the satellite's position so that its camera was aiming at a theoretical target on Earth. "We'd tell it to move, and the motors would whir up, and it was superloud," Boshuizen said. "It would turn, and then it would overcorrect and bounce around, and then it would send us a photo."

The project made the cover of a *Make* magazine issue dedicated to do-it-yourself space projects. Marshall and Boshuizen were photographed with big smiles as they held the curious-looking creation aloft with one hand. Once again, Pete's Kids from Ames had managed to think up a new and different approach to space that caught the public's attention. More important than press coverage, though, the project had sparked a whole wave of ideas in the heads of Marshall and Boshuizen. What had started out as something of a ho-hum project for students made the Ames researchers realize how good everyday consumer electronics products really were. "Will and I ended up driving to a conference in Long Beach the weekend after we built the Lego thing, and we showed it to some people and said that it was the future of space," Boshuizen said. "They all thought we were full of shit and that it couldn't do anything, but at that point I was a true believer."

Pete Klupar, the aerospace veteran Worden had hired as director of engineering at Ames back in 2006, was tight with Marshall, Boshuizen, and the whole Rainbow Mansion crew. For a couple of years, he'd been ending meetings at Ames by holding up his new smartphone, waving it around, and urging the scientists and engineers to think about what it represented. Apple and the Android phone makers had ushered in a sea change regarding what a small computing device could do. Their phones had tons of computing horsepower, lots of data storage, accelerometers for measuring speed, gyroscopes for picking up on movement, GPS for location, powerful cameras, and radios for communication. In many ways, they were more capable than the expensive computing and sensor devices being produced by NASA and others in the aerospace industry.

Up to that point, aerospace traditionalists had been convinced that any hardware going into orbit had to be made rugged to survive the extreme conditions of space. People in the industry would buy only specialized computing and communications systems and other components that had been tested to the extreme and preferably been flown to space on previous missions. No company mass-produced that type of gear, and as a result, it always ended up being expensive and bulky.

Klupar's thesis was that the aerospace industry had blinded itself to the advances taking place across the consumer electronics industry. Companies such as Apple and Samsung were spending far more on research and development and factories than governments and aerospace-focused companies were. They had perfected the art of cramming tons of horsepower into a small case and having the devices withstand the rigors of daily use. As Klupar saw it, there was reason to believe that everyday electronics might survive just fine in space. And if that was the case, NASA could build cheaper, more powerful systems than it had ever imagined. He said, "If you went to NASA headquarters and tried to explain this to them, they would just tell you, 'The Xbox and cell phones don't do what we do. We need these exquisite instruments.' They did not understand the course that things had taken. I was banging on the doors and telling everyone at NASA, at the air force, at Space Command, but no one would listen."

The student project, though, had primed Marshall and Boshuizen to be receptive to Klupar's message. The Lego satellite felt like a gimmick at first but had actually been pretty functional. It seemed that the next logical step would be a literal interpretation of Klupar's words. Why not just send a phone to space and see how it behaved? And so, in 2009, Marshall and Boshuizen began the NASA PhoneSat project.

The major goals for PhoneSat were pretty simple. Marshall and Boshuizen wanted to buy an off-the-shelf smartphone, blast it into space, and see if it would stay on long enough to snap some pictures and send them back to Earth. They would also gather data from all of the sensors built into the smartphone and more or less determine how many useful things the device could do in space.

At Worden's urging, the PhoneSat team kept a low profile and tried to stay well outside NASA oversight. No one involved in the project could tell if it would be just a playful experiment or something more meaningful. What they did know was that if NASA labeled PhoneSat as a "mission," it would come with all of the committees, reviews, and other baggage that would inevitably end up slowing progress down and making things cost more. Best to lurk in a corner somewhere at Ames and not draw any attention to the stab at doing things differently.

Marshall and Boshuizen found a small out-of-the-way office at Ames and furnished it with three big mahogany desks, a coffee table, a sofa, and a rug. Two of the desks were for day-to-day computing work, while the third was where the first PhoneSat would come to life. The budget for the first PhoneSat came in at $3,000. The figure was low enough that Marshall and Boshuizen could purchase the items they needed without requiring official spending approvals. To round out their team, the two men brought in a group of interns who were willing and able to exchange cheap labor for a chance to work on a real space program.

In July 2010, the PhoneSat team made it very clear how different they were from the usual NASA operation. Their project was still in its infancy, and they wanted to assess just how crazy their underlying idea really was. One obvious first step would be to slap a smartphone inside a rocket and see if it survived the vibrations and forces of a launch and the ascent into space. Hitching a ride on a real rocket would require millions of dollars and months of planning. Since they were in a hurry and had no money, the PhoneSat team opted for the next best thing.

Marshall and Boshuizen and their small crew picked up a couple smartphones and drove out to the Black Rock Desert in northwestern Nevada. That patch of desert is best known for being the annual home to Burning Man. For rocket enthusiasts, however, it's a place to test out serious, homemade machines. Each year, dozens of people head to Black Rock for an event called Balls, in which the very best amateur rocketeers turn up with their own propellants and twenty-foot-tall vehicles and shoot them as high as three hundred thousand feet into the air.

Marshall and Boshuizen wanted to hitch a ride on one of those homemade rockets and invaded the amateurs' event. Soon enough, they ingratiated themselves with Tom Atchison, a talkative rocket maker who often helped student groups on the playa. "I'd heard they wanted to put a smartphone on a rocket and send it to orbit," Atchison said. "But they were complaining that it would take eighteen months to get on one. I said, 'Fuck that. We're flying now. It'll be a rough ride, but we'll be able to put your stuff through the paces.'"

For their first experiment, the PhoneSat team broke off a panel on one of the hobbyist rockets, inserted a smartphone, and then drilled a hole in the panel that the camera could see out of. The rocket, called the Intimidator 5, launched and did great, flying to twenty-eight thousand feet under one thousand pounds of thrust. It allowed the team to check how parts like the accelerometers functioned under heavy gravitational forces and to gather pictures taken as the rocket went up and then came back to Earth under a parachute. A second launch with a second phone seemed to go less well. The rocket took off and flew, but a malfunction prevented its parachute from opening. The rocket slammed into the ground, compacting everything inside it.

Though the crash had the makings of a disaster, it turned out to be a blessing. The PhoneSat team dug into the wreckage and pulled what they could out of the rocket. Their phone had been squished and shattered, but it still had a working storage drive full of valuable data. The two launches gave the engineers more confidence that they were onto a big idea. "They proved what they needed to prove," Atchison said. "Which was that consumer electronics components could handle a real-world rocket ride."

The Marshall unreality field also activated during that trip to the desert. One of Silicon Valley's best-known amateur rocket enthusiasts is the venture capitalist Steve Jurvetson. A longtime friend of Atchison, Jurvetson had been testing his own rockets when he caught wind of what the NASA Ames engineers were up to. Jurvetson had been one of the earliest investors in SpaceX and quickly began to see the PhoneSat effort as perhaps the start of the next major shift in commercial space. He befriended Marshall and Boshuizen on the spot and began following their work with his checkbook at the ready.

Marshall functioned as the PhoneSat chief from the start. He managed the project, set deadlines, and used his well of impatience to try and move things along. Boshuizen took on far more of the technical work in conjunction with the rotating cast of interns. The Ames interns were often tasked with writing PhoneSat code, which Boshuizen would check and fix as needed.

With the launches behind them, the PhoneSat team shifted from putting a smartphone through its paces to producing an actual satellite. The engineers opted to mimic something called a CubeSat, which was a four-inch-by-four-inch-by-four-inch cube of metal scaffolding that could be packed full of electronics. The CubeSat concept had been pioneered by universities* looking for a way to simplify and standardize the construction of small satellites in the hope that more students would have a chance to work on and launch real spacecraft. By having a common satellite design to work from, students could trade information on which solar panels, electronics, sensors, and other components worked well in the device and not have to repeat the efforts from scratch at all their respective schools.

The PhoneSat design centered on a $300 Nexus One smartphone built by HTC. It would be packed into the CubeSat scaffolding and surrounded by other sensors and electronics needed to take measurements and keep the phone alive for about ten days. Some of the key pieces of technology included twelve lithium-ion batteries, an off-the-shelf radio transmitter, and a separate computer chip that would watch over the phone's behavior and, if needed, send it a signal to reboot. The phone's accelerometer and magnetometer provided plenty of movement data, and additional temperature sensors were added for more precise readings. The PhoneSat team also wrote some software to schedule the picture-taking operations and to have the phone pick only the best pictures to send back via the radio.

It took about eighteen months to go from a sketch on paper to a working device. During the initial prototyping period, the engineers had to test the PhoneSat components to see how they would cope under major pressure and temperature swings and how the radio would perform over long distances. Much of the work took place in NASA's laboratories, but the young engineers also had to journey into the field for some experiments. On occasion, they would divide into two teams, climb to the top of two mountains in the area, and

* California Polytechnic State University, San Luis Obispo, and Stanford University starting in 1999.

see if they could communicate between the PhoneSat device in one spot and receivers in the other.

"We did some tests where we put the phone in a vacuum," Boshuizen said. "It kept working, which seems obvious now. But back then, we had no reason to suspect the phone would keep running in some of these conditions. The assumption was that it would die."

By the middle of 2011, the PhoneSat team designed another experiment, in which they would hang their device from a balloon that would carry it twenty miles into the air. During that test, they ran into major problems for the first time. The frigid temperatures caused the smartphone to turn itself off. It's actually warmer in low Earth orbit than in our windy upper atmosphere, but the engineers decided to take precautions anyway and built an insulated case for the phone.*

While the PhoneSat engineers refined their satellite over several months, the march of consumer electronics continued. Newer and newer smartphones arrived with faster chips and better sensors. Boshuizen made a couple of friends in Google's Android phone division, and the Google employees sometimes showed at up Ames with bags of the latest phones for the PhoneSat team to try out.† In the background, lithium-ion batteries were improving, and so, too, were solar cells. The PhoneSat engineers realized that their small devices could likely perform longer missions than first expected and that smartphones themselves were becoming more and more capable of running sophisticated flight software.

There were not many people building small satellites at the time. Those who were, however, would usually take three of the CubeSat scaffolds and fuse them together. One cube might be dedicated to batteries, another to electronics, and the third to whatever scientific

* According to Marshall, "There was a time when one of the test balloons landed in a field in the Central Valley and was picked up by a policeman, who was then chased by some bulls out of the field. This was all caught on camera, and the commentary is utterly hilarious with the officer discussing whether or not the device was a UFO before concluding 'It's probably something from one of those scientist types.'"

† The Google engineers were soon drawn deeper into the project and began helping with the balloon tests and providing the PhoneSat team access to special software tools.

task the satellite was supposed to accomplish. Even though those small satellites were often built by university teams trying to do things on the cheap, the satellites' cost tended to rise quickly. People still made assumptions that they needed to use "space-grade" circuit boards and electronics that had been in orbit before, and the "space-grade" hardware tended to be a decade or more behind the times in terms of performance and price.

The PhoneSat team, meanwhile, remained fully committed to its thesis and kept pushing for extremes. For their second satellite prototype, the engineers removed the smartphone casing and plucked out the main circuit board. Instead of having an entire cube filled with computing systems, the PhoneSat had the equivalent amount of horsepower in a single thin board. That, in turn, allowed the engineers to fill the rest of the cube with the requisite supporting parts and to create the most powerful small satellite ever constructed.

The objective with the first PhoneSat design had been to create a device that could stay alive in space for a few days, report back on its health, and snap a few photos. But the advances with the second iteration had opened up new opportunities. The cube had room for solar panels, a stronger radio, and components that would let the satellite adjust its position better while in orbit. Where the first PhoneSat had been mostly a proof of concept, the second device was expected to do real work and wow people with its abilities.

After putting their new design through the usual battery of tests, Marshall and Boshuizen realized the time had come to stop hiding in their small office and make the PhoneSat project official. If they wanted to fly the PhoneSats into low Earth orbit on a real rocket, they would need real money, and the only way to get that would be to upgrade their effort from a skunkworks operation to a NASA Mission with a capital M.

Pete Worden had, of course, been supporting PhoneSat behind the scenes all along. He'd approved the project without much questioning and helped the team hide in their lab away from prying eyes. He had no idea if PhoneSat would turn into something major, but it looked very much like a step toward the low-cost satellites he'd been urging the government to build for decades. No NASA veteran would ever bother with an effort that appeared so trivial at first blush and that came with such a minuscule budget. Pete's Kids, however, were viewing the technology with a fresh perspective. It was exactly the type of thing Worden had brought the youngsters to NASA Ames to build.

When Marshall and Boshuizen told Worden they were ready to fly, he began making phone calls to try and find an upcoming rocket launch with spare room for three of the three-pound PhoneSat devices and to free up some funds for the launch. Through his contacts, he learned that an Antares rocket due to fly in late 2012 could fit the satellites in and launch them into orbit for $210,000. NASA approved the funding, and the satellites were given names: Alexander, Graham, and Bell.

Marshall and Boshuizen were thrilled by the prospect of the launch and began making the rounds of various NASA centers to fill people in on their mission. They wanted to spread the good word of cheap space throughout the agency. Only their public relations campaign did not go as well as they had hoped.

During a visit to NASA headquarters, Marshall and Boshuizen were in a lobby while waiting for a meeting with a high-up official. The walls in the room had signs detailing some future scientific missions NASA hoped to fly. One of the placards noted that NASA wanted to put up a constellation of weather satellites that would keep an eye on the sun's solar flare activity and its effects on Earth's atmosphere. The NASA posters noted that the mission would likely cost about $350 million. Marshall and Boshuizen did some quick calculations and estimated they could pull it off for $35 million with their new technology.

When their meeting began, Marshall and Boshuizen told the official all about the PhoneSat program and then presented her with more good news: they would be happy to take on the solar flare project next and to do it cheaper than NASA could have ever imagined. "We were so excited," Boshuizen said. "We told her we could help with all kinds of things that she knew would never really get funding otherwise because they were too expensive." The official, however, did not return the enthusiasm. "She laughed us out of the room and told us that our ideas were not credible," Boshuizen said. "That's around the point where we started thinking that maybe we should do some of this stuff ourselves."

The launch of the three PhoneSats did take place, although in April 2013 instead of late 2012 as originally planned. NASA hailed the launch as a huge success and told anyone who would listen about its newfound ability to do space on the cheap. Curiously, another small satellite had made its way onto the rocket, too. It was called a Dove and had been built by a California start-up named Cosmogia.

CHAPTER SIX

THE BIRTH OF A PLANET

From the earliest days of the PhoneSat project, Marshall and Boshuizen felt as though they were onto something big. The satellite industry had bogged itself down by decades of tradition and staid thinking and almost completely ignored the huge advances in consumer computing technology. The big question for the two scientists was what exactly could be done with a new type of satellite. What sorts of jobs could tiny satellites perform better than large ones? And, given their ideological leanings, what tasks could a new breed of satellites perform that would most benefit mankind?

During the day, Marshall and Boshuizen pursued the PhoneSat effort at Ames, but at night they pondered those questions with friends at the Rainbow Mansion. Marshall, being Marshall, tried to add some rigor to the pondering by creating a spreadsheet that collected and ranked people's ideas. The spreadsheet boiled down to the twenty or so things that small satellites could do that would aid society, earn some money, and be novel from a technology standpoint. The group discussed things such as gathering images, creating a new GPS location system, doing science experiments, and setting up a new communications system in space. As the deliberations went on, the imagery concept gained the most traction for a variety of reasons.

Almost all of the imaging satellites in space were controlled by governments, research bodies, or a handful of companies. It was a small, not terribly forward-thinking club. The satellites cost anywhere from $250 million to $1 billion each. They tended to be large—about the size of a van or a small school bus. They also usually took several years to design and launch and were meant to keep working in space for as many as twenty years. Because of their high cost, they were relatively rare commodities. Not even the US government had all the spy satellites it desired. The imaging industry, for the most part, could snap pictures only now and again and only of spots of known interest.

The PhoneSat experiments had pushed Marshall and Boshuizen to rethink how an organization might go about photographing the earth. Instead of building a handful of superpowerful, expensive satellites, someone could build lots of cheap satellites and completely surround the planet with cameras. Marshall did some calculations and figured that about a hundred such machines would be needed to snap a photo of every spot on Earth every day. The trick would be to mass-produce the satellites at a reasonable enough cost that the idea would feel feasible to investors. "People were spending about a billion dollars for a single imagery satellite," Marshall said. "We could put up hundreds of ours for way less cost than one of theirs. It would be a lot of money but not an extraordinary amount of money. It would be something a venture capitalist could fund."

The satellites Marshall and Boshuizen dreamed about would basically be disposable. Instead of having a life span of twenty years, the small, cheap satellites would be designed to orbit Earth for three to five years. Then they would tumble back into the atmosphere and burn up on reentry. It would require a lot of rocket launches to keep putting new satellites up into space, but that was also kind of a benefit of the model Marshall and Boshuizen were chasing. The fresh satellites would have the latest computing and electronics components, so the constellation of imaging machines would keep getting better all the time.

Just as SpaceX had changed the economics of rocket launches, Marshall and Boshuizen hoped to change the economics of satellites. Launching one would no longer be an all-or-nothing affair where $1 billion had to be invested in an object that would need to last decades and simply could not fail. Instead, devices could be thrown into orbit that would be good enough for the task at hand and then they could be improved over time. If components in a few of the satellites failed, so be it; their more modern replacements would follow soon enough. The same thinking went for the rocket launches: if a $1 billion satellite blows up on the pad, people's careers end and companies die. If a few cheap satellites blow up, they just move on and build some more.

Beyond all of that, the satellite constellation Marshall and Boshuizen were proposing would change what's possible when it comes to understanding the earth. Rare photos from odd moments in time would be replaced by a constantly updating record of, well, everything. It would be possible to keep much better track of the health of the oceans, forests, and farms. It would be possible to monitor much of humanity's economic activity, including the movement of cargo, the construction of roads and buildings, and how active people are in one region versus another. Rather dramatically, this could be provided as a service outside government control: the images would be fed into a database that anyone could search. Marshall and Boshuizen realized that they could build a Google-like analysis system for the entire planet.

Their work at Ames had put Marshall and Boshuizen into a unique position. They had built lunar landers and spacecraft and been part of complex missions and the data analysis that had followed. That gave them firsthand experience with the very real complexities of space. At the same time, Worden's overall push to think differently and try cheaper approaches had put them into the right mindset to see opportunities others had missed. "We realized we could make the capability per kilogram put into space somewhere between a hundred times and ten thousand times more effective than what others were doing," Marshall said. "It's an unusual situation where the industry had been so lackadaisical and sitting on its heels that you could improve it that much. There was something deeply wrong with the industry."

By the end of 2010, Marshall and Boshuizen had committed to forming a company. They wanted to see the PhoneSat project through a bit longer at Ames but were also busy putting things in place for their new venture. Reluctantly, they informed Pete Worden of their plan to leave Ames. Initially, he bristled at the idea, but then he accepted the situation. He urged them to meticulously log all of the hours they spent working on projects at Ames so they could prove that only their evenings and weekends were going to the new venture. That would hopefully prevent the higher-ups at NASA from grumbling—or worse—when word of their start-up eventually became public.

"We waited and waited and waited to tell Pete," Boshuizen said. "He'd become such a dear and close friend, and we were probably some of the best examples of what he'd been trying to achieve at Ames. We knew he'd get upset, and he did get upset for a couple of days. He didn't like the idea of us leaving. But then he came to the conclusion that our company was emblematic of what he'd set up. Us leaving was a sign of his own success."

Marshall and Boshuizen spent a couple of hours at the Rainbow Mansion brainstorming names for the new venture. Marshall wanted "Gaia" in the name to celebrate Mother Earth, and both men wanted something with a spacey flair. They typed variations of those themes into the internet to see which names were already taken and eventually settled on the befuddling "Cosmogia," a merger of "cosmos" and "Gaia." "I remember us high-fiving each other and feeling so accomplished that we'd found the perfect name," Boshuizen said. "It turned out, though, that the name was lost on everyone. It was a terrible name. But in our fever dream, it seemed fantastic."

Cosmogia soon morphed into Planet Labs, and about six months into 2011, the company started to assemble its team. Along with Marshall and Boshuizen, Robbie Schingler joined as a cofounder. Schingler had done a four-year stint as Pete Worden's special assistant and spearheaded numerous programs to open up more of NASA's data and technology to the public. He had also helped lead a handful of satellite and small-spacecraft missions. Just before starting Planet, Schingler worked at NASA headquarters as the right-hand man to the agency's chief technologist. The three men had all spent countless nights at the Rainbow Mansion discussing their ideas regarding the technology and plans for the venture. For the longtime housemates Marshall and Schingler in particular, it was a chance to create a company together with one of their best friends.

Each person brought his unique skills to the endeavor. Boshuizen would run the technology side of the operation. Marshall would be the CEO, championing Planet's vision and keeping things moving along quickly. Schingler had a gift for dealing with the more practical business matters, such as setting and executing a strategy and hiring, and would focus on nurturing relationships between Planet and its customers. A handful of people from Ames also agreed to risk the stability of their NASA jobs and go on the start-up adventure, including Vincent Beukelaers, Matthew Ferraro, Ben Howard, James Mason, and Mike Safyan.

Like any good Silicon Valley start-up, Planet Labs sprang to life in a garage. The team would discuss their concepts in the comfort of the Rainbow Mansion and then head to their garage to try and turn their ideas into working hardware.

The basic concept was to build the smallest, cheapest satellite capable of taking decent pictures of Earth. It would basically be a box with a telescope inside it plus computing and communications systems for storing and sending photos. The satellite would also need systems to control its orientation in space and to stay alive for years. Inside the garage, the small group of engineers began laying out all the components on tables to see how much physical space they would require. Over the course of a few months, they whittled the number of components down and concocted artful ways to pair different elements. Once they reached the stage where a prototype needed to be built, the Planet founding team knew it was time to move out of the garage and find a real office.

The PhoneSat program lived on at Ames under Worden's supervision. Meanwhile, Planet set up shop in the center of San Francisco. The Planet founders had been paying for their experiments at the Rainbow Mansion out of their pocket but now needed to raise money as their expenses mounted. Remembering their time in the Black Rock Desert, Marshall and Boshuizen decided to call the venture capitalist Steve Jurvetson. To their great surprise and joy, he agreed to cut Planet its first check. "We raised three million dollars at the start, and Steve put in two million of that," Marshall said. "To his credit, he saw it, and he placed the bet."*

Along with the new job, Marshall and the Schinglers changed their living conditions. They wanted to be closer to Planet's office and live in San Francisco. They traded one mansion for another, finding a 7,500-square-foot, eight-bedroom Victorian house near Alamo Square.† Once owned by a shoe magnate, the house had a bowling alley in the basement, a library, and plenty of sitting rooms. The new digs were as inspiring as the old, and Jessy Kate again turned the home into a thriving communal scene with about a dozen residents. People in the tech industry often turned up at the house to sit in on a salon or strategize about their next start-up.

Satellites are typically built in clean-room settings to prevent particulates and other contaminants from reaching electronics and mechanical parts. Anything with a lens demands a particularly clean environment to ensure that the resulting images come out pristine. Sending an imaging satellite all the way to space only to end up with pictures marred by fluff off someone's clothes or dust would be beyond self-defeating. At that point, however, Planet did

---

* James Mason, one of the first Planet employees, had put together a detailed plan regarding Planet's technology and business case to show investors. "In the end, it was completely useless," he said. "All it took was Will meeting up with Jurvetson at Burning Man." The other investors were Capricorn Investment Group and O'Reilly AlphaTech Ventures.

† Chris Kemp opted to live on his own and did not make the move to the new house.

not have money for a state-of-the-art facility and had to do things on the cheap.

To create a makeshift satellite laboratory, the company first bought some greenhouses and air filters on Amazon. The greenhouses were large enough for people to fit inside and do their work. Planet had hired around thirty people by that point, and a good portion of them would don lab coats, venture into the greenhouse, and start trying to make satellites from scratch. "It was clearly a greenhouse that you were supposed to put in your backyard, but it worked great," said Ben Howard, who had come to Planet from Ames.

It took Planet about a year and a half to craft its first Dove. It was a rectangular cuboid shape measuring ten by ten by thirty centimeters, making the Dove roughly three times as large as the early PhoneSat cubes. Inside, there was a cylindrical telescope wrapped in gold-plated tape to provide thermal insulation. Surrounding the telescope were several lithium-ion batteries, with individual heaters for each one, and a handful of circuit boards. The machine also had solar panels attached to its sides and an antenna. Instead of $1 billion, the Dove cost less than $1 million to produce.

I met the Planet founders for the first time in the middle of 2012 as they were bringing their initial Doves to life. The greenhouses had been replaced by a slightly more formal setting. Planet had created mini-manufacturing centers that were separated from the rest of the office by sheets of plastic. Though the office had impressive testing equipment and all manner of other gadgets, it felt more like aerospace underground than professional satellite factory.

Marshall and Schingler gave me a tour of the place and told me the backstory of how Planet had originated from the NASA experiments. Passion poured out of them. They were going to put up more satellites than any organization in history to "monitor deforestation in Africa, track illegal fishing, and monitor the ice caps melting." It was not clear to me how the company would make a ton of money from any of that, but Marshall and Schingler sure seemed like nice, very idealistic guys. "We want to bring about a new phase change in human consciousness and in understanding our planet," Marshall said. "We care a lot about making this information available to those who need it most," said Schingler.*

In April 2013, Planet launched its first two Doves. By sheer coincidence, one of the machines ended up on the same Antares rocket as the first PhoneSats. A second Dove flew aboard a Russian Soyuz rocket. A few months later, Planet launched two more satellites and began gathering significant amounts of data for the first time.

The successful flights came with all the usual trappings of a start-up having its initial taste of success. The engineer who made contact with the first Dove charged out of the satellite-tracking station where he was working and ran circles around the antenna while screaming. Inside the tracking station, other engineers cracked open bottles of booze while they performed their communication tests with the machines. Later, when one of the satellites began its return to Earth and disintegrated in the atmosphere, the Planet team held a combination party and wake for their lost Dove.

The first image to come back from one of the satellites showed a forested area, but no one on the team knew the exact location. Planet's tools for pinpointing pictures were rudimentary at the time. After a few hours, someone determined that the satellite had snapped photos of Oregon. "Robbie came in with this picture on his phone," Marshall said. "It was fucking amazing. It was beautiful. I was so amazed that you could see individual trees. The fact that it worked just as we had planned was shocking to me. Robbie still carries the picture on his phone."

As Planet gathered more images over the course of several months, its employees began to be struck by what astronauts describe as the "overview effect." This is the experience of seeing the earth from above and gaining a new insight about the fragility of this relatively small object hanging out in the void with the thinnest

* "Jurvetson says that he likes to fund people that don't have making a ton of money as their main ambition," Marshall told me. "If money is your main ambition, you tend to think more near term. If you have a longer-term goal like settling Mars like Elon or saving Earth, like us, you are going to take much bigger leaps to change things dramatically. Larry and Sergey did not have a business plan at Google. They simply wanted to make the internet useful. You create something that has immense value and then strap a business model to it later."

of atmospheres keeping it viable for human existence. The Planet engineers watched forests change colors through the seasons. "You could see somewhere like Africa essentially breathing," said James Mason, who joined Planet after working on PhoneSat and other missions at Ames. "We saw the planet evolving in real time."

In those early days, Planet's engineers showed a level of quick thinking and an ability to adapt on the fly that were uncommon in the satellite industry. For example, a radio on one of the Doves had a software error that erased the device's memory. To bring the radio back to life, a couple of people rewrote the core code that made the device function and then beamed it up to the satellite. They could "talk" to the satellite only when it was over a ground station, so it took several passes to get all the new code installed on the radio, which did eventually begin working again.

This early trial reinforced Planet's idea that satellites could be thought of as flexible rather than fixed devices. They did not have to sit in space unchanged for years and years. They could be improved and updated just like consumer computers and smartphones. Once again, Planet helped erode years of traditional thinking that satellites were fragile objects that should be left alone once they were in space.

The first couple of launches boosted investors' confidence in the young company. In mid-2013, Planet raised $13 million more in funding. Steve Jurvetson once again led the investment round, which was joined by the venture capitalist Peter Thiel, Eric Schmidt, and others. Planet used the money to build a fleet of twenty-eight new Doves. It put those into space in 2014 by hitching a ride on a cargo rocket bound for the International Space Station. Once on the ISS, astronauts ejected the Doves into orbit. In 2015, investors really bought into Planet's vision and pumped another $170 million into the company. Soon Planet was in the process of building more than a hundred satellites and trying to find rides on any available rockets.

As Planet began pumping out satellites and conducting more launches, it discovered that it had misjudged the performance of its early satellites. The first Doves had been in space for only a short time before they deorbited. As new Doves began multimonth journeys in low Earth orbit, the machines overheated, and their batteries died. Managing dozens of fast-moving satellites from a few ground stations also proved very difficult. "We basically ended up needing to design a system that could just babysit these horribly performing satellites in a very detailed way," Howard said. "We were a small, inexperienced team, and it was very hard. We'd been thinking that we just needed to hit 'Print' and manufacture a hundred of them and put them in space and print data. That turned out not to be the case."

There were times when the Doves received too much persistent sunlight, causing their temperatures to soar. It then became too risky to perform operations such as charging the batteries or really to do much at all. So Planet's engineers would shut the satellites down for days at a time to let them cool off.

The lenses had issues, too. In a typical imaging satellite, pains are taken to make sure that the lens does not experience temperature fluctuations. In the Doves, however, the lenses were heating up and then cooling down, and those drastic shifts caused them to shift out of focus. "There was basically nothing we could do about that after the fact," Howard said. "We should have probably had an optics engineer." On top of all that, the radios were not powerful enough to send out all the images the satellites were collecting. "We went through a huge number of launches without having a constellation that could actually produce data that was at the quality we needed it to be and at the rate we needed to be profitable," Howard said. "There was a real concern that we would not be able to pull it off."

That period was a rude awakening for Planet. People in the aerospace industry had scoffed at the start-up and its new-kid-on-the-block attitude. Planet had tried to bring some of Silicon Valley's fake-it-till-you-make-it spirit to satellites and had been rushing at tech industry speed to get its first hundred machines into orbit. Some aerospace veterans felt that Planet was getting what it deserved. Satellites were very hard to build, and the renegades from NASA had overestimated their abilities. They'd championed vision over engineering.

Planet declined to disclose many of its issues to the wider public. Marshall could be found at technology conferences talking up Planet's world-saving tools as if the company had already accomplished most of its goals. Behind the scenes, however, the Planet team was busy tackling one problem after another and developing the set of skills it would take to place hundreds of satellites into orbit and orchestrate the flow of information between them and the earth.

The first job was to figure out how to navigate the world of rocket launches. Planet wanted to put up lots of satellites and do so as cheaply as possible, but at the time rocket launches were hard to come by. The experiment with the ISS had been nice because the ride to space had cost less than a typical launch. The problem, however, was that the ISS's orbit was not ideal for Planet's goals. The satellites had to spend months making their way to the perfect spots for taking photos.

In the years that followed, Planet put Mike Safyan, one of its early employees, in charge of forming relationships with rocket companies around the globe and negotiating contracts. Planet became the ultimate rocket vagabond, sending its devices to Russia, India, and SpaceX in the United States. Planet also began courting some rocket start-ups that were starting to appear on the scene. These companies had yet to prove that their rockets worked, but they were promising cheap, frequent rides to space, flying small satellites as their main cargo instead of as secondary payloads, and paying extra attention to placing the satellites into ideal orbits. To help foster the industry, Planet bought one of the first rides from a New Zealand outfit called Rocket Lab, which had an intriguing rocket called Electron.

No company had ever needed to negotiate so many launches before, and Safyan emerged as one of the best-connected people in the launch industry. He learned to haggle and navigate the ever-present launch delays to figure out the most efficient means of getting the satellites into space in a timely fashion.

To communicate with its satellites, Planet also had to build a vast network of ground stations, small buildings that could be managed by a couple of people and had powerful antennas and radios for shuttling data back and forth between the earth and the satellites zipping by overhead. Often, the ground stations were placed in remote locations by the poles and the equator, and engineers learned how to coordinate the data transfer among dozens of the ground stations and hundreds of satellites. Each day, many petabytes of encrypted information flowed across the network.

Planet also had to master the art of building tons of satellites. As it raised more money and became a more mature company, the greenhouses and plastic protectors were replaced by a factory that took up much of the bottom floor of its office. Satellite companies usually built one or two machines at a time, but Planet's factory had to knock out dozens on relatively short notice if the company found an opportunity to snag a ride on an upcoming rocket launch.

The man in charge of Planet's factory was Chester Gillmore, a bow tie–wearing, frenetic ball of energy who seemed to enjoy his job as much as anyone has ever enjoyed a job. The key to his operation was to keep the manufacturing process flexible enough that Planet could swap out existing components for the latest computing systems or sensors while still maintaining a high level of quality with each satellite. Whereas the first PhoneSats had a handful of components, Planet's satellites eventually evolved to have two thousand different parts. Almost every piece had a unique barcode that would be used to track its arrival at the factory, its placement in a satellite, and its performance in outer space.

No company had ever tried to make that many satellites before, and Planet's methods were very different from the status quo. "If the CIA wanted to build a big spy satellite, here's what would happen," Gillmore said. "They would send out a document detailing the technical specifications of what they need. Various companies would spend six months putting together cost proposals. The CIA would pick the one they like, and then they'd start designing the satellite, and that would take four months. Maybe a year after that, someone has built some prototypes that have to go through a series of approvals. Then, eighteen months later, there might be the final satellite that's done. After that, it's another six to nine months to launch it. By the time you're finished, the technology that you started with is about five years old."

At Planet, it took around a dozen people to knock out as many as thirty satellites per week. The company typically hired people from outside the aerospace industry and trained them on the factory floor. One guy used to be a paralegal. Another had been a bike mechanic. They would move among forty-two different stations to handle both manufacturing and testing tasks. The whole idea of this team was to remain fluid and roll with the ever-changing components Planet's engineers would need. Generation by generation, improvements were made to field of view, resolution, image quality, battery life, amount of storage, computing power, location tracking, and the solar panels.*

During a typical launch, Planet now sends up anywhere between twenty and ninety Doves. To place the Doves into space, a rocket's upper stage tips forward and then slowly rotates, releasing a satellite or two every couple of degrees. It takes about five minutes for all the Doves to be ejected. Once floating in space, each satellite's solar panels unfurl, a lid at one of its ends pops open, and an antenna extends.

New Doves will join existing satellites in an orbit that goes around and around the poles. They're spread out so that each satellite is responsible for photographing a certain swath of land below it. The satellites function, in effect, as a line scanner with the earth rotating below them, while they take a near-constant stream of photos. To get the satellites into optimal position, Planet uses its differential drag technique to slow a chosen Dove down with respect to the others.

Once it's in the right spot, a satellite's attitude, determination, and control system set the machine's orientation. Gyroscopes and sensors on the Dove look for magnetic fields and seek out the earth's horizon, the sun, and other stars. Magnetorquers and reaction wheels

---

* From 2013 to 2021, the Doves increased their performance, in this case the amount of data collected per satellite each day, ten thousand times.

then adjust the satellite's movements until it reaches the desired alignment.

The path the Doves follow is known as a sun-synchronous orbit. An individual satellite passes over the same spots each day at the same times each day, and this orbit helps ensure that the pictures have uniform light and shadows. It takes each Dove about ninety minutes to complete an orbit, which equals sixteen orbits per day. As the earth rotates below a satellite, it might be over New York at 9:00 a.m. local time, then over St. Louis at 9:00 a.m. local time, then over San Francisco at the same time, and so on.

Each Dove collects thousands of images covering around eight hundred thousand square miles per day, an area the size of Mexico. The pictures are relayed during ten daily eight-minute sessions on custom-built radios between the satellite and its ground stations. Once the images reach Earth, Planet's software compiles them, cleans them up, and deletes photos marred by clouds and shadows. Customers can then log on to an application and browse the pictures as they please. Planet charges corporations and governments a certain price for its photos and offers discounted access to reporters, nonprofits, researchers, environmental groups, and the like.

Since its founding, Planet has faced criticism that the quality of its photos is not good enough. The images produced by the Doves have three-meter resolution. That means that each tiny pixel in a photo on your computer screen is equal to a three-meter-square bit of land. You can see buildings, cars, and landmarks, but it's hard to pluck fine detail out of the photos. Though Planet has been able to make many improvements to its satellites over the years, the resolution is a hard thing to increase because of the physics of the satellites. A telescope of a certain size placed at a certain height above the earth can take pictures of only a certain quality.

Planet's argument against such critiques has been that there is immense value in photographing all the earth, all the time instead of just points of known interest. By creating a massive record of the earth over a long period of time, Planet's satellites spot trends and changes that others would miss. It's only by making the satellites small, and thus more affordable, that the company could assemble such a large network and build a data set that no other company or government has.

In 2017, Planet moved to increase the quality of its photos by acquiring a company called Terra Bella* that made imaging satellites known as SkySats. Like Planet, Terra Bella was a start-up that applied modern techniques to manufacturing its machines. The SkySats, though, were larger: refrigerator-sized units compared to Planet's shoebox-sized devices. The beefier satellites allowed Terra Bella to have larger lenses and a resolution of fifty centimeters. Those satellites went into different orbits around the earth, too, which enabled Planet to gather more pictures of various locations at different times of day.

The two satellite systems working in concert gave Planet a major technological edge over what had come before. The Doves were always there, always looking, and powerful enough to spot changes on Earth, be it a forest being cut down, a new building going up, or a missile being fired. There were not enough Terra Bella machines to see everywhere, but they could be adjusted to zero in on a location if the Doves detected something interesting taking place. In the years following the acquisition, Planet applied its manufacturing and launching expertise to build more of the large satellites and place them into orbit.

As you read this, there are hundreds of Planet satellites traveling overhead, mostly Doves and about two dozen of the big boys. The technology has improved enough that Planet can snap at least twelve photos of a single location in one day. The whole network takes well over 4 million photos per day, and Planet's archive has an average of two thousand pictures of every spot on the earth's landmass.

---

* Terra Bella used to be called Skybox Imaging and was acquired by Google for $500 million in 2014. It operated as a stand-alone division within Google. Will Marshall later convinced his good friend Sergey Brin to sell the company to Planet.

CHAPTER SEVEN

THE GREAT COMPUTER IN THE SKY

The rumors had been floating around military circles in Washington, DC, throughout the early part of 2021. China was said to be in the midst of expanding its nuclear arsenal and constructing missile silos in remote parts of the country. No public sources of information existed to prove the theory, but the people doing the whispering were convinced that a massive weapons buildup was underway. If someone could confirm the presence of the missile silos, the disclosure would point to an increasingly aggressive People's Republic of China and heighten tensions between it and the United States.

Decker Eveleth heard about the weapons rumors from a couple of mentors in what's known as the open-source intelligence community. Open-source intelligence analysts are people who scour publicly available information to try and discern things about military and economic activities. Their insights might come from trawling through tax records or military contracts or analyzing satellite imagery. With these tools, the analysts can sometimes uncover details about a North Korean missile test or an illegal shipment of oil to a country subject to sanctions. In general, the work is done to bring information to light that governments and bad actors want to hide, allowing the public to know what's happening in their world and to talk about it in open forums.

An undergraduate at Reed College, Eveleth had gotten close to the open-source intelligence community through his hobby as an information hunter. While other students were funneling beer and modifying their gravity bongs, Eveleth sat in front of his computer digging through databases and analyzing satellite imagery. And it turned out he was really good at those tasks. He could spot patterns that far more seasoned analysts missed. He'd often post his findings on Twitter, and he produced enough solid discoveries that veterans in the open-source community began chatting him up.

In mid-May 2021, Eveleth decided to take a crack at looking for the rumored missile silos in China. He assumed that the sites would have an appearance similar to that of a previously discovered group of silos over which the Chinese military had built inflatable dome coverings to hide their work. Analysts had come to refer to the white domes as "bouncy houses of death" because they resembled the semipermanent structures you might find on a sports field or in a bouncy castle at a kid's party. Eveleth also assumed that the domes might be in the desert of northern China because the military had been particularly active in that region, and there were vast areas of flat land available there.

Through his hobbyist work, Eveleth had set up a Planet Labs account, and he began pulling up images and dividing thousands of miles of desert into grids that he searched one by one. It took more than a month, but in late June, he made a major discovery: he spotted about 120 of the white "bouncy houses" of death. In the previously discovered sites, China had a couple of dozen structures at most. If Eveleth had really found 120 new silos, his intel would reverberate around the globe and signal that a new arms race might well be in progress.

At 8:00 a.m. on June 27, Eveleth contacted Planet and notified the company about the possible find. The Dove satellites had taken tons of pictures of the area in question over the course of many months, and Eveleth could use the photo timeline to recreate the silos' construction process. To get even better, more recent pictures of the site, Eveleth asked if Planet could aim its higher-resolution SkySats at the area. Planet agreed to help.

Over the next day, Planet's engineers sent radio signals from their ground stations on Earth to their satellite constellations. The computers on board the machines received the signals, and the satellites turned on their reaction wheels to change their positions and better orient themselves toward the target. Traveling at 4.7 miles per second, the satellites took rapid-fire shots of the desert. Radios transmitted the images back to Earth, where they were decrypted and then processed by Planet's software. At 8:46 a.m. on the twenty-eighth, Eveleth logged in to Planet's service and saw not just the domes but also trenches for communications cables leading out from underground facilities where the military likely had its launch operation centers. He presented the collection of images to open-source intel veterans, and they all agreed that he had indeed located the missile silos people had been whispering about. "We knew that it was a big deal," he said. "There's a very special kind of excitement that goes into this, knowing that you're the first to find something."

After Eveleth showed the images to reporters, stories about China's nuclear buildup hit the front pages of many newspapers. The State Department categorized the find as "concerning." Chinese publications tried to downplay the pictures by dismissing Eveleth as an amateur sleuth who had merely stumbled upon the construction of a wind farm. Though a worthy effort, the Chinese misdirection proved laughable because the images had so many telltale signs of a nuclear weapons site.

The reporters covering the story were naturally obsessed with the political ramifications of the discovery. None of them, however, stepped back to look at the other major part of the tale: that an undergraduate student poking around on his laptop had uncovered a major Chinese military operation. He'd done so by tapping into a fleet of hundreds of satellites built by a private company rather than by a military or government body. And anybody else could do the same. "It used to be that the government had satellites, and we didn't," said Jeffrey Lewis, an expert in nuclear arms control and Eveleth's mentor. "Now they have slightly better satellites. Okay, that's nice for you, but it doesn't really matter."

Going back to the 1940s, people within the US military establishment had theorized about putting image-snapping satellites around the earth. The attack on Pearl Harbor had highlighted major gaps in US intelligence gathering, and the spooks in Washington, DC, felt it would be nice to have some all-seeing eyes in the sky at all times. The only things stopping the military from pursuing a spy satellite at the time were technological limitations concerning how to get a camera into space and then retrieve the images it produced.

In the 1950s, the desire for spy satellites became even more pressing. In 1957, the Soviet Union launched *Sputnik 1*, triggering concerns that the United States had fallen behind its rival in space technology. Beyond that, the US military worried that it did not have a clear picture of the Soviet Union's missile arsenal. Spy planes could capture useful pictures, but they mostly had to be used over known sites because of the risks tied to flying them through Soviet aerospace and the relatively short durations of their flights. The United States lacked a way to scour huge swaths of land and look for previously undiscovered silos and military installations. Without an accurate assessment of what the Soviets were building, the United States really had no idea what it was up against and whether it was ahead in an arms race or woefully behind.

By 1958, the US government had come up with a secret plan called CORONA that called for the development of a raft of new technologies to make its spy satellite program work. Rockets would be built that could ferry satellites into orbit. The satellites would have cameras specially designed to deal with things such as atmospheric distortion and vibrations of the spacecraft so that they could take clear photos of the earth. As if those jobs were not hard enough, the United States also had to figure out how to get the photographs taken in space back to Earth. The data transmission systems at the time were nowhere near quick enough to send bulky photo files from orbit to a ground station. Rather fantastically, a group of engineers decided that CORONA would work by having the satellites eject physical canisters of film from space that would fall back toward the ground under parachutes. The capsules holding the film would have heat shields to protect them from burning up upon reentry into the atmosphere and were to be grabbed in midair by a plane that would fly a hook on its belly into the parachutes. No big deal.

To disguise all that activity, the government concocted a public program named DISCOVERER, which it billed as a series of scientific space missions. So if someone happened to notice a bunch of rockets going off, they were to be explained away as research endeavors that would improve all of humanity's understanding of Earth rather than supersophisticated spy operations. The CIA and air force were put in charge of developing CORONA's technology, and a new office was formed to support the effort and other crazy space shit like it, which became known as the Advanced Research Projects Agency (ARPA) and later as the Defense Advanced Research Projects Agency (DARPA).

CORONA was a monumental undertaking. The government had many of its best engineers working on the project and also recruited people from various industries to help with the photographic elements of the operation. All of them were sworn to secrecy and pushed to work as fast as possible. But things did not go well at the start. In the first eighteen months of CORONA, the United States sent up a dozen rockets, and the missions failed every time, either because a rocket exploded or a capsule could not be recovered or because the cameras did not work right. But launch by launch, the operation started to hit its stride, and the first images from space began arriving in 1960.

The results were spectacular. The initial batches of photos covered huge swaths of the Soviet Union. A single capsule recovery could return more photos than four years' worth of flights by spy planes. The United States hired hundreds of people to work in a new top secret office called the National Photographic Interpretation Center, where they unspooled six-hundred-foot-long rolls of film and analyzed each frame under a microscope.

The first major revelation to come from CORONA was that the Soviet Union appeared to have a much smaller nuclear arsenal than the United States had feared. That was a major, if temporary, relief and proved the value of the program. The images provided a level of truth that the United States could leverage in military planning and politicking against the Soviets. In addition, the analysts discovered myriad previously unknown sites across the Soviet Union that seemed to be engaged in military activity.

In the years that followed, rockets often blew up and cameras often failed, but the United States pursued CORONA with unrelenting vigor. Almost twenty CORONA missions were launched in 1961, and they kept right on coming at a similar pace through the 1960s. Each year, the image analysts would be asked to go through around two hundred miles of film. Since they did not have great computing systems to keep track of the information gleaned from the pictures, they often created stories around important images and packed details into their narratives. They retold the stories to new analysts, creating a type of institutional memory that catalogued the photos.*

Technology advances, of course, led to major changes in gathering satellite imagery. In relatively no time at all, other nations joined the United States in orbit with powerful satellites of their own. The machines became more capable, as they could be aimed via remote commands to zero in on certain targets, and the resolution of their images improved again and again. While some measure of secrecy still surrounds those programs, it's thought that

* Thanks here to Jack O'Connor for his superb book *NPIC: Seeing the Secrets and Growing the Leaders: A Cultural History of the National Photographic Interpretation Center*, which documents much of this history.

dozens of spy satellites circle the earth with the ability to spot objects just a few centimeters in size from space. The pictures the satellites produce no longer have to tumble back to Earth through a series of engineering miracles but are simply transmitted directly into computer databases.

In the 1970s, the satellite imagery field moved beyond the spooks. Organizations such as NASA began launching satellites to survey the earth and monitor geological changes. Thanks to billions of dollars in taxpayer funds, we now have millions of images that cover fifty years of time, and people are free to peruse them. In the 1990s, the US government began to allow private companies to start flying their own imaging satellites and sell the pictures. The government placed restrictions on the quality of the images produced by the commercial satellites by limiting their resolution. Even so, the pictures were of use to the military, corporations, and researchers, and a handful of companies entered the market with their own satellite imaging fleets.

Over the past sixty years, it has been humans that made sense of all the satellite images. The US military, for example, has a tradition of finding bright youngsters and putting them through demanding training to learn how to spot items of interest in a photo. These analysts must memorize the size and shape of every kind of tank, truck, airplane, aircraft carrier, missile silo, and nuclear reactor of the militaries of numerous nations. If you can't remember that a Russian T-64 tank has two equipment boxes on its body while a T-64B has three equipment boxes, you're failed out of the program and given a different job. Typically, only about 10 percent of the people who start the image intelligence training end up passing the course.

Once done with the rote memorization, an image analyst has to hone his or her craft. A single analyst might look over the same fifty-square-mile chunk of territory for six weeks trying to find subtle changes in the landscape or new additions to a facility. Much of the work is monotonous, and the breakthroughs come from spotting subtle clues. Maybe different-colored vehicles start arriving at a building under watch, indicating that a new group has taken over the facility. Or, on the grimmest of notes, the texture of the ground in a field might change from one image to the next because a group of militants has dug a mass grave.*

In the hierarchy of who gets what images when, the military has always enjoyed top billing. It has the best satellites and the highest-resolution pictures. Most often, its satellites are pointed at known areas of interest. You can, for example, be sure that a machine in space has its eyes on North Korea and that thousands of images are flowing back to analysts every day. North Korea knows this and sometimes goes out of its way to hide military machinations from the satellites. On other occasions, it tries to make use of the fact that it's being spied upon. Its missile tests are often timed to take place when a US satellite is overhead so that there's a record of the event and an accompanying projection of power.†

The military, though, does not have enough satellites to see everything all the time, and neither do the traditional commercial satellite operators. The highest-resolution satellites are simply too expensive to build and launch en masse. Historically, this has led to large gaps in information and the inability of a company or analyst to obtain a photo of an area of interest on demand unless it happened to be a spot that was already under surveillance. Before Planet Labs arrived, that meant that a company or individual might submit a request for an image and then wait months for the picture to arrive. A satellite would have to be "tasked," in imagery industry speak, to aim at the right location, and that request would fall into line behind thousands of other requests. After the picture was finally captured, the company or individual had the pleasure of paying many thousands of dollars for it.

"You have to call the salespeople at one of these imaging

* People on the autism spectrum sometimes excel at this work and can find patterns and spot changes better than others. "I'm autistic, and we are very good at visual tasks," said the undergraduate Eveleth. Along these lines, the Israel Defense Forces have a large number of autistic soldiers in their geospatial intelligence unit.

† One rumor going around Washington, DC, is that the North Koreans short South Korean stocks ahead of their missile tests, banking on the fact that images of the country's tests will appear and rattle investors.

companies and tell them what you want," said Jeffrey Lewis, the open-source intel expert. "They'll quote you a price and look at their schedule and tell you where you fit in. There are different pricing tiers depending on how urgent your request is. But even then, their satellites sometimes miss your spot, or maybe there are clouds in the image that block what you want to see. It's just this very involved process of negotiating with them to put a satellite in place and then finding the right match for what you're willing to pay."

Because of Planet, people such as Lewis are now awash in imagery of the earth. The pictures produced by the Doves might not have the best resolution, but they are abundant and tell new stories. For the first time, we can see what image analysts call "patterns of life" for spots all around the globe. These are the daily machinations of humans and industries that reveal detailed truths about what's happening in a given locale.

Human analysts still play a key role in assessing the patterns of life, but increasingly the work is done by computers and artificial intelligence (AI) software. Many thousands of images are fed into AI systems to teach them how to spot items of interest on Earth. The AIs learn about cars, trees, buildings, roads, cargo ships, oil wells, and houses. Once they know what these objects look like, the AIs watch them all the time and make note anytime a road changes, a house is torn down, or a ship leaves port. This global analysis system never stops running and serves as a sentry system for the human analysts. When something interesting on Earth changes, an alert is sent to a human, who then takes a closer look at what's happening.

You are no doubt familiar with products such as Google Earth and Google Maps that seem to do similar work. Much of the imagery in those systems comes from commercial satellite systems, and Google has certainly done a remarkable job at using the pictures to catalogue the world. The pictures, though, are often dated and tend to be much better in highly populated areas. What Planet and other companies with novel AI tools have built can make Google's products seem like toys.

In 2019, Planet revealed that it had used its images and AI software to create the first complete map of every road and every building on Earth. To make the map easier to parse, the software color coded all of the buildings in blue and all of the roads in red, and the resulting images are almost like anatomical drawings. A city such as San Francisco has grids of blue boxes with veiny, winding red lines running through it. Were it a mere momentary snapshot of the earth's infrastructure, it would be useful enough. Planet's images, though, update as roads change and new buildings rise.

Similar systems have been built to map and count all the world's trees. The AI software can not only tabulate the number of trees but also tell what types of trees are present. It can then calculate their biomass and come up with a reasonable estimate of their carbon dioxide consumption.

These images and calculations add precision to previously opaque problems. In South America, Planet's technology has been used to monitor the state of the Amazon rain forest. On a base level, we can, all too depressingly, measure how much the rain forest shrinks each year. But we can also now hold people accountable for their actions. Numerous lawsuits in South America have been filed, and sometimes won, in which the satellite imagery has been the key evidence demonstrating that a company illegally cut down trees. Planet's images are adding muscle to carbon-offsetting programs as well. Auditors can use the company's software to check that a company that promised to plant a certain number of trees on someone's behalf actually did so.

The commercial uses for this type of technology pay Planet's bills. The US government shells out tens of millions of dollars per year to analyze the company's images in a variety of ways, stretching from intelligence gathering to environmental work. Other countries, which have not typically created satellite fleets of their own, have similar deals in place and have gained access to state-of-the-art space technology without needing to build satellite or rocket programs. Some of Planet's largest customers are farmers, who use special sensors on the satellites to monitor their crops in almost magical ways. The satellites can measure the amount of chlorophyll the crops produce to check on their health and the ideal times for harvest.

Start-ups such as Orbital Insight have appeared that take images purchased from Planet and free ones from public databases and then apply yet more sophisticated analysis to the pictures. Orbital can count the cars in Walmart parking lots during the holiday shopping season to see how busy the stores are. It then sells the data to hedge funds and others on Wall Street, who seek to profit from the privileged information. The company can also look across all the cornfields in the United States and track their health to predict what their yield will be when it comes time to harvest. Commodities traders on Wall Street purchase this forecast, which has proven incredibly accurate, as they place their bets on where the price of corn is heading. Yet more AI systems can estimate the gross domestic products of countries all over the world by counting how many lights are on at night, follow the movement of every ship at sea, and tally how much coal comes out of mines each day. It would require a thousand human analysts to do any one of these jobs, but the AI software performs them all without tiring.

Orbital's most impressive technology comes from the way in which it measures the global oil supply. The company analyzes images of oil storage containers, which have floating tops that move up and down depending on the amount of oil inside. By measuring the size of the shadow cast on a container's side when the roof is depressed, Orbital can tell just how full each container is and come up with a figure for the total amount of oil a country has stored at any given time. On numerous occasions, Orbital has run its image algorithms across thousands of storage containers in China and found that the country has far more oil than it publicly discloses to analysts and economists. As Orbital's founder, James Crawford, put it, "What we are selling is truths about the world."

The number of truths Planet's satellites reveal increases each year and adds much-needed context and detail to the machinations of the earth and its denizens. Just in California, the images produced by Planet help scientists monitor drought by measuring the amount of water in reservoirs. Other scientists survey forests and pinpoint the areas most at risk for wildfires, which helps them suggest places to thin or burn trees. Meanwhile, the people charged with protecting California's public lands turn to the satellites to locate illegal drug-growing operations.

Open-source analysts have presented reports on the construction of China's first homemade aircraft carrier, its takeover of islands in the South China Sea, and its expansion of Uighur reeducation centers. In such cases, Planet's images typically end up on the front pages of newspapers such as the *Wall Street Journal* and the *New York Times*, where they bolster the stories and make them more visceral for readers. Similar work has led to stories on the discovery of remote missile facilities in Iran, the construction of Tesla's enormous battery factory in Nevada, and attacks on Saudi Arabian oil refineries. When an explosion rocked Beirut in 2020, Planet quickly had images that depicted the extent of the destruction. And as the covid pandemic took hold, Planet's pictures documented the emptied cityscapes that accompanied a global economy grinding to a halt.

The truth, of course, is not always welcome. In 2019, Planet found itself in the middle of a dispute between India and Pakistan. Indian prime minister Narendra Modi's government claimed to have successfully bombed an Islamist group's training camp in northeastern Pakistan in retaliation for a suicide bombing that had taken place earlier in Kashmir. In the midst of an election year, Modi hoped to use the bombing as a show of strength. Officials in Pakistan argued that Indian jet fighters had missed their target and been bested by Pakistan's own military aircraft. Modi's government denied the claims and insisted that its strike had wiped out "a very large number" of terrorists.

In the past, people in the region were left on their own to try to discern which government was telling the truth. Each side would push its version of events and accuse the other of spreading disinformation. Reporters would do their best to go to the site and speak with witnesses to the bombing raid, but their accountings of events would still be surrounded by doubts and doubters.

Planet, however, had satellite images that clearly showed that India had missed its target. The bombs dropped by the Indian jets had fallen in empty fields. Despite having significant business in India, Planet opted to share its images with reporters who requested them, and their subsequent stories embarrassed Modi during a delicate political moment. As Will Marshall put it, "Pictures don't lie."

"Every couple of weeks, someone comes to me from inside the company and asks if we should release an image," Marshall said. "I don't think I've ever said 'No.' But there are situations where we would say 'No' if there's good reason to believe that the images would put civilians in harm's way or something like that. That's the kind of thing we would watch for. But if it's for reasons of embarrassment, that's a different matter."

In the forty-eight hours after Planet released its images, the news networks in India and Pakistan discussed the photos nonstop. Planet's customers in India complained about what the company had done, and people bombarded Marshall with messages via Twitter. A short while later, Planet ran into difficulties as it sought to purchase a ride for its future satellites on an Indian rocket. Someone in Modi's government had told the Indian space agency to make life more difficult for the start-up.

"It was really stupid and had to do with this election cycle and Modi showing he's powerful," Marshall said. "The good thing, in general, is that it showed that governments aren't going to be able to just do things and get away with lying about them anymore. This is part of a transformation toward worldwide transparency. We are trying to be careful about how we bring this out responsibly. But it is going to change the way that governments have to deal with the world. They can't hide."

The US government has funneled vast amounts of money into commercial satellite imaging companies over the years. The assumption has long been that the government could ask the companies to keep sensitive images secret and to point their satellites away from things the United States does not want others to see. With the advent of Planet, it became inevitable that images will eventually come out. Too many people have access to its all-seeing eyes for interesting things to be missed. Like its predecessors, Planet does a lot of business with the United States and its military and has to keep them sweet. Marshall, however, has bet that the days of secrecy requests have passed and given way to a new era and a new reality. "We think the data is far more useful for an open, democratic society," he said. "The more countries get on top of it and get used to it, the better they will be positioned. We will have issues with all governments on some level until they get used to this new, transparent regime."

Most people have no idea that these types of imaging satellites exist or that artificial intelligence is out there observing their patterns of life from space. We regular humans can take solace in the fact that the satellites cannot see our faces and that the analysis work focuses on broad trends rather than individual actions. Certainly, however, we're in the same position as governments and must come to terms with "this new, transparent regime" as well. A vast network of computing and observation systems has been placed over our heads, and they never stop watching or analyzing what we do. Although such technology seems sophisticated, it's in its relative infancy. The cameras will improve. The amount of data will increase. The algorithms will get better and better. The sum total of human activity will be turned into an extraordinary database, and certain people will find ways to mine that information in unexpected and perhaps unwanted fashion.

Innovative analysts and software engineers have already found ways to pair the satellite imagery databases with databases tied to individual behaviors. Orbital Insight, for example, has started using geolocation data gathered from smartphones to complement its imagery analysis. The apps on your smartphone constantly monitor your location, and the app makers sell the data to companies that anonymize it—honest—and use it to track the flow of humans around a city. Orbital can request data on how many people are going into and out of a Tesla manufacturing plant to see if the carmaker is running double or triple shifts to pump out vehicles or its manufacturing lines have slowed.

A military analyst told me that she has automated alerts set up for different shipping ports around the globe. If a satellite detects an unusual amount of activity at a given port, she begins analyzing the images to try to find out what's happening. On one occasion, she received an alert to examine Puerto Cabello on Venezuela's northern coast. The analyst tapped into the satellite imagery and saw that a large oil tanker had pulled into port. She then took the port's geographic coordinates and fed them into various social networking search systems, where the coordinates were matched up to the location metadata of photos people had posted online. Sure enough, the analyst spotted several Russian seamen documenting their travels near Puerto Cabello. Taken together, the data points suggested that a Russian oil company was in the midst of delivering crude to Venezuela in violation of US sanctions.

When many people hear about Planet's abilities for the first time, their minds tend to race toward all the ways the technology could be put to ill use, such as spying on the activities of regular citizens or nefarious governments gaining access to powerful technologies they could not otherwise build on their own.

Marshall does not dismiss the tensions that Planet's images cause or the issues they pose. Planet goes to great lengths to secure its satellites and monitor how its pictures are used. Still, its images come with the usual unsatisfying trade-offs of most new technologies between the mix of powers and perils that they unleash.

Marshall's hope, of course, is that the good will far outweigh the bad and that Planet can contribute to solving some of our most pressing problems. "We are producing a data set that can significantly help us face the global challenges that we're facing as a species today," he said. "Tracking and stopping deforestation, tracking and stopping illegal fishing, protecting the coral reefs, helping humans get better quality of life via better access to water, better food productivity, and better transportation efficiency. We can help humanity take better care of our resources."

INVENTION IS OFTEN SYMBOLIZED BY a picture of a light bulb going off over someone's head. This leads us to believe that an invention is a big idea that comes to you in a flash. And, yes, this does happen. Of course, celebrating the notion of a moment of genius does a disservice to all the mess that precedes the grand insight. Invention is not luck or a clever thought that arrives in the shower; it's a process—and a weird, inexplicable one at that.

Marshall had the bright idea to have satellites work together as a constellation. It's doubtful, however, that he would have hit upon that insight had it not been for a complex set of circumstances. It took an odd astrophysicist general, an idealistic kid fond of gazing at the stars, a communal home of tech hippies, and all of them being at the right place at the right time in terms of the technology for an Earth-enveloping network of space cameras to make sense.

Another thing about the brightest of ideas is that once they move from arrival to fully formed, they can feel so obvious. People once doubted that tiny satellites could do much of anything useful. No one knew if they could be maneuvered in space by using solar panels like sails. No one had thought much about mass-producing satellites as if they were any other technology product. But then, once Planet pulled it off, new satellite companies started to appear by the dozens.

When I first started reporting this book, there were about two thousand functioning satellites in orbit. With its more than two hundred satellites, Planet alone accounted for about 10 percent of that total. And it would continue to do so if the situation were static—which it most certainly is not.

By the end of 2021, there were five thousand satellites in orbit. About two thousand of them were built and launched by SpaceX. These satellites do not take photos but rather are part of SpaceX's

Starlink internet system. The machines orbit around the earth and beam down high-speed internet to antennas on the ground. The major near-term objective of Starlink is to create the first truly global internet service. Anyone with a Starlink antenna can tap into the Web from wherever they are. For roughly 3.5 billion people who cannot get a high-speed internet connection, this could be a godsend. They will be able to join the modern world. Meanwhile, people on planes, on boats, and in cars or in remote spots will have the same luxury. The internet will be inescapable. For the first time, the earth will have the information network that we have created moving around it like an unstoppable force.

The two thousand satellites, though, are enough to fulfill part only of this vision. They cover a limited swath of the planet. SpaceX expects to put up as many as forty thousand satellites in the sky to complete its giant network.

SpaceX mimicked Planet's philosophy of satellite construction: it created smaller, more modern communication satellites than had ever existed before. It learned how to mass-produce them after much trial and error. It flies the satellites relatively close to the planet in low Earth orbit to ensure that their signal strength remains high. And it treats the satellites like disposable objects: they go up, orbit the earth for a few years, burn up on their way back down, and are replaced by newer, better models.

As it happens, SpaceX is not alone in its desire to make a worldwide space internet. In recent years, companies such as Apple, Facebook, Amazon, Samsung, and Boeing have looked into flying fleets of thousands of satellites, too, as have nations such as China and Russia. For the moment, Amazon appears to be SpaceX's main rival among this brand-name bunch, and it wants to put up around 3,500 satellites as quickly as possible.

But the company best positioned to challenge SpaceX is a startup called OneWeb. Like SpaceX, it was inspired by Planet's technology and began planning a massive space internet system at the same time as Elon Musk. It has already put up hundreds of satellites with the help of European and Russian rockets. None of this is cheap.

Through the early part of 2022, OneWeb had raised a staggering $4.7 billion from investors such as the UK government, Coca-Cola, SoftBank, and Richard Branson's Virgin Group. Though SpaceX has the advantage of owning its own rockets, it, too, has needed to raise billions of dollars to fund Starlink.

Outside these major players, there are yet more companies of various shapes and sizes hoping to build space internet systems. They're all competing for the communications spectrum that will carry their signals from orbit down to the ground and for space in space. The satellites have to be arranged so that they do not interfere with one another's signals and do not run into one another.

In the United States, government agencies monitor these issues, and international bodies, including the United Nations, overlook the heavens as well. They try to ensure that rocket and satellite companies know what they're doing and are putting machines into orbit safely. The bureaucrats also try to parcel out spectrum in a fair manner and divvy up the territory above us in an equitable fashion.

What's become clear in recent years, however, is that regulators cannot keep pace with the launches or the wills of the people leading the various companies. The regulatory bodies spent decades operating under a regime in which a handful of rockets went up every couple of months and the number of satellites increased by twenty to fifty per year. Now the number of satellites going into space has hit an exponential curve, and companies aim to put up tens of thousands of them annually. Musk and others keep launching their rockets and satellites while people on the ground below spend months or years debating the regulatory merits of the various constellations.

No one knows if the business case of the space internet makes any fiscal sense. In the late 1990s, a company called Iridium spent $5 billion to put up eighty satellites for an early crack at a space internet system. Creating a network in space at a time when the consumer internet and cell phones were taking off proved a terrible idea. The company's subsequent bankruptcy scared all comers off of dreaming up anything so ambitious until Planet came along twenty years later and offered evidence that the times had changed.

The 3.5 billion people without access to high-speed internet tend to live in the poorer parts of the world. How much money SpaceX or Amazon could make from those customers remains to be seen. Businesses and wealthier individuals will pay for the convenience of a quick connection wherever they go, but, again, nobody knows how large the audience will be. As things stand, SpaceX is valued by its investors at more than $100 billion, and the vast majority of that figure is premised on Starlink creating a major revenue stream. Even for a company as efficient as SpaceX, there are not huge profits in rocket launches. It's much better to be a worldwide telecommunications company with subscribers paying monthly fees.

The politics of space internet systems come with plenty of ambiguity, too. SpaceX and others must apply for licenses to provide their services in most countries. Countries such as China and Russia that control what sorts of information can flow across their networks despise the idea of anyone being able to buy a Starlink antenna and evade their draconian firewalls. Still, just about any country that cares about its data infrastructure and has the money to invest will want a space internet. This makes the oncoming onslaught of satellites an inevitability.

Either comically or tragically, the average Earthling is not paying any attention to what's happening above. You would be hard pressed to find someone who knows we're destined to go from five thousand to fifty thousand satellites in a few years and then much more after that. Even astronomers, who had every reason to be concerned about those objects blocking their views and have heard Elon Musk promising to fill the skies with Starlink systems for years, did not start objecting to the space internet idea in earnest until it was already being built. By then, a few complaints from academics sitting behind their telescopes were never going to counter the ambitions of billionaires and nations.

Outside these high-speed internet systems, there are dozens more constellations being built for imaging and low-speed data services. Some companies have developed imaging satellites that use a specialized type of radar to see through clouds and snap pictures at night. Others have created devices designed to take precise measurements of methane coming out of gas wells and the health of the oceans. One start-up called Swarm Technologies has managed to make satellites that are no bigger than a pack of cards. Regulators were worried that tracking systems on Earth would not be able to spot the devices and that they would thus pose a risk to everything else in orbit. Swarm was told not to launch its satellites by US officials but did so anyway in 2018 by sneaking them onto an Indian rocket. It was the first illegal satellite launch that anyone could recall and also a sign of how frenetic and borderline out of control the aerospace industry had become. Swarm received a stern talking-to and a $900,000 fine from the Federal Communications Commission—and then kept right on launching satellites.*

People, of course, fear that all these satellites could bash into each other in orbit, which would be catastrophic for our modern way of life. There's a phenomenon known as the Kessler syndrome that predicts that low Earth orbit could turn into an absolute shitshow as the result of a relatively small number of collisions. One satellite would bash into another at high speed, and the crash would result in thousands of pieces of debris. Each piece would then morph into a high-speed missile that could collide with other satellites and create a cascading effect. If enough debris were to end up in low Earth orbit, it would be difficult to send new rockets and satellites through the mess. What's more, existing technologies such as GPS and our communications systems could be ripped to shreds, sending life on Earth back to a different era.

Naturally, there are now start-ups that track all of the satellites and the existing debris and offer their services to the likes of Planet and SpaceX. They can advise the companies of pending collisions and where to move their satellites for a spell to avoid danger. Other start-ups are appearing on the scene proposing to become space debris garbage collectors.†

---

* After the initial launch, it turned out that radar could, in fact, see Swarm's satellites. The subsequent launches were done with the blessing of regulators.

† Will Marshall, ironically, worked for a while on exactly this sort of thing at NASA Ames.

Because we are humans, we will ignore whether or not these satellite constellations are sound businesses worthy of all the risk. Low Earth orbit has emerged as the most exciting and untapped real estate market imaginable. Regulators and governments can control things to an extent while the rockets and satellites are still on Earth. But once something is actually in space, there's little that a paper pusher or politician can do. Companies now largely have the wherewithal to go into space as they please and put whatever they want up there.

We are in the early days of the next great infrastructure build-out. A communication system is being constructed that will surround the earth with a digital heartbeat. Our computers and cell phones will never be outside the reach of an internet connection. What's more interesting, though, is that neither will our self-flying planes, autonomous cars, or drones. Almost every sci-fi contraption that you've been promised over the past twenty to fifty years will come to depend on this ever-present information network.

What's more, so will a host of new computing devices that we're only now seeing glimpses of. Farmers will put moisture sensors all over their land and have the devices report back on what they detect to the computer in the sky. The same will go for tiny sensors on shipping containers and the objects inside them. The internet will come from space, be ubiquitous, and change life as we know it. Unless, of course, everything goes to shit first.

SpaceX may have kicked off this charge back in 2008 with the Falcon 1. But you don't have to work very hard to argue that Planet is equally, if not more, responsible for the brave new world ahead. Elon Musk lowered the cost of launching a rocket by tens of millions of dollars. That said, $60 million for a ride to space is still a lot. Planet's satellites were not just a fraction better than the status quo; they were a thousand to ten thousand times better. Cheaper. Smaller. More powerful. They changed not only how we get to space but what we can do once we're there.

On Earth, the global economy and productivity have boomed over the past six decades in large part because of Moore's Law. This is the technology industry's dictum that computers will run twice as fast every couple of years while also becoming cheaper and smaller. It's the force of unyielding progress that has resulted in the modern world as we know it.

This same push had never quite made it to space. The computers and related technology in low Earth orbit were always well behind the times. Space was still dialing into AOL on a modem, while Earth was consuming TikToks on smartphones.

Planet altered the equation. Put simply, it brought Moore's Law to space. The Doves were the first step toward aligning the pace of innovation between the earth and space and putting our terrestrial and orbital economies onto the same clock.

The only real gating factor preventing the space economy from taking full advantage of this new reality and exploding at internet speed has been a lack of rockets to put up all the new satellites. We needed supercheap rockets that were flying all the time and venture capitalists to fund their creation.

For people who already fancied themselves as the next Elon Musk in their dreams, the call to action was loud and clear: Get yourself a team and some money. Let the great rocket race begin.

THE PETER BECK PROJECT

CHAPTER EIGHT

BIG, IF TRUE

Elon Musk called in the early part of the evening. Or at least my evening.

It was November 2018, and I was staying in Auckland, New Zealand, for a couple of weeks, renting a house in a nice suburban neighborhood. My day had been spent hanging out at the main factory of Rocket Lab, a maker of small rockets, and my thoughts had been on the company and its founder, Peter Beck. One of Musk's assistants, however, had contacted me after my Rocket Lab visit to say that Elon would be calling any minute, which required a change of focus.

Ahead of the call, I ate a pot gummy smuggled into the country by a friend* and knocked back a beer. Those actions could be considered self-care, you see, because I'd not talked to Musk in any meaningful way since the publication of my biography of him three years earlier, mostly because he had not appreciated some of the things I had written in the book and had threatened to sue me, which had made our relationship considerably less amicable than before. Attempting to deaden years of emotional baggage and smother the

---

* Dear New Zealand immigration authorities: this is not a hypothetical friend but an actual friend. Please let me back into your beautiful country.

tense anticipation thumping my skull, I'd turned to the THC-laden gelatin and a pilsner. We've all been there.

I figured we'd address the multiyear cold shoulder right at the start of the call and talk through some of our issues, but Musk had other ideas. He knew I was in New Zealand and fixated on that. "Aren't there just like a lot of sheep over there?" he asked. "That's what I heard. A lot of sheep. And Kim Dotcom."

For those who don't know, Kim Dotcom ran an internet service called Megaupload that allowed people to swap large media files. Authorities in New Zealand, where Dotcom lived, and the United States were appalled by the amount of copyrighted material being traded on Megaupload and raided Dotcom's mansion in 2012. It was one of those no-joke raids with gun-toting police helicoptering into Dotcom's compound and bringing their perp to justice after a punch in the face and a few swift kicks to the ribs. "If I go to New Zealand, I want to see Peter Jackson's house, which is basically like visiting *Lord of the Rings*, and Kim Dotcom," Musk said. "Those are the two things. We could reenact the raid."

It turned out that Musk wanted to talk with me about Tesla and how the company had navigated a disastrous year and come out the other side alive. We did that for a while, and then I steered the conversation back to Rocket Lab. Peter Beck's company had recently joined SpaceX in the ranks of successful private rocket companies, flying one of its machines to orbit from its own spaceport. I wanted to know what Musk made of the upstart. "It is impressive that they managed to reach orbit," he said. "It's fucking hard. Bezos has spent a shitload of money, and he hasn't made it."

I told Musk that Beck would like to have dinner with him some time, which Musk found amusing. "I'll take you out on a steak dinner date," he said with a jokey voice. "There better be some flowers."

The two men would eventually have a meeting and a curious one at that. But during the call, Musk mostly blew off Rocket Lab and Beck as things that were not of much interest or concern to him.

Back in 2018, Musk was not alone in those sentiments. Rocket Lab had created the second coming of SpaceX without many people noticing. Its relative anonymity could be traced to the company's roots in New Zealand, which is so far from the rest of the world that its goings-on are sometimes ignored. Beck also did not arrive with the usual trappings of a space mogul. He was not a billionaire, nor did he have any track record of building a previous blockbuster technology company. He did not make controversial statements or do flashy things. Quite the opposite: though enthusiastic about his work, Beck mostly put his head down and got on with rocket building.

Out of a mix of luck and curiosity, I'd discovered Rocket Lab in 2016 during a visit to New Zealand. People in the space trade press had written some stories about an Auckland-based outfit that was trying to build a small rocket called Electron. While the stories had caught my attention and convinced me to see the company, my expectations had been low. For one, Electron was basically a glorified version of the Falcon 1 rocket that SpaceX had made and flown successfully way back in 2008. Rocket Lab was making Electron out of more modern materials and applying some new technological twists, but the concept remained the same. It was to be a small rocket that could be flown cheaply and carry a few satellites into space on each journey.

What really made me skeptical of Rocket Lab, though, was Peter Beck, and the fact that he was trying to build a rocket in New Zealand. Beck, various stories noted, had not studied aerospace engineering formally. In fact, he had not attended college at all. His work experience consisted of a stint at a dishwasher manufacturer and some time in a government research lab. All the while, rocketry had been Beck's hobby, which he explored late at night and on weekends. Somehow this man had convinced venture capitalists to fund his pastime.

The story being peddled made no sense. One did not simply will a rocket company into existence. The United States, with tons of resources and knowledge at its disposal, had produced only a single successful rocket start-up, SpaceX. New Zealand, by contrast, could not even be described as an aerospace backwater. It was basically void of all the things needed to make a rocket, such as well-trained aerospace engineers, the right materials, and good infrastructure. Beck, an amateur rocketeer, would have to solve all those issues along with the complexities of rocket science while on a literal and metaphorical island. Surely the investors had made a terrible mistake.

Back in 2016, Rocket Lab operated out of a large building near Auckland Airport. It had the standard rocket maker setup, with engineers doing their thing behind computers, a couple of rooms for crafting and testing electronics, and a large factory floor where three Electrons were coming to life. The company had another spot out in a pasture where it could test engines. Those arrangements were, in many ways, ideal because it was doing its engineering and testing near the heart of a major city instead of being cast off in a desert, as was the case with many a rocket start-up.

The things Beck had accomplished shocked me. A sixty-foot-tall, four-foot-wide Electron was in the final stages of construction, with the other two well on their way to completion. Rocket Lab had made the machines of carbon fiber instead of aluminum or stainless steel in a bid to improve the rocket's strength while keeping it lightweight. The black material gave the rocket a punk, polished look. It put the sexy back into space phalluses. Electron would be powered with the help of nine of Rocket Lab's Rutherford engines, named after the famous Kiwi scientist Ernest Rutherford.* The engines were works of art. Pieces of twisted metal were arranged around electronics with precision and care. The factory itself was organized in the same exacting way, with all the tools categorized and placed on workbenches just so.

Beck, who was thirty-nine at the time, materialized in the factory with minimal pomp and circumstance. Of average height and build, his most striking feature was a dome of curly brown hair. He seemed pleased that an American reporter had traveled all the way to New Zealand to check out his rocket palace and gave me the grand

* "Rutherford has a really famous saying, 'We have no money, so we have to think,' which resonates with us well," Beck said. "This is about solving a problem and really thinking about different ways of approaching a really complicated problem."

tour, spewing out technical details on Electron and the Rutherford engines all the while. His lack of pretense almost undermined what Rocket Lab had created. As a rule, rocket moguls love to boast and preen. Beck came off more as a normal guy who just happened to have built a rocket.

The thesis of Rocket Lab, as laid out by Beck, was clear: Rocket Lab would complete the mission that SpaceX had begun and then forgotten. It would deliver the world's first cheap, reliable rocket ready to fly into space at a moment's notice.

Not long after the initial successful flight of the Falcon 1 in 2008, SpaceX had abandoned work on the rocket and shifted toward the much larger Falcon 9. The company's main mission was to create a flourishing human colony on Mars, and only big rockets would be useful for the task. Even in the nearer term, SpaceX did not think the Falcon 1 served a practical purpose. In 2008, companies still made large satellites that required large rockets. Devices like those built by Planet Labs were several years away from becoming mainstream. SpaceX might have crafted a perfectly fine small rocket, but no one needed it at the time.

Founded in 2006, Rocket Lab had bounced around for a few years pursuing various projects before realizing that the Falcon 1's demise coupled with the rise of small satellites had opened the perfect opportunity for the company. It would make a small rocket that could carry about five hundred pounds of cargo to orbit for $5 million per launch. Instead of launching the rocket at the typical cadence of once a month, Rocket Lab would seek to launch one every month, then every week, and then perhaps every three days.

The launch price and the number of launches would be a revolution for the industry. Rockets generally came in two sizes: medium and large. The going rate to hitch a ride aboard one of the machines started around $30 million and went up to $300 million. The customers paying for those launches had large satellites that cost anywhere from $100 million to $1 billion to make.

Companies such as Planet Labs with smaller satellites and less money typically paid to be excess cargo on other companies' rockets.

Their machines were tucked into spare nooks and crannies around the larger satellites that were the primary concern. SpaceX, for example, would fly a large satellite to its desired orbit first as the main payload and then let the small satellites trickle out into space like afterthoughts. Whereas the large satellites were placed into orbit with great precision, the small satellites often had to maneuver their way into the right orbits over the course of several months. The small satellites were treated as second-class citizens.

Rocket Lab, by contrast, would cater to small-satellite makers' every need. Companies like Planet Labs would no longer have to wait around for a lucky moment when some extra cargo room on a big rocket freed up. They could order an Electron of their very own and have it fly exactly where they wanted it to go. To a start-up trying to get its business going, the ability to put something into space in a timely fashion felt invaluable.

By lowering the price of getting satellites into space and making rocket launches a regular occurrence, Rocket Lab would become like a rapid shipping service to low Earth orbit. And by 2016, it certainly looked as though the world might need exactly such a thing. Companies like SpaceX and Samsung were revealing plans to fly thousands upon thousands of satellites into orbit to power new space-based internet systems. Numerous other companies both large and small had similar goals and wanted to build huge satellite constellations of their own. If such systems came to fruition, there would not be enough rockets to meet the demand for all the launches. In addition, those constellations would need to be serviced with rockets flying up to replace satellites that had broken or been decommissioned and burned up after deorbiting. Rockets would have to be blasting off all the time, but the existing large rocket makers launched only one of their vehicles per month on average.

"If you put the internet into space, you've put an infrastructure in space that's no different to any other utility," Beck told me. "It's like water or power. Your internet can't go down. You might use a large rocket to send up a lot of satellites in one go and build the initial constellation. But if you have a couple of satellites go down, within hours you need to have replacement satellites in there to get your infrastructure back up. That's where we can play as well; the turn time for us to get something on orbit will be a matter of hours."

Looking farther out, Beck predicted that the presence of a dependable, relatively cheap ride to space would have a dramatic knock-on effect on the number of companies willing to try out new ideas. More and more satellite companies would appear as a voyage to space shifted from a rare, expensive happening to an expected everyday occurrence. Rocket Lab would help change people's mindset around accessing space. It would help the space economy boom.

Where others saw Rocket Lab's place in New Zealand as an issue, Beck contended that it was an advantage. Most spacefaring nations had started their spacefaring decades earlier and had become bureaucratic and bogged down over time. Their spaceports tended to be dominated by military contractors or government bodies. New Zealand, by contrast, offered Rocket Lab a clean slate. The country had no legislation governing space activities, which presented Beck with the unique opportunity to tell an entire nation how it should regulate a rocket company.

From a pure logistics standpoint, Rocket Lab would need to contend with fewer people, planes, and boats when launching rockets. The company intended to build a private spaceport in a remote part of New Zealand's South Island. From there, it could theoretically launch rockets at will without anyone really minding. Unlike other rocket makers, Rocket Lab would not need to wait for lulls in air traffic or for boats to clear out of the way. It could simply stick a rocket on the pad, press a button, and let it go. And since Rocket Lab would own its spaceport, it would not need to pay the $1-million-per-launch fee typically charged by NASA and others that ran launchpads.

Though there were plenty of reasons to think that Rocket Lab would not accomplish its goals, the company's very existence spoke to a massive shift starting to take place in the aerospace industry. Governments had built rockets. Then Musk had come along and put his fortune toward building a rocket. Now we had a company making a go of it based on venture capital money alone. Investors had decided that space could be a real business like any other. Did the world need a rocket that could fly every week? Would it be a profitable business? Would our entire psychology around flying things to space change if a regular, cheap ride to the heavens existed? People were willing to pay to find out.

Tradition had dictated that new rocket programs required the efforts of thousands of the brightest scientists and engineers and billions of dollars in funding. SpaceX had altered those long-held assumptions by lowering the cost of rocket production and recruiting young, inexperienced engineers. Much of its most important technology, though, came from the brains of Old Space veterans who had done stints at places such as NASA, Boeing, and Lockheed Martin. They knew what mistakes to avoid even as they tried out fresh approaches. Rocket Lab lacked the ability to tap into any such stores of wisdom.

Beck had never built a real rocket before, and neither had his employees. Because rockets are intercontinental ballistic missiles by another name, the United States placed tight restrictions around people working on the machines and the associated technology. That prevented Rocket Lab from hiring veterans of past US rocket programs and left Beck to assemble his team by recruiting fresh-faced kids from universities in New Zealand, Australia, and Europe.

By rights, the combination of an inexperienced leader with inexperienced staff trying their hands at rocket science should not have worked. Beck, though, had the makings of an engineering savant who seemed to possess an innate feel for physics and how machines wanted to operate. He bet that his talent, coupled with the amazing advances that had occurred in the fields of computing and materials, meant that rockets could now be made by new groups of people and under new conditions. It was still rocket science, and it was hard. But the technology no longer had to live in a mythical realm populated entirely by geniuses. Rockets had become approachable.

That thesis did not carry the same sex appeal as endeavors during the Apollo era or even Musk's risky gambit with SpaceX. Rocket

Lab and its investors were basically saying that technology had come far enough to allow any sufficiently creative and capable group of humans with a decent amount of money to reach space. If they were right, space would feel less magical and more pragmatic. That was perhaps why people were not rushing to fawn over Beck and his company. To me, however, Rocket Lab was that much more fantastic for what it represented.

The earliest rocket pioneers—Konstantin Tsiolkovsky in Russia, Hermann Oberth in Germany, and Robert Goddard in the United States—had all hit on the idea of building machines to explore space at about the same time. Inspired by the science fiction of Jules Verne and H. G. Wells and the breakthroughs of the Industrial Revolution, those men began proposing in the 1920s that liquid fuels could be combined to propel a projectile into orbit. Goddard did the best of the bunch and managed the first such rocket launch in 1926, albeit only to a height of forty-one feet. "It would cost a fortune to make a rocket hit the moon," he wrote. "But wouldn't it be worth a fortune?"

Over the next two decades, Goddard and others received money from individuals and military bodies to pursue their work. Though rockets did not have obvious commercial potential, they were viewed by some observers as objects of scientific prestige and accomplishment. Wealthy people had been funding the development of ever more sophisticated telescopes for centuries, and there was a chance that rockets and later satellites could have gone the same route.* It was the Cold War and the Space Race that came with it that pulled aerospace technology in a new direction, far, far away from private capital and toward a reliance on nation-states. In the United States, it became an assumption that NASA did space, and that was that. Rockets and satellites became intertwined with the will of the country and the power it wanted to project to the rest of the world. The entrepreneurial spirit and the individual passion

* Alex MacDonald, the chief economist at NASA and one of Pete's Kids, made a strong argument along those lines in his insightful book *The Long Space Age: The Economic Origins of Space Exploration from Colonial America to the Cold War*.

of the dreamer had to fade into the background. While the Apollo missions inspired so many, they also calcified people's impression of who can try to get into space and how they can try to get there.

Other hobbyists and wealthy individuals attempted, on occasion, to reinvigorate the century-old spirit of Goddard and largely failed. It took Musk and his merry band of engineers to make commercial space feel real. Beck and Rocket Lab, though, had the potential to be the next dramatic step. "Rocket Hobbyist in New Zealand Reaches Orbit" would be exactly the sort of headline that would have warmed Goddard's soul, only he would not have expected to wait until 2016 to read it.

Rocket Lab certainly was not looking to make the most complex of spacecraft. Large rockets still required huge investments and a certain level of technical expertise. Beck, however, did want to make the most elegant and precise small rockets. He did not dream of sending people to the moon or Mars; he dreamed of making a tool that would unlock the potential of space for others. "I'm here to commercialize space," he said. "That's what's important to us. We don't need to have a big song and dance. We've got a job to do; let's just get on with it."

Beck told me that in January 2016 and vowed that Rocket Lab would kick off its plan for low Earth orbit domination by the middle of the year, when the first Electron was due to take flight. He said it with such enthusiasm and confidence that I wanted to believe him and to believe that Rocket Lab would start a new, exciting era in commercial space. But all I could really do was smile politely, while chuckling to myself on the inside. Hadn't Peter Beck read about what happens next? First come the delays and then come the explosions and then the money runs out. It's always the same story.

CHAPTER NINE

A BOY AND HIS SHED

Peter Beck grew up at the end of the world.

To find his hometown of Invercargill, you must travel as far south as New Zealand will allow. To the north and east of Invercargill, farmers have taken over verdant, flat lands and filled them with cows and sheep. To the west, there's Fiordland National Park, which ancient glaciers, or perhaps a divine hand, carved up into an obscenely gorgeous mix of forests, lakes, and mountains. Invercargill slides into the spectacular surroundings with some measure of embarrassment. It's a sleepy city of sixty thousand people known to the rest of the country as an all-too-windy and all-too-cold place.

The buildings that make up the modest city center of Invercargill appear to have been shipped direct from Scotland and England in the 1850s. The same goes for the street signs: Dee Street, Tyne Street, Pork Pie Lane. Tourists visit Invercargill to take in its throwback charms, and how could they resist? A billboard at one of the main hotels notes that the major local attractions include a redbrick water tower built in 1889 that's an astonishing one hundred feet high, a park (it's quite nice), and Bill Richardson's

Classic Motorcycle Mecca, which is "Australasia's premier motorcycle museum," which, okay, if you say so.*

The enthusiasm for motorcycles is not Bill Richardson's alone. Like much of New Zealand, Invercargill is populated by gearheads who enjoy modifying and racing machines. The city's most famous figure is probably Burt Munro, who customized motorcycles with such skill that he set land speed records and whom Anthony Hopkins played in a movie. Peter Beck is also one of those people who likes to meddle with metal and seemed somewhat destined to call Invercargill home.

If you take a stroll around Invercargill and ask people if they know Peter Beck, some people will say "Yes" and some will say "No," and this is remarkable.† Outside of Elon Musk, no one has achieved as much success in the private rocket business as Beck. As a result, he's one of the country's wealthiest people and would likely be a household name just about anywhere else. New Zealand, though, is a peculiar place.

The Beck whom most Invercargill residents do know is Russell, Peter's father, who passed away in 2018 at the age of seventy-six.

Russell was a Renaissance man. He spent two decades as the director of the Southland Museum and Art Gallery, where people could learn about things ranging from Māori traditions to rare animal species and the artistic movements of the area. Though the museum building predates Russell's arrival in 1965, its most appealing architectural elements owe themselves to Russell. He raised funds in the 1990s to erect a white pyramid over the building; it is the largest pyramid not only in Australasia but in the entire Southern Hemisphere.‡ To the left of the pyramid is an observatory with a twelve-inch-diameter telescope that Russell built as a teenager. It became the regular meeting place of the Southland Astronomical

* The billboard makes no mention of Invercargill's lingering eau de livestock or its unusually abundant number of advertisements for free bowel screening, but billboards are not known for their honesty.

† Such was the case in 2019, although surely it cannot remain true for long.

‡ So claims the internet.

Society and provided the first brush with astronomy for many local schoolchildren, including Peter.*

Outside of his museum work, Russell Beck pursued a passion for jade, which the Māori call *pounamu*. He studied where to find *pounamu* on the South Island and throughout the world; the stone's history and relation to the Māori culture; and how to work with it to create jewelry and other artistic pieces. Russell authored several books on jade and in fact became one of the top experts on the greenstone. At the end of his life, he donated a collection of 1,500 pieces gathered from trips around the globe to a New Zealand research institute, gifting it what was considered the most extensive jade collection ever assembled.

Russell Beck also left his mark all around Invercargill and New Zealand's South Island via large sculptures that often have playful scientific themes. In the city center of Invercargill, he built a large metallic umbrella that functions as a sundial and also has various constellations built into its see-through canopy. The handle of the umbrella resembles a *koru*, the spiral shape of an unfurling fern frond often found in Māori art, and is a nod to the Māori study of the stars. Not far away is his *Cube of Learning* sculpture, which looks like a cube at a distance but reveals itself to be a rhombohedron as you approach it. Beck said the optical illusion was meant to teach people that "not everything is what it seems and one must question and investigate." Near the coast, he built an enormous anchor chain sculpture that juts out from the water's edge and locks onto the land.

After Russell Beck died, the local paper wrote an obituary that stated, "There seemed nothing he could not do." And that's really the point of all this. The Becks had moved to New Zealand from

---

* Here's Beck recalling one of the earliest space memories he shared with his dad. "My father would tell me the story, but when I was very young, he took me outside, and he showed me a shooting star and explained to me that that was a satellite and that that satellite was doing stuff. I said to him, 'All these other stars, are they satellites, too? Are they made by men, by humans?' He told me those were suns and they have planets and they could have people on them. That was probably the first moment where space was relevant to me, so where the concept of a satellite was not a foreign thing to me. I had books and books, but this knowledge of satellites was something supercool. They had the ability to impact a lot of people on the planet."

Scotland and immediately established themselves as a very capable, imaginative people. Russell's forebearers came from a long line of blacksmiths and upon immigrating to New Zealand set to work building machines for local farms, eventually turning their creations into a thriving business. Russell shared the ability to craft just about anything with his own hands, and so, too, would Peter.

Invercargill may be cloudy and cold, but it's a pretty nice place to raise children, and Russell and his wife, Ann, a teacher, created an idyllic existence for Peter, who was born in 1977, and his two older brothers. The family lived in a redbrick 1950s-era house with three bedrooms and a large workshop, situated in a well-manicured part of town.

The focal point of the Beck boys' childhood energy was the workshop and building things in it. The workshop, a garage with an olive green door, started out as a parking spot for Russell's car and a place for him to store research materials. As the years went by, the car was left outside, and the research materials were replaced by milling machines, lathes, welders, and other tools, and the building was transformed into a research-and-development facility.

If the family bought a new stereo, one of the brothers would soon nick it from the living room, take it to the workshop, rip off the back, and try to see how the machine worked. Instead of freaking out at the possible destruction of his new high-end equipment, Russell would join in with the boys, saying "Oh, well, let's go ahead and have a look and see what's in this thing." He sought to encourage experimentation and turn his boys into what he called "industrious makers." "I think the quickest way to learn is to have a desire to make something and then the ability and facility to make it," he said.*

Russell would teach the boys how to use various tools and then try to get out of their way. "We never had someone standing over the top of us, saying 'Be careful of that' or 'Do that,'" Peter said. "We'd be using a power drill and see Dad out of the corner of our

* Courtesy of an interview conducted by the Southland Oral History Project.

eye and knew he was watching, but he would never intervene unless there was an absolute certainty that we were going to hurt ourselves in a bad way. It would probably be seen these days as a bit hippie in that we had these resources and would just go for it. It was probably a different childhood than most people have."

As teenagers, the boys spent as many of their waking hours as possible working on cars. The older brothers went first, buying cheap, beat-up vehicles and bringing them to the workshop for fine-tuning. They would strip away just about every part of a car and then retool it based on its skeletal frame. They were not out to make regular, purely functional cars; they sought to make fast cars that could be tested during weekend races. After that, they would sell the cars and use the proceeds to buy more machinery for the workshop and more vehicles and repeat the process. "Nobody just bought a car," Peter said. "You *built* your car. You bought a clapped-out rig, and then you hotted it up, and that was what you did."

One of the first major projects Peter undertook was the construction of an aluminum bicycle. Mountain bikes were the craze at the time, but, fourteen-year-old Peter figured, no one had a superlight, supersturdy aluminum mountain bike. "I wanted the Ferrari of bikes, not a normal bike," he said. He announced his plan to the family, and his father offered a few words of caution: "That's a very complicated project. Make sure you finish it."

His words were serious. In the Beck household, it was unacceptable not to complete a project. Peter relished the challenge.

Since Peter had no money, he found most of the aluminum he needed for the project by going to metalworking shops and scrounging through their scrap bins. In the school machine shop, he taught himself how to create things such as the headset bearings that let you turn the handlebars and steer the bike. Sometimes he would find old bikes and chop them into pieces to use as molds for his new aluminum parts. Part of the process involved heating some metal in his mother's oven and cooling other metal bits in her freezer and then bashing the parts together in a process known as a "press fit" that is an all-or-nothing, instantaneous affair. "I'd run out of the house with a stinking hot piece and then run out with a freezing cold bit and then slam that with a hammer to make sure it went in just so," he said.

After much toil, Beck had done exactly what he set out to do and created a bike like no other. It had an ultramodern design with a single aluminum tube that functioned as the heart of the bike and a seat post that arched over the back tire and seemed to defy physics. Upon reflection, he viewed that early experiment as indicative of how his mind worked; he would aim for an unlikely, ambitious end point right from the start rather than trying to edge his way to a goal through a series of more modest constructions. "It rode fantastic," he said. "I was on the front page of the paper as this kid who had built a bike and all that crap, but I guess the point is that I didn't want to do all the steps in between. I didn't want to build a normal bike or modify a bike. I just wanted to go straight there. If there was a school science fair or something, I didn't turn up with a beaker, I turned up with a flamethrower. So it was not uncommon behavior for me."

At fifteen, Peter asked his parents for a $300 loan and acquired his first Morris Mini. He could drive it home but only barely because rust had eaten away a huge hole in the car's floor and he had to make sure his feet did not slip and touch the road flying by underneath. Over the course of the next six months, he just about lived in the workshop, trying to upgrade the vehicle. His schoolwork suffered as a result. Still, Beck's parents let him beaver away without so much as a grumble. "I would get home from school and go right into the shop," Peter said. "Mum would bring dinner down and set it on a bench, and it would get cold. Eventually, I would get hungry and find it. She never said, 'You have to stop angle grinding and get to bed.'"*

Peter first fixed all the rust damage. Then he rebuilt the car's engine and upgraded it with a turbo boost, fixed the suspension, and so on until he'd moved all the way from the front of the car to

* It must be asked—has any other mum ever said such a thing?

the back, retooling every inch. "I learned how to do most of it from books and talking to people," he said. "An engine is not that complicated when you boil it down to the basics."

To make some extra cash on the side, Beck held a couple of after-school jobs. He worked for a spell at Thwaites Aluminium, which manufactured a variety of aluminum products. He also worked at the local hardware store, E. Hayes & Sons, where he assembled mills and lathes and cleaned toilets. A metal retailer and a hardware store were natural fits for Beck, and the jobs aligned with the family philosophy of "Nothing given, everything earned."

In the background of all the tinkering and odd jobs, Peter had been pursuing a love of space at his father's side. The telescope Russell had constructed as a teenager was not your average hobbyist effort. He'd made all the metal casings by hand and learned how to grind lenses and mirrors. He had also fabricated a wooden building around the telescope to create a makeshift observatory in his parents' backyard. It was all a bit ridiculous for a home astronomy setup and made much more sense once Russell transported the contraption to the museum. It was there that, as a six-year-old, Peter began using his father's telescope to peer up into the night sky. "I have very fond memories of being up in the observatory and it's freezing cold and you're locked onto something cool like Jupiter," he said.

At around the same age, Peter began tagging along with his father to the monthly meetings of the Southland Astronomical Society at the observatory. The group attracted a wide range of curious, intellectual people, and Peter had a chance to listen to their discussions and probe them for knowledge. The average age of the society's members was about fifty, but there was tiny Peter right beside them, reveling in the chance to stay up late and have a bonus round of tea and biscuits. "They all had their own large telescopes and would be talking about their latest breakthroughs and discoveries," he said. "Most of the time I had no idea what anyone was talking about, but it all seemed very cool and interesting."

In 1986, when Peter was nine, Halley's Comet was due to make a run past Earth. Peter became fascinated with the comet and threw his energy into researching it. He put together a book on the subject for a school project and learned enough about the comet to become the resident expert. "The teachers would come to me for any information they needed or updates on it," he said.

The experience with the comet research was indicative of much of Beck's scholastic career. If he liked a topic or needed to know something to help with one of his projects, he dug into the work and often ended up with a school prize or two. For the most part, however, he had little interest in pursuing academics just for the sake of it and did not stand out as the brightest kid in class. "I didn't slack off," Beck said. "I have a need to do well. It drives me insane to present something that's not done well. But if I couldn't see the relevance to something, it just didn't seem important."

Beck's oldest brother, John, left home at sixteen and went on to pursue his passion of modifying and selling cars and then racing motorbikes. The middle brother, Andrew, also departed and took a job at the local aluminum smelter instead of attending college. Even though his older brothers had been protective of their tools, they had helped teach Peter and push his pursuits along. "I was surrounded by good engineers," Peter said. "If I was milling something the wrong way, one of my brothers would tell me that I was fucking it up." Once the brothers left home, Peter had the run of the workshop, which only intensified his interest in machinery and engineering.

At sixteen, Beck decided to leave home as well, ending his parents' dream of one of their sons attending college. He took a standardized test to secure the equivalent of a high school diploma and then thumbed through brochures about engineering trade programs. "I needed to go into a trade and just never felt that university was right for me," Beck said. The trade he settled on—tool and die making—was perhaps the toughest of the bunch in that it would require him to learn most of the skills needed to mass-produce the major objects used in industry and everyday life. In 1995, Fisher & Paykel, a local appliance maker known for its well-engineered products, offered Beck an apprenticeship, which required that he move to Dunedin.

Leaving home at such a young age did not frighten Beck at all. He'd been very independent and decided early on that either you stay in Invercargill forever and hang out with your mates or you head off and try to see what you can do in the world. He felt drawn to the challenge of experiencing the unknown and seeing how far he could advance his skills.

"The only reason, really, that I was interested in the die-making trade was because it's hard," he said. "It's precision engineering and the next level beyond everything else I was looking at. If someone says something is difficult, that's like petrol on a fire for me. I needed the hand skills and abilities that that trade was going to give me to do the things that I wanted to do."

It takes two and half hours to drive from Invercargill to Dunedin, a city located on the southeastern side of New Zealand's South Island. Though hardly a booming metropolis, Dunedin has about 150,000 people, making it three times the size of Beck's hometown, and it was a vibrant community of college-aged residents thanks to the University of Otago, the country's oldest university, and Otago Polytechnic.

Beck threw himself into the work at Fisher & Paykel, studying under a pair of old-school machinists who hailed from England and the Netherlands. They were obsessive about the art of tool and die making and expected Beck to meet their high standards. One of the youngster's first projects involved hand filing a square so that it could be mated with another square and have no more than one-thousandth of an inch of clearance at every point. More and more tests followed, and Beck excelled at them to the point that he won one apprenticeship award after another. It was a mix of his natural skill, his experience with tools, and his willingness to work long hours that made that possible.

Beck could not believe his good fortune. Fisher & Paykel had the best manufacturing equipment money could buy, and he was able to use the machines all day. Instead of putting on layers of warm clothes to while away the hours in his garage workshop, he turned up in jeans and a T-shirt at a state-of-the-art facility with computer-guided cutting tools that cost $250,000 each and giant $1 million machines that created molds for some of the best appliances in the world. Best of all, since he'd proved that he had a high level of skill, Fisher & Paykel soon put him in charge of running some of those machines. And when Beck made suggestions for how products could be improved, people actually listened to him and followed the advice. He appeared to have a gift for engineering that eluded not only the other apprentices but even many of the veterans. He zipped through the four-year apprenticeship in record time, finishing in less than three years.

After demonstrating his skills on the floor, Beck was elevated to the design office, where he had more direct influence on the form and function of products. Fisher & Paykel specialized in making things like high-end dishwashers and washing machines. The company, however, had been struggling with the issue of how the dishwashers dispensed detergent. Most appliance manufacturers bought their dispensers from the same supplier, but since Fisher & Paykel wanted to be better than other companies, it paired Beck with an experienced engineer and asked them to come up with an original design. Without delving too deeply into the sublime intricacies of dispenserology here, it can be said that Beck thought up a novel convergent-to-divergent nozzling system that really let the detergent's water softener do its thing.

Beck enjoyed his job so much that Fisher & Paykel could not keep him out of the building. He'd clock in early in the morning and often not leave until 3:00 a.m. The human resources department and his unionized colleagues did not always appreciate Beck's zest for labor. To throw them off his scent, he made sure to punch his time card at five o'clock each afternoon even though Beck would remain in the building trying to complete a job. "I didn't want anyone to get their nose out of joint that it looked like I was working longer than I should," he said. "I didn't want to show anyone up. There was the odd occasion when HR found out what I was doing and said,

'Look, you can't work those kinds of hours.' I would have to be inconspicuous for a while."

Away from the day job, Beck had still been working on his Mini and other cars, only the modifications no longer held the same appeal for him. He'd been adding superchargers to the vehicles and fuel injectors, but he wanted more power, more speed, more everything. Beck decided that the limiting factor of the vehicles lay at their core: the internal combustion engine. "It just didn't seem like the right way to go," he said. "So that's when I started building jet engines. But they still didn't produce enough power. And, you know, that's when I moved into rockets, because it's where I really needed to be."

CHAPTER TEN

YOU JUST STICK IT BETWEEN YOUR LEGS AND PRAY

New Zealand demands resourcefulness. First off, it's an island nation situated at the nether regions of the world. Your next stop after leaving Invercargill, for example, is Antarctica. Kiwis have long had to learn to make do with what they have on hand. This is doubly true for the large portion of the people who are farmers and live outside major cities. New Zealand has only 5 million inhabitants, and you can drive for long stretches on both the North Island and South Island without encountering much of anything beyond sheep, which are a New Zealand cliché for a reason. This sort of isolation breeds self-reliance and its own flavor of inventiveness. You must often think and build your way out of problems.

To celebrate this part of its nature, New Zealand champions what it calls a "Number 8 wire" mentality. This refers to the

0.16-inch-diameter gauge of wire that became popular during the mid-1800s as a cheap material to build fences for enclosing flocks of sheep. Since the farmers had lots of the wire on hand, they often used it to fix various things on their property. It became the New Zealand version of duct tape. Over time, the Number 8 wire lore grew to symbolize the Kiwi spirit of fixing a fence, a microwave, a car, a whatever with the bits and pieces lying around combined with guile and resolve.

Peter Beck detests the national infatuation with the Number 8 wire mentality. It's not that he doesn't think that Kiwis are clever or dislikes the idea of being resourceful. It's more that he finds the Number 8 wire philosophy self-limiting. New Zealand has come a long way from being a rural nation off on its own just trying to get by and survive. It's full of smart people and plentiful resources now. As he sees it, perhaps it's time to aim higher and make fantastic things the rest of the world wants rather than being content with a nifty do-it-yourself job that solves an isolated problem on your farm.

The amusing part about his stance is that the Peter Beck working at Fisher & Paykel applied the Number 8 wire mentality in a fashion so extreme that it has not been topped before or since.

At seventeen, in the early days of his apprenticeship, he began to nurture his fascination with rocket engines and rockets. He'd found a house in a Dunedin suburb and built an R&D shed in the backyard. That's when his *Breaking Bad* period kicked off in earnest. He checked out a few books from the library to read about propellants and rocket engine designs. It did not take long for Beck to settle on building an engine that ran on hydrogen peroxide, a compound that's safe enough to swish around in your mouth in diluted form but incredibly volatile in its pure state. It was dishwashers by day, explosives by night.

The hydrogen peroxide you can buy at a store is about 3 percent proof. To make the compound do what Beck needed it to do, he had to ratchet the concentration up to run-for-your-life 90 percent proof. After doing some research, he learned that he could get a head start on his refinement effort by purchasing 50 percent proof hydrogen peroxide from a local chemical company. He ordered the product and received word that a specialist would deliver it to his home. "I remember being at work, and I just couldn't wait for the day to be finished," Beck said. "I knew this peroxide was waiting at my door. It was a lovely sunny day, and I drove home flat out." After he parked and rounded the corner of his house, he could see a five-gallon vat of hydrogen peroxide on the porch. Only it didn't look quite right.

The delivery specialist had not been that special. He'd picked a bad container for the chemical, which was decomposing and letting off oxygen. The once cylindrical drum had turned spherical. "I was sitting on my front porch, thinking, 'What am I going to do with this?'" Beck said. "It was in the doorway, and I had to get past it in order to get any kind of protective gear on so that I could move it. So I had to creep over to it and vent off some of the gas. That was the first time I realized that I was dealing with a bit more than I'd expected."

After venting the gas, Beck hauled the drum into a spare bedroom. He built a bubble column reactor in the bedroom and proceeded to try to distill the hydrogen peroxide through a series of experiments. One morning, Beck awoke with his brain throbbing because the fumes from his mad scientist laboratory had circulated through the house. He then moved the operation to the backyard shed. But the safety conditions there were not ideal, either.

Typically, someone dealing with the souped-up version of hydrogen peroxide wears a hazmat suit. The chemical reacts with any organic matter. In the case of humans, for example, it will turn your skin white as it burns your flesh and you scream in agony. Like any teenager convinced that the rules of life and death didn't apply to him, Beck chose to deal with the safety issues by putting on a welding helmet and flameproof overalls and wrapping his arms and torso with garbage bags.* Occasionally, the hydrogen peroxide still found its way to a piece of Beck's clothing and spontaneously combusted,

* No doubt some Number 8 wire held the bags in place around his arms.

causing it to catch on fire. "You could call the shed a garden shed, but it looked like a crack shed," Beck said. "It had a light in it and a compressor that ran all day, every day and all these valves and vents hissing and roaring as I refined the peroxide."

Beck persisted with the experiments and built an early engine to run on his homemade fuel. Eventually, however, he decided that another, safer approach might be needed. "I still see people talking sometimes about using hydrogen peroxide for low-cost rockets and shake my head," he said. "I'm glad I had that experience and got that out of my system. It looks benign and simple, but it's really nasty stuff. I abandoned the research. It was superdodgy."*

Although he had given up on hydrogen peroxide, Beck continued to experiment with other propellants and began to work more on rocket engine designs. Though he still often stayed late at Fisher & Paykel to complete company projects, he also spent many, many late nights using the high-end workshop tools for his side projects. Just about everyone at the office knew what Beck was up to, and some people provided help. One day, for example, Beck asked the Fisher & Paykel supplies manager how much a large hunk of aluminum would cost. The supplies chief reported back that it would run Beck about $2,000, a figure well out of the apprentice's budget. "I told him that I'd find another way to do the project, and then, a couple of days later, this lump of aluminum turned up on my desk," he said.

The rest of the work, though, took place in Beck's shed. His home was located in the Kaikorai Valley of Dunedin, where many of the residences sit atop a hill at the eastern edge of the valley and look out upon a vast open expanse of lush rolling land to the west.

---

* As Beck said, "The problem you've got with peroxide is that it's highly susceptible to organic contamination, so that means if you sneeze into the top of an open jar of ninety-two percent proof hydrogen peroxide, it's a bad day. It's also prone to thermal runaway. And you're distilling it right up to that point where the water vapor comes off but not the peroxide, so, you know, it was a little bit touch and go.

"Once I got it distilled out to the kind needed for useful propulsion purposes, I began doing other experiments. I had petri dishes out on a rock and put other liquids in syringes and injected them into the peroxide. There was a dead patch of foliage near the rock that I am sure is still dead."

As Beck built his first engines and then began testing them late at night, the booms from the controlled explosions echoed throughout the valley, making dogs yelp and neighbors come out to see what was happening. "It's amazing that no one ever turned up at my door," he said. "It was sort of out in the bush, so it was hard to pinpoint exactly where it was all going down. I'm not sure you'd get away with it these days."

Looking to perfect his rocket engine technology, Beck read the classics of the field, like *Rocket Propulsion Elements* by George P. Sutton and Oscar Biblarz. He also mined the internet for scientific papers and took advantage of NASA's rather generous archives of technical documentation and manuals. The more he read and the more he experimented with the propellants, the more comfortable Beck became with the idea of making something real. At eighteen years of age, he committed to placing himself atop a rocket-powered bicycle.

The contraption Beck built looked like a cross between a racing motorbike and a bicycle. The bright yellow body of the bike was stretched out, with the handlebars right over the front wheel, meaning that Beck did not sit upright to ride it but lay prone over the body with his chest almost touching the frame and his legs extending back on either side of the back wheel. Though the body looked sturdy and like that of a proper racing machine, the rest of the vehicle appeared somewhat comical. Just below the frame was the rocket assembly made of a pair of propellant cylinders that looked as if they had been hastily wrapped in tinfoil and were connected to an engine via a mess of tubes. An exhaust pipe extended out from the rear of the bike right behind Beck's ass and above the back wheel. All of that machinery was supported by two twenty-inch BMX bicycle tires.

Late one night, after much labor, Beck decided to put his life into the hands of his engineering. Clad in work overalls and a standard-issue bicycle helmet, he walked out to the company parking lot and hopped onto his invention. "That first time I'm lying down on it and the start button is millimeters from my finger, and I'm thinking 'Hmm. What's going to happen here?'

"The first run was just phenomenal. Even though it was a short little burst, it was that first taste of what it feels like to ride something that's rocket propelled. It's an impossible feeling to describe. You just need more.

"It's this one little push of a button, and the millisecond that you push that button, you're off and basically along for the ride at that point. The dodgiest bit is when you punch off the line, because if you're not lined up in the perfect direction, then you're in trouble. You're going whichever way you're pointing. That first half second is a bit terrifying.

"As the propellant tanks empty, the bike gets lighter and lighter, and although the air resistance is increasing, the reduction in mass means that the acceleration is increasing. You've got the acceleration. You've got the noise. You've got the instantaneousness of the whole thing. Overall, it's sensory overload."

Let's step back for a moment and appreciate the sheer stupidity of this all. The five-foot, ten-inch Peter Beck weighed 140 pounds or so back then. His rail-thin frame was enough to support his curls and little else. Beck talked in an excited, supremely confident manner about his engineering prowess and had the volatile enthusiasm of a mad scientist. Think a young Doc Brown with a perm. He'd been cooking up homemade fuels in a homemade garden shed and making rocket engines from spare parts gifted to him by kind souls. It was that person who opted of his own volition to place a bomb between his legs and ride it.

After performing a number of tests, Beck held a more official demonstration in the Fisher & Paykel parking lot for the company brass and its employees. Not long after that, he unveiled the bike at Dunedin's Southern Festival of Speed, an annual event that took place on the city streets and included a variety of races and drag competitions.

Even among a group of people used to modified hot rod vehicles, Beck's bike stood out as a freak show and raised serious safety questions for the event organizers. The main races ran down a one-eighth-mile-long stretch of road in which spectators were separated from the vehicles by stacks of tires and prayers. Would the curious bystanders be safe should something go awry with the built-in-a-shed missile being ridden by an apprentice dishwasher detergent dispenser engineer? To find out, the event's "chief scrutineer" assessed Beck's bike in the most laid-back New Zealander ways, asking the young man if the vehicle had good brakes. Beck assured him that they were "great brakes" and was given the all clear to race with one major precaution: Beck would be followed down the track by an ambulance.

Before racing anyone else, the organizers wanted to see Beck conduct a "slow" run on his own. Only, in this case, "slow" was not an option. The bike had two modes: stopped or "everyone hope for the best." Adding to the spectacle, he could not just place the bike at the starting line and let it rip when the light turned green. That would not give him enough time to place his legs on the foot pegs at the rear of the vehicle. Instead, Beck needed to mount the bike and employ a "gangly motion" with his legs in which he kicked and scraped at the ground to get the bike up to stable speed *and then* placed his feet on the pegs *and then* lined the machine up straight *and then* hit the magic button on the handlebars that made it take off. "Nobody knew what was going on," he said. "Here's this strange yellow thing with this guy running down the street and an ambulance following behind him."

Beck at least looked the part. He'd put the usual grimy overalls back into his work locker and replaced them with a red-and-black racing suit, yellow gloves, and a black motorbike helmet. It was while he was wearing that getup and his mother was wondering what she'd done wrong that Beck managed to blast himself down the street successfully. His bike was clocked at ninety miles per hour at the end of the five-second run, although by that time he'd already started to slow down from a top speed well over a hundred miles per hour. To bring the bike to a halt, Beck could not use the brakes right away because the pads would melt and turn to liquid at such a high speed. He'd devised a technique of sitting up on the bike and using his torso to create enough wind resistance to lower the speed to sixty miles per hour, at which point the brakes could be applied.

After the crowd erupted in pure joy, they demanded that Beck do it again. This time they wanted to see him in a proper race, and, wouldn't you know it, someone just happened to have a Dodge Viper—zero to sixty in four seconds—on hand. Beck got his running start, and the Viper's driver punched his gas pedal when the two vehicles were even. Beck won. "It was a good time," Beck said. "Afterwards the organizer said to me, 'Boy, we're really glad that didn't go bad.'"

Beck would end up working at the appliance maker for seven years. He perfected his ability to use more tools and built up his reputation as someone who got stuff done. He also rose through the design division, taking on more and more demanding projects and becoming an ace at refining Fisher & Paykel's flash hardware. By the time his run ended, Beck had been assigned to designing the huge stamps and presses used to create the appliances. He was building the machines that built the machines.

An introvert by nature, Beck did not have much of a social life while at Fisher & Paykel, although he did succeed in meeting his future wife, Kerryn Morris, in the design office, where she also worked as an engineer. The pair had teamed up on a project, with Peter crafting some tools for an appliance design cooked up by Kerryn. They'd also been hanging out as a result of the "Friday Night Challenge," a weekly event that Peter started in which one Fisher & Paykel employee would propose some kind of task and the rest of the crew would try to accomplish it. The missions included things like cliff diving and crossing rivers by shimmying across suspension cables—anything to get people out of their comfort zones. It was a very Peter Beck way to spend your free time.

The real romance, though, sparked as a result of Peter's rocket obsession. Kerryn and some other women in the design office took pity on the lad who worked endless hours on his hobby and never seemed to eat. They began inviting him over to their flats for the odd meal. With Kerryn, the odd meal turned into more frequent visits. Soon she became the first person to lend some safety support to Peter's rocketry adventures. "I would ring her up before an engine test and tell her that if I didn't ring back within a certain period of time, that she should call the ambulance and fire brigade," Peter said. "The engines were getting a lot more powerful, and things could go bad." Later, Kerryn would simply turn up at Peter's place to help out with the tests.

Following the rocket bike, Beck built a rocket-powered scooter and then a jet pack, which he wore while on roller skates. Kerryn witnessed most of these experiments and stayed with Beck anyway. "There was one test with the rocket scooter where I went to hit the igniter for the engine and I got a massive wallop of an electric shock," Beck said. "It was huge, and I jumped into the air screaming. Once I settled down and came to, I could see Kerryn on the ground absolutely pissing herself laughing. I honestly felt there was a good chance I was going to die during that experiment, but she thought it was incredibly funny."

Kerryn grew up on the outskirts of Dunedin on an oceanside dairy and sheep farm that the Morris family had started cultivating in the 1860s. The breathtaking property, which has visits from sea lions and yellow-eyed penguins, demanded that Kerryn embrace the Number 8 wire spirit, too, and she proved adept at fixing things around the farm and aiding the family business before obtaining a degree in mechanical engineering. In 2002, while dating Peter, she took a new job with the oil field services company Schlumberger in New Plymouth, a city on the western edge of New Zealand's North Island. Peter quit Fisher & Paykel and followed.

In New Plymouth, Peter found a job working for a manufacturer of custom yachts for the superrich. The gig had its appeal in that the boats were made of carbon fiber and titanium and presented tough engineering challenges. The owner's cabin, for example, was always located at the back of the 124-foot yacht, which put it close to the propellers and other noisy machinery. People like Beck were recruited to try and find ways to dampen the sound. The boatbuilding industry, however, valued experience above all else and had a built-in resistance to new ideas. Beck would make suggestions and run computer simulations to prove that his suggestions would likely work, but his ideas were usually nixed. "Their thinking was that if something had worked in the past, then that's how they do it," Beck said. "It drove me totally insane."

Peter's home life was not that much better. After moving, he'd lost access to a shed and had little time to work on the rocket side projects. Meanwhile, Kerryn's job required that she travel frequently throughout the Middle East and the United States to analyze the performance of oil and gas wells. Peter tried to make the best of things. He flew to Cairo to meet Kerryn for a vacation, the first time he'd gone solo on a big overseas trip. He also kept at the engineering problems on the yachts, even though his boss didn't really care what he discovered. As luck or persistence or a combination of the two would have it, that work soon pushed Beck toward the foundation of his rocket empire.

CHAPTER ELEVEN

I EXPECTED MORE FROM YOU, AMERICA

The owner of the yacht company demanded that Beck make titanium doorknobs for the boat's kitchen by hand. That meant grinding away at metal pieces for hours and forming the knobs perfectly even though something very similar could be purchased with ease. The whole purpose of the exercise was to make a slightly lighter doorknob and shave a few ounces of weight off the yacht. Yet when it came to muffling the roars and clangs of the engines and propellers, the owner opted for brute force: he commanded the team to fill the back of the boat with sand. Tons and tons of sand.

Beck could not let such an affront to his engineering sensibilities go unchallenged. He'd studied the propellers and analyzed their acoustic output. He knew that the noise they produced could be canceled out with baffles and resonators tuned to dampen the right frequencies. Beck had built computer models to check his theories, but he wanted to be absolutely sure before creating a fuss with his boss. So he found an acoustic engineering expert at a New Zealand government-backed research lab called Industrial Research Limited (IRL) and arranged a meeting.

The organization had three offices in New Zealand filled with some of the country's top scientists and engineers. They were asked to come up with big ideas and to help businesses solve tricky problems in the name of fostering industry in the country. Beck turned up at IRL's office in Auckland and had his meeting about the yacht noise. While there, however, he noticed that the lab had lots of high-end equipment and some job openings. Beck applied for an engineering position, got the job, and left the yacht project after a year in New Plymouth.

As Kerryn headed off for an extended posting in Libya, Peter moved to Auckland, New Zealand's largest city. The four-story IRL office seemed somewhat out of place in what was mostly a residential neighborhood with well-to-do homes and some nice restaurants and coffee shops nearby. Still, it had served its purpose pretty well for a while. From the early 1990s on, government researchers had shared the office with technology start-ups, divvying up the use of laboratories and equipment and swapping ideas. By the time Beck arrived in 2004, however, the government side of the operations had lost funding and started to wane, while the start-up side became more vibrant.

Beck, then twenty-four, started out working with the government researchers. They threw him onto a team studying composite materials such as carbon fiber. New Zealand's proud sailing tradition and success in the America's Cup had turned the country into a hotbed of carbon-fiber experts who were looking to push the material into new fields beyond boats and planes. Under the tutelage of a couple of scientists, Beck learned new and very precise techniques for testing how much stress certain materials could endure. He became a strong member of the team right away, merging his practical abilities with the theoretical backgrounds of his colleagues.

For one project, he built a testing rig that enabled the researchers to slam panels of carbon fiber into a pool of water. The process mimicked a boat hull crashing into a wave. Measurements were taken, and the team would try to make lighter and lighter panels that would stay strong and resilient against such harsh forces. It was the type of knowledge that could come in handy if, say, you wanted to jam a carbon-fiber rocket against the atmosphere later in life.

On another project, Beck and some partners went down into the IRL basement to put a sixty-five-foot wooden wind turbine blade through its paces. Beck built a frame that held the blade in place and then a mechanical mechanism that made the blade wobble up and down, the point being to shake the thing until it broke. When the machine was turned on and the wood bent back and forth, it generated a thunderous thumping sound—*pwooomp, pwooomp*—that ricocheted throughout the basement. "He's such a clever bugger," said Doug Carter, who worked with Beck at the lab. "Pete designed and built this whole thing himself. He spread the load among the different columns in the basement and transferred the force of the frame back into the building. The machine was so solid that cracks formed in the walls, and we started to worry we would break the building and not the blade. He was an absolute natural scientist."

In its glory days, IRL had done pure research funded by the government, pushing the limits of science in various fields. The IRL Beck inherited worked more as a consultancy in which companies would request help on projects and receive scientific guidance for a fee. Although IRL had lost some of its luster, it retained hundreds of top scientists, almost all of whom had advanced degrees if not PhDs. Beck could hold his own with the researchers, although his lack of any sort of a degree stood out. "It did seem quite odd that he didn't have any qualifications," Carter said. "But he certainly could have gone to university and gotten any degree. He's just a sponge. I think he's one in a billion."

The degree issue came up once when IRL asked Beck to serve as an expert witness in a court case that revolved around a series of carbon-fiber composite failures in a major project. Over the years, he had taught occasional classes at local universities and even advised PhD students, but IRL, upon remembering Beck's lack of credentials, decided that he could not go in front of a judge.

Beck did not appreciate the slight, and it added to his suspicion that organizations and companies often made the mistake of placing paper credentials above experience when looking for talent. "There are two ways you can learn," Beck said. "You can go to a university and be taught about a shaft breaking, or you can go to industry and have a shaft break. At university, the major consequence is a poor grade. The consequences in the factory are the production lines stopping, so they're much bigger."

For the first couple of years at IRL, Beck threw himself into the research tasks. Where he'd learned the ins and outs of industry at Fisher & Paykel, he now had a new set of tools with which to probe the behavior of materials at the atomic level. "I had all the Crown's assets at my disposal," he said. "I had vibration-testing machines, shock-testing machines, and the best computer codes. I didn't build a lot of stuff, but from a research and knowledge perspective and forming a deeper understanding of physics, there was a lot gained during this period."

The people at the lab liked Beck. He did not lack confidence, but he also did not boast or belittle people when pointing out the errors of their approaches to problems. His appearance helped do some of the disarming. Still skinny and with his mushroom cap of curly hair, Beck looked like a surfer who liked to dress up in a white lab coat for fun. His hobbyist vibe made him seem more approachable, too. After work, he could sometimes be found in the parking lot underneath a jacked-up ancient Corvette, digging into his bags of tools and covered in grease.

The PhDs did not make a ton of money, but they earned far more than Beck, who started at an annual salary of $40,000. To help support his side projects, he took on a second job at a tool-and-die-making shop across town. He'd work there and do some consulting projects from about 6:00 to 9:00 p.m. and then drive home and work on his propellants and rocket engines from ten at night to two or three in the morning. Those late-night sessions paid off, as his expertise with propellants expanded right alongside the size and power of his engines. Beck planned to build a rocket car that would set the land speed record in New Zealand and had started to manufacture the vehicle's propulsion systems and frame. That's when he took a vacation that spun his life in a new direction.

Around 2006, Kerryn's job required that she spend a month in the United States. For Peter, that month was the chance to take the space nerd's pilgrimage of a lifetime. He had been writing to people at NASA, large aerospace companies, and hobbyists in the United States for advice on his various projects and struck up online relationships. He thought that the vacation would allow him to meet some of those people and check out the latest in aerospace technology. Not only would he get to visit a space superpower, but he could also show folks pictures of what he'd been working on and tell them about his hands-on background. Maybe, just maybe, someone would offer him a job. While Peter had been struggling to make ends meet, he saved up enough money for the journey and asked for a month off work.

Although meticulous in most things, Beck did not plan this trip out in exacting detail. He knew that he wanted to split his time between California and Florida, the United States' two main aerospace hubs. Beck, however, made very few firm appointments ahead of time. He mostly planned to turn up at the headquarters of companies and NASA centers and ask to speak with an online chum or to see some research-and-development work. "This was a way to meet people and insert myself into the industry in this country," he said. "That was the prime objective."

Upon landing in the United States, Beck felt overwhelmed by a sense of awe and possibility. Los Angeles International Airport happens to be in the heart of LA's aerospace scene. Drive a couple miles south from LAX, and you'll see the drab, boxy offices of Boeing Satellite Systems, Northrop Grumman, and Raytheon Space and Airborne Systems—the aerospace old guard. Drive a few more miles, and you run right into the headquarters of Elon Musk's SpaceX and its massive rocket factory. "There were those police guys on the motorbikes, and it's like the TV show *Chips*," he said. "I thought that was made up, but it turned out it was real. I was pressed hard against the window of the taxi because here I was in America. I was driving past all these aerospace buildings, and it was mind blowing."

Beck's first stop in Los Angeles was Norton Sales, a space geek mecca. From the outside, the establishment looked like a pawnshop in a low-rent neighborhood. On the inside, it looked like an aerospace buff hoarder's paradise. Over the years, the shop owner had collected engines, turbopumps, valves, and even satellites from defunct space programs and plopped them down pretty much wherever. Beck figured that either he could go to an aerospace museum and peer through a glass case or he could experience space junkyard nirvana and actually hold the artifacts in his hand. Shelf by shelf, valve by valve, and hour after hour, he made his way through the store as the bemused owner watched. From there Beck drove to Boeing, Pratt & Whitney, Lockheed Martin, and Aerojet Rocketdyne. All the while, he carried a scrapbook with pictures of his rocket bike, jet pack, and other weird machines, hoping, rather naively, to prove to whomever he might meet that he was not a crackpot. Sometimes he actually made it past the reception desk and had a chat with an engineer; far more often the security guards told him to get back into his car and come back after making an appointment.

All rocket enthusiasts who find themselves in Southern California must make their way to the Mojave Desert, so that was what Beck did next. It's in the Mojave that the United States has spent decades testing its most advanced aircraft and space start-ups have appeared in more recent years trying to find cheaper, faster, better ways into orbit. Much of the officially sanctioned government action takes place at Edwards Air Force Base, which was Beck's first stop. "I saw the signs for it and am like 'Holy shit! This is the Holy Grail!'" he said. "I was so excited. I jumped out of my car and started taking photos by the guardhouse. Then this guy with a gun comes up to me and asks for some ID. I show him my passport, which was full of stamps from the Middle East. It didn't look very good. But then he realized I wasn't much of a threat, and something clicked in his head, and he started asking about *Lord of the Rings*. I reckon those movies saved my ass. He told me never to return."

Once he reached the town of Mojave, Beck found the dozen or so start-ups trying to build rockets and space planes and lunar landers inside their hangars at the local airport. Those outfits were more welcoming than the traditional aerospace set and accepted Beck's rocketry scrapbook as solid proof that he belonged among them.

As he went from one gritty workshop to another, Beck took great pleasure in discovering that the Mojave people were building similar things to him and suffering from similar problems. The only real difference between the Mojave crowd and Beck was that they had US government research grants or space dreamer millionaires backing them, while Beck had whatever was left over each month after paying his rent and buying food. "Back in New Zealand, I had no idea where I sat in the context of things," he said. "You have no idea if you're at the bottom of the pile, the middle, or the top. They were using the same equipment, building the same engines, and having the same struggles. They were doing exactly what I was doing in my garage. That was great, great context. It was hugely inspiring."

From Mojave, Beck hopped on a plane and arrived at Cape Canaveral in Florida. Despite being an absolute rocket maniac, he had never seen a large rocket or launchpad up close. He made the rounds at Kennedy Space Center just like the rest of the tourists except that each stop placed Beck in a near-orgiastic state. "I was freaking out because my camera started to run out of batteries," he said. "It was the first time I got to touch real rockets. It was like taking an alcoholic and putting them in a swimming pool of vodka." At night, he would call Kerryn and pepper her with tales of his discoveries and insights, and she would politely acknowledge his zeal.

The most significant visit during Beck's trip was, in many ways, the worst. He'd returned to Los Angeles and made an appointment for a tour of the famed Jet Propulsion Laboratory, which has produced some of NASA's most advanced space exploration vehicles. At first, Beck had his usual space enthusiast glow. He'd arrived at one of NASA's sacred sites, the very same NASA that had built huge rockets and put people on the moon. As the tour progressed, however, the glow faded. JPL appeared stuck in the 1960s. The research facilities had old computer equipment and old scientific instruments. Making matters worse, the tour guide kept telling the group how disenchanting NASA had become.

Beck would spot an engineer working in a lab and try and break away from the crowd for a private conversation. He wanted to get a look at the good stuff. But the guide would pull him back in and shut his curiosity down. Beck thought that JPL would have the frenetic buzz of a start-up with some people running around trying to hit their deadlines and others sleeping in the corridors after pulling all-nighters. Instead, he found bureaucracy and malaise.

"I'd held NASA to this massive level of esteem," he said. "I wanted to see the supercrazy metals and ceramics and the mind-blowing hardware. NASA was meant to be so many steps beyond anything else, and it just wasn't. People were complaining that they didn't have decent computers. It was a relic of the past. I walked out of there, and I felt that it was not how it was supposed to be. It was as depressing as hell. That day at JPL destroyed me."

Not long after visiting JPL, Beck boarded a plane for the thirteen-hour flight back to Auckland. His mind raced back to all the hardware he'd seen at Norton Sales. He remained impressed at how well the aerospace gear had been engineered in the 1960s. Then he thought about the guys in Mojave who were tinkering away just like him in their dusty hangars and the sadness of the visit to JPL. Beck did not really know about companies like SpaceX and Blue Origin at the time or the opportunities they might present. He'd gone to the United States to try and secure a job at a prestigious place such as Northrop Grumman or Lockheed Martin and to work at the cutting edge of aerospace. The problem was, however, that the cutting edge didn't seem to exist. Whether at a company or even NASA, everyone did things the same old way, and these people doing things the same old way didn't seem to have much passion. No one really appeared to have new ideas or want to advance aerospace technology.

As the flight took off, Beck edged forward in his seat and looked out the window. The last thing he saw of the United States was the glowing blue sign on the Northrop Grumman building. "That was the real kick in the guts," he said. "I'd come with this whole vision, and there's the final reminder of this wild, emotional journey. I thought the blue letters might just flicker out."

Despondent, Beck didn't know what else to do other than think. And that was what he did. As the others around him ate their crappy meals, watched their tiny TV screens, or fell asleep, Beck thought for a solid eight or nine hours straight, full of inner turmoil. "It was like everything you had believed and held true was now wrong. There was zero desire to go out and do the things that I thought were important to achieve. I remember one discussion with one of the Lockheed guys. I was proposing some ideas. I can still hear his words: 'We don't do anything unless the government tells us to and they pay for it.' But engineers are supposed to solve problems. I decided that I could either wallow in my self-pity, which I did for a very short period of time, or figure out a solution. I decided that, okay, I'll do it myself."

In that moment, Beck experienced the engineer's equivalent of a religious revelation. The people who were supposed to be building cool things were not building them. They had, in fact, given up. What they should be doing, as he saw it, was changing the way humans reached and interacted with space. Someone needed to make a cheap rocket that could fly cheap satellites to space and do so on an almost daily basis. If such a rocket existed and companies and scientists knew they could count on it, humans' relationship with space would alter in a fundamental way. They would think of access to space as a given, and all sorts of possibilities would arise. Beck had found his calling.

Ever the engineer, he broke down what he needed to do next into a series of steps: Start a company. Raise some money. Build something small first. Gain confidence. Then build something bigger. Raise more money. Just having a plan erased Beck's anguish and pumped him to the brim with enthusiasm. "There are times when something is incredibly stressful for you, and then a solution comes and it just washes over you," he said. "And then I became filled with adrenaline. It was 'Off you go!' I told my wife that I was going to own that Northrop Grumman building someday."

After landing in Auckland, Beck raced home and then dashed to his downstairs workshop. He brainstormed for a while and came up with the name for his new venture: Rocket Lab. Then he drew up a simple logo on his computer with the name placed at the center of a black rocket with a red flame shooting out the back of the machine. He printed out the logo and taped it to the outside of his basement door. That basic gesture suddenly made the whole thing feel real.

Rather than being ecstatic, however, Beck felt frustrated and overwhelmed. It was the middle of 2006, and he would soon turn thirty. Beck did not doubt himself or his abilities. He did not fear the challenge ahead. He simply felt as though time was working against him. There was too much to do.

CHAPTER TWELVE

"YOU FUCKING BEAUTY!"

Peter Beck knew he wanted to run a rocket company, but he did not know exactly what kind of rocket company. Since he'd been working two jobs and did not have much money, he decided to start small—really small. He would build what are known as sounding rockets. These are machines that have all the major rocket features, including an engine and an aerodynamic design, only they're not as ambitious as the big boys. They fly beyond the earth's atmosphere and into space but lack the speed needed to head sideways and venture into orbit.

While the sounding rockets are limited, they can do useful things. They're able to carry payloads into areas higher than those occupied by balloons and lower than those occupied by satellites. Since they are small and less complex, they're also relatively cheap to build. This makes them appealing to scientists who want to send probes into space to take measurements or test the properties of chemicals and molecules in zero gravity. Not many sounding rocket manufacturers exist, so Beck thought he could find customers among universities and other scientifically minded organizations if he could build the machines quickly and cheaply.

Even though he'd experienced a magic revelation, Beck soon encountered the self-doubt that accompanies any risky, bold endeavor. He went to IRL and pulled a couple of people aside to ask them if they thought starting a rocket company seemed like a good idea. "One of the guys I went to was Doug Carter, who'd been doing business development for the lab," he said. "I told him my plans. And looking back on it, it was quite a formative discussion. He could have told me it was a terrible idea, but he didn't."

The funny thing is that Doug Carter remembered the conversation with Beck quite differently. "He'd been to America or something and had met all these people from NASA," Carter said. "I couldn't believe it. He told me he was really into rockets and wanted to start a business. I remember thinking 'Well, that doesn't sound very realistic.' He looked like some surfy dude with his curly hair. My main feeling was that he didn't have a chance. I told him to maybe get into making rocket parts." Carter, however, did have an idea that might help Beck. He'd been reading a magazine story about a rich New Zealander who was named, if you'd believe it, Mark Rocket, and who professed a love for all things space. "I said, 'You should give him a ring and see if you can get some money.'"

Mark Rocket's birth name was Mark Stevens. During the early days of the consumer internet boom, he'd started a New Zealand tourism website and directory. It had listings of things such as rental car locations, hotels, and activities. The site had become popular enough to attract the attention of the local yellow pages publisher, which acquired Stevens's company and turned him into a multimillionaire.

Stevens had been fascinated by space since his childhood and had longed to take a trip into the heavens. "I was really disappointed that I wasn't born in America, where they had an astronaut and space program," he said. Flush with cash, he decided to let his inner space nerd loose. He legally changed his name to Mark Rocket and paid around $250,000 for a reservation on Virgin Galactic's planned spaceplane, which would take rich tourists to the edge of space for a few minutes. "With the name, it was one of those ideas that grabs hold of you,"

Rocket said. "Words are really powerful, and I think they can resonate in many different ways. You can create your life as if it's kind of like an artwork. As for the trip, I decided to put my money where my mouth was."

Beck called Rocket at an ideal time in 2006.* It pained Rocket that the Southern Hemisphere had little to no activity in terms of aerospace technology or commercial space projects. Rocket had been hunting for something to invest in but to date had found only people "who were pretty flaky." Beck, by contrast, talked like a real engineer on the phone, and his sounding rocket plan struck Rocket as feasible. If nothing else, it was on brand for the investor.

Rocket asked Beck for some additional information and began making calls to check on Beck's background. The proposal that Beck sent outlined plans for a rocket that could fly 175 pounds of cargo into space for a few minutes. Universities and their science departments were expected to be the main customers for the vehicle. Rocket felt that even 175 pounds of cargo was ambitious. He suggested that the company start out trying to fly just 5 pounds of material. "I thought we could build a demonstrator vehicle first, so that people could see we were capable of making engines and the overall system," Rocket said. "But Peter was adamant that the bigger rocket was the way to go."

Despite the early disagreement, Rocket needed only a couple of weeks to decide to go ahead and fund Beck. He cut Rocket Lab a check for $300,000 and took ownership of 50 percent of the company. "It seemed like an easy way to get rid of a bunch of money, but it was exactly what I wanted to get involved with, and it didn't seem like anyone else was thinking along the same lines," he said.

Beck already had his full vision for the future of Rocket Lab in mind but held back on dishing it all out to Rocket. The plan to build a space company from nothing in New Zealand with $300,000 sounded audacious enough; Beck feared that he might scare off the

---

* According to Beck's recollection, he did not learn about Rocket from Carter. Beck recalled driving in his car one day and hearing Rocket giving an interview on the radio and talking about his purchase of the Virgin Galactic trip.

only person in the Southern Hemisphere willing to back his hopes and dreams. He also appeared somewhat desperate and inexperienced in the art of starting a company, as he gave away half of the venture right from the get-go. "I was terrified that all of my plans would just sound too crazy," Beck said.

Money in hand, Beck marched back to IRL and asked if he could take over some space to build his new company. The lab granted Rocket Lab an office area on one floor and another area in the basement where the company could conduct its more serious experiments with the shielding of concrete walls as a modicum of protection to others in the building. Best of all, IRL gave Beck access to everything for free.

Like a man possessed, Beck began building his first rocket. The project required that he once again experiment with propellants. This time around it was not so much about refining the chemicals as finding the right recipe of explosives to combine. Knowing almost nothing about chemistry, he read books on the subject and sought advice from anyone who could help, including scientists working down the hallway. After running many experiments over several months, Beck settled on a combination of ammonium perchlorate, aluminum, and hydroxyl-terminated polybutadiene, which were fused together into a bar of solid rocket fuel. He could hold the hunk of fuel in his hand, and he would place it in a safe each night before he left the office.

Crucially, Beck had to learn how to move beyond making a stand-alone engine and toward making an elegant, complete machine. The engines he'd built in the past had always been slapped onto something like a bike or a pair of roller skates. Rockets, even small ones, required more holistic thinking. The machines simply had no room for error or spots for superfluous parts. Each piece of the machine, including its body and electronics, had to be engineered with precision and careful planning. "It was a wonderful time," Beck said. "It was a very simple time. It was just me, and you start to realize how much you can get done with one person who doesn't know much, and, well, it's not much."

Initially, Beck gave himself a year to create and fly the sounding rocket. He enjoyed running a real company and took pride in the Rocket Lab name and company logo on the office door. The "Lab" part of the moniker had been picked to add some gravitas to the operation. "If it was just Rocket Inc. or Rocket Company, it could mean anything," he said. "But if you say 'Rocket Lab,' people understand that it's a laboratory, and then all of a sudden you gain some credibility." He wore a white lab coat around the office for further seriousness.

As a solo inventor, Beck tried to design the rocket engine, its body, and its electronics all at the same time. He bounced from one project to the next as quickly as possible. Even though he was Rocket Lab's sole employee, he had to learn the mechanics of running a business, too. "When it's a hobby, you don't need to worry about insurance," he said. "If you burn the house down, well, that was poor planning. But as a business, if you burn someone else's building down, then it's not poor planning; it's a problem."

In his mind's eye, Beck could envision what the final product would look like with absolute certainty. "It's not like it's conceptual," he said. "I see the finished thing in my head." The distance between his starting point and the final product, then, was simply a matter of long hours at work. The blessing was that Beck had such a clear vision and was not intimidated by the number of things he had to accomplish. "Once I decide on doing something, it's a fact in my mind," he said. "I don't wonder if it's going to happen." The problem was that Beck often underestimated how long it would take to make things in terms of both time and money. "I've since learned that I should times everything by pi," he said. "The amount of time and cost is usually about 3.14 times what you think it will be."

In 2007, Beck began to recruit a couple of people to help with his rocket quest. He'd been friendly with an electrical engineer named Shaun O'Donnell, who hailed from the coastal city Napier. O'Donnell had worked at a start-up and run his own consulting business, and he'd done a couple stints at IRL, where, among other things, he'd helped build a database tracking system for New Zealand meat exports. Beck spotted the obvious aerospace potential in the meat cataloguer and made his pitch one evening. "I was leaving the building, and Peter followed me out to the footpath," said O'Donnell. "It was weird because he'd never done anything like that before. Pete stopped me and said that he wondered if I might do some work for him at his rocket company. I thought, 'Aw, yeah, that sounds great.' But it seemed a little bit nuts."

Beck hired O'Donnell on a part-time basis to work on the rocket's avionics, or electrical systems. O'Donnell settled into one of the two desks in the office and created an area in which to perform electronics tests. Like Beck, he learned a lot about what the rocket would require from books and by searching on NASA's website.

Not long after O'Donnell arrived, Beck hired Nikhil Raghu as his first full-time employee. Raghu had immigrated to New Zealand from India with his family when he was ten years old. He came from a family of engineers, had studied mechanical engineering at the University of Auckland, and joined Rocket Lab right after finishing his master's degree. Beck gave him the title of engineering and operations lead but could have slapped "Everything Engineer" on Raghu's business card. Raghu would need to help with building the rocket, creating financial models and materials for potential investors, managing budgets, and generally doing whatever Beck asked of him.

Neither Beck nor the company's main investor, Rocket, had attended university, but that general lack of pedigree did not bother Raghu. The young engineer found that Beck had a knack for picking up engineering theory from books and then translating it into practical applications. Beck's hands-on approach struck Raghu as a refreshing change from his more abstract university courses. "I never sat down and asked myself, 'Is this a hoax?'" Raghu said. "I believed Pete could do it. With him, you just kind of get sucked into the vortex. He's very motivated and always goes a hundred miles per hour. He just never gives up on solving problems. It's very easy as a young engineer to get focused on theory. But with Pete, we'd be working on a problem, and he would go home at night and machine something in his home workshop and prove the theory wrong or right."

The three-person Rocket Lab team must have looked comical to outsiders. Their office was the size of a small studio apartment. They carved up the little space they had into various work areas by putting up temporary dividers: computers and desks over here; electronics assembly and testing over there. Beck had scrounged all the furniture from other groups in the building, so nothing matched. The office had a window, but since the building abutted a hill, only a sliver of blue sky appeared at the top of the opening. One visitor described the space as "a dungeon."

The basement offered more space where "you could do cool and dangerous things," Raghu said. He and Beck bought a ten-by-ten-foot shipping container and installed it in the basement as their rocket engine testing center.* They also built a shed and made sure that it was completely sealed and electrostatically grounded, so that they could work on propellants in a controlled environment. Such precautions were necessary as they fiddled with substances like ammonium perchlorate, grinding the chemical's crystals into just the right size.† "Not only does it like to set things on fire, but you don't want to breathe it in because it can do all kinds of nasty things to human health," Raghu said. Predictably, Rocket Lab scared the hell out of everyone else in the IRL building with constant, wall-rattling explosions.

Some of the trickiest work took place outdoors as Rocket Lab tried to perfect the art of having its rocket return to Earth at the end of a parachute. To perform tests, the company replicated the weight of the rocket and its cargo by building 150-pound hunks of metal equipped with all kinds of sensors and accelerometers. Next, it located a pilot who was willing to fly over a bombing range and let

* Later on, Rocket Lab also did engine tests at a model rocketry range and at a facility owned by Air New Zealand.

† One day, someone at IRL was throwing away an old, broken microscope. "It was so complicated that no one could fix it," Carter said. Beck saw the microscope heading for the bin and offered to buy it for $1. He fixed the device and used it to examine the crystals.

someone toss the metal blobs out of the plane's cargo door. As the pilot flew, one member of Team Rocket Lab, who was harnessed into the plane's scaffolding, hurled the metal onto the pine tree–sized sand dunes below. Another member of the team then had to try to follow the falling object and communicate with it via radio signals. Things often did not work as planned. "If you think rocket fuel is crazy, wait until you get to parachutes," Raghu said. "It's a whole scientific field unto itself."

As the Rocket Lab employees did their research, they became more encouraged by the opportunities ahead. People working in the aerospace industry during the 1950s and 1960s had accomplished so much in a relatively short period of time. Beck felt more and more convinced that the aerospace industry had become complacent and fearful as the decades had passed. A small team with limited resources could accomplish an awful lot if it just thought about problems in inventive ways.

That said, the company had to operate under constant financial pressure. The investment from Rocket helped but took things only so far. Beck had hoped to drum up business from university customers who would pay some money in advance for future launches. The schools, though, did not take Rocket Lab all that seriously due to its nonexistent track record, and it proved difficult to figure out how to get the right approvals for an American or European payload to fly on a New Zealand rocket. Beck would have liked to hire more people to push the project along faster, but no money existed to make that possible.

"It was so frustrating for Pete and for me as well," Raghu said. "We kept saying 'Why can't we find someone who has money? Why is it so easy for all these companies in Silicon Valley to go in with an idea on a napkin and a PowerPoint deck and walk out with millions of dollars, and yet here we are? We can do this for a fraction of the money these other guys have raised.'"

To find extra cash, Beck and Raghu constantly chased grants. Their space start-up counterparts in the United States could usually talk their way into a few million dollars from NASA or DARPA. Rocket

Lab, though, had to get by on much smaller contracts tied to work outside the aerospace field. The company secured a couple deals to perform ultrasound testing on carbon-fiber hulls for superyachts. Another contract required the Rocket Lab team to probe the strength of a new type of piping by detonating small explosions inside the pipes. The experiments both excited and scared Raghu. He mostly took comfort in the fact that Beck still had all his fingers and toes after fifteen years or so of blowing things up.

In April 2008, Rocket Lab received some of its first major media attention when a reporter from the local magazine *Metro* popped by the IRL offices to interview Beck and Rocket. The prompt for the story appeared to have been a $99,000 grant doled out to Rocket Lab from the government to create "New Zealand's first space programme."

The story had a skeptical but sympathetic and curious tone. Right from the start, it noted that no government scientist had vetted Rocket Lab's technology or claims as part of the grant process, and it wondered whether Beck was "daft as a brush or brave as a lion." It recounted parts of Beck's youthful exploits, describing the rocket-powered scooter and rocket jet pack displayed in his office. It also managed to get Beck on the record about the company's first mission.

Beck revealed that the initial rocket would be dubbed the Ātea-1, borrowing the Māori word for "space." Rocket Lab had mocked up a demo of what the rocket might look like, showing the reporters a model eighteen feet tall and eight inches wide, although Beck said the final product would be a bit larger. It would fly fifty-five pounds of cargo up to 150 miles into space for $80,000 a launch. After being deployed, the payloads would slowly travel back to Earth in an arc with a parachute easing them to the ground.

Over those first eighteen months of trying to build the rocket, Beck had clearly tempered his vision of how much weight the machine could carry. He'd also become more of an entrepreneur. He did not let on that university customers were proving tough to acquire and continued to present science-focused organizations as

Rocket Lab's primary business targets. One breath later, though, he proposed that Rocket Lab would also be open to the idea of flying things such as trinkets and dead people's ashes into space if families, you know, wanted their loved ones to be "certified as 'space travelers.'"*

Beck suggested that people could pay $5,000 to send things like photos, business cards, and locks of hair to space on one of his rockets. However, he did not fully commit to the sales pitch. When the magazine asked Beck why someone would want to do such a thing, he replied, "You're asking the wrong person. The ashes one I can see a little bit more. 'Cause if you want to go to space you spend $20 million and go in a Russian rocket, or $250,000 and go on [Virgin Galactic], or wait till you're dead and go on Rocket Lab for a couple of grand. You still get to space. But it's just you're not alive." With his money on the line, Mark Rocket embraced the revenue opportunity with more gusto, explaining that it would be nice to put a relative's spacefaring ashes over the fireplace. "You know, part of Uncle Bob that has been to space," he said.

In a series of statements about an issue that would later come to affect their relationship, Beck and Rocket promised that the one thing they would never do was take money from the defense industry. "We said right from the beginning if it's involved in the military we don't want anything to do with it," Beck said in the article. "The military can be quite a tempting cherry because a lot of money gets poured into it but we're about science, we're not about killing people. . . . No weapons." It would eventually prove ironic that Rocket then tried to soften the rhetoric to leave room open for defense funds under the right circumstances. "I mean, we wouldn't be interested in winning a NASA contract to build weapons but if it's to do with research communications kind of stuff then we're a bit more open-minded on that," he said. ". . . To be clear, we're not anti-military."

The date for the big first launch was September 2008. At that

* "If you want to do something a bit funny then we can do it," Beck told the magazine.

time, Rocket Lab would send up six—yes, six!—Ātea-1 rockets and turn New Zealand into a commercial space powerhouse. T-shirts were being prepared for the event—get your merch!—and new investors were welcome. "This is something for New Zealand," Beck said. "We want all New Zealand to be proud."

A few months after the interview, SpaceX flew the Falcon 1. Even though Elon Musk's outfit appeared to be years ahead of Rocket Lab, Beck and his cohorts were inspired by the launch. SpaceX had more money and more people, but it had built a rocket in the same gritty fashion adopted by Rocket Lab. "We heard these stories about them having beds in the manufacturing building and sleeping by the rocket at night," Raghu said. "They managed to come through it and pull it off. We felt like we were in a similar boat, and for us, it was a huge validation."

The comedy in Raghu's reflection is that most people in the space industry back in 2008 considered SpaceX something of a joke even after its long-awaited first success. The company had almost gone bankrupt trying to produce a rocket so small that serious rocket makers judged it a toy. People figured that the company would implode as it tried to manufacture larger machines and doubted that it had achieved any real breakthroughs in terms of making rockets that would be cheaper and easier to fly. Rocket Lab's ambitions were less grandiose by miles. It had two and a half people working on a tiny rocket that was not even meant to reach orbit. If SpaceX was a joke, Rocket Lab was a ninety-minute comedy special.

Rocket Lab was, however, similar to SpaceX in the same way that all rocket companies are similar: it was forever behind schedule.

In the months after the *Metro* interview, Rocket Lab scaled back its design goals to more or less what Mark Rocket had envisioned from the start. The Ātea-1 that Rocket Lab put all its effort into building would be twenty feet tall and half a foot wide and carry five pounds of cargo to a height of ninety miles. Rocket Lab would now aim to fly the machine in late 2009, almost three years from when Beck had started the company.

Though many people in the aerospace industry would view the Ātea-1 as a toy, Beck looked at it as an opportunity to showcase his engineering chops. In an interview during that period, he explained that the main challenge was solving something he called "the spiral of doom." The physics of rocketry dictated that for every gram of mass added to the rocket (for something like the propellant tank or fins to keep the rocket from spinning), Rocket Lab would need to add ten grams of fuel. "Say I want to add ten grams of screws to the front of the rocket," he said.* "All of a sudden, now I need to carry a hundred grams of propellant to lift those ten grams of screws. Because I have put more propellant in there, I need a bigger fuel tank. So I need to add another ten grams of tank to carry that propellant. Now I've added more inert mass, and I need another hundred grams of fuel. But then I need a bigger tank."

Rocket Lab had decided to make the "spiral of doom" work in its favor. It would try to cut the weight of the vehicle as dramatically as possible in a bid to produce the most efficient sounding rocket anyone had seen. True to Beck's goals, the company did pioneering work around using carbon fiber for the rocket's body and came up with a powerful proprietary fuel recipe that mixed both solid and liquid propellants. It even designed its own heat shield materials to help the rocket withstand the 1,600 degrees Fahrenheit caused by friction as it pushed against the atmosphere. It may have taken Beck 3.14 times longer than he'd expected to make the Ātea-1, but his small team had proven their inventiveness and skill.

By November 2009, Rocket Lab had started to run out of money. Mark Rocket made some additional loans to the company to keep it going, but it was clear that Rocket Lab needed to do this thing and show the world its technology if it wanted to remain a going concern. No more engineering. No more refining. Let's see if the rocket works. "There had been so much pressure building up," Raghu said. "We'd been working nonstop very long days and nights for weeks and weeks."

* This interview took place on local television.

Finding somewhere that would let Beck have his moment of truth also required imaginative engineering. Beck first reached out to people in Australia to see if he could borrow a chunk of uninhabited land in the desert, but no one would take him seriously. After more investigating, he found some land owned by the navy near the Mercury Islands on New Zealand's eastern coast. It would be just a short hop from Auckland to the chain of seven islands, and the spot Beck had located was considered a weapons-testing range, which made it easier to justify sending up a small missile.

As he looked into the location more, Beck realized that a wealthy banker named Michael Fay co-owned one of the Mercury Islands, Great Mercury Island, outright. Beck figured that dealing with a rich guy who could do whatever he wanted on his own island might be easier than messing with the military. Since it was New Zealand and everyone knew one another, Beck called a guy who knew a guy who knew Fay and asked if he'd like to host a historic rocket launch. "I'm a very private person, which is why I live on an island," Fay said. "But my friend called and said that some guy wanted to do a rocket launch. He offered to tell Peter to fuck off on my behalf, but I said, 'No way. Give me his number.'"

Before agreeing to the launch, Fay asked to visit Beck's lab in Auckland. He saw the pictures of Beck's rocket bike and other inventions. "I went in skeptical," he said, "but it struck me that there was good design and sophistication to what he had done. The workmanship was outstanding. He explained very clearly to me what he had made and why it would work."

In late November, Beck, Raghu, and O'Donnell turned up on Great Mercury Island. Fay had really taken to the event and put helicopters, barges, and boats at Team Rocket Lab's disposal, as they spent about a week transporting the necessary gear to the remote location. "Michael told us, 'You do nothing else but launch this rocket, and I'll make sure everything else happens,'" Beck said. Used to hosting celebrities such as Bono on the island, Fay decided to turn the launch into a party, sent out invites to friends and the media, and hired a chef to feed them all.

As word of the impending launch spread, the various layers of the New Zealand government offered surprisingly few objections. The local council that advised matters on the islands noted that nothing in the district plan said that a company could launch rockets but also that there was nothing saying it couldn't. Through Fay's connections, it took all of two phone calls to clear the airspace over Great Mercury Island on the appointed launch day, with the airlines agreeing to detour all their flights. "The hardest bit was that a customs officer had come up with the idea that the rocket would be leaving New Zealand and then reentering it and would require some kind of import paperwork," Fay said. It took a couple more conversations to convince the customs office that it would look pretty dumb asking Beck to fill out forms on the spot after he'd blasted the first New Zealand–made rocket into space.

Fay decided that he would oversee the retrieval of the rocket from the water upon its return. He bought an anchor from a local boat shop and planned to use it as a hook that he would dangle out the side of a helicopter and then scoop the rocket up. He performed a number of test runs with some spare rocket bodies Beck had on hand. "We tied the anchor to the pilot's seat," Fay said. "It worked pretty well except the rocket kept filling with water. We practiced lifting it more slowly and letting all the water run out before bringing it inside."

On November 30, Beck and his crew prepared to launch the Ātea-1 early in the morning. They used a small bulldozer to cut a chunk out of a hill, creating a bunker between them and the launchpad. Into the bunker went a garden shed, which would function as their mission control. The setup looked very much like what a low-budget Bilbo Baggins might concoct if he were into launching rockets. "We were quite close to where we were going to be launching the rocket and put some wooden poles across the top of the bunker as some kind of protection," O'Donnell said. "I remember Pete saying that if anything went wrong the liquid fuel would probably catch fire and flow toward this hole we'd created in the hill anyway."

Inside the rocket's payload bay, Michael Fay placed a homemade lamb sausage that had been carefully wrapped in tinfoil. The observers, including a couple news crews, camped out on the grassy hills surrounding the makeshift launchpad. A Māori ceremony was held to bless the rocket, as everyone prepared for a good show.

No first rocket launch goes off without a hitch, and Rocket Lab met its nemesis in the form of a defective small fitting that controlled part of the fueling process. The part cost $5, but it was threatening to derail the whole operation. Fay took care of the visitors by plying them with food and drink, while Beck jumped into a helicopter to fly to a hardware store back on the North Island. He could not find the exact part but got something suitable enough and rushed back to the island. In his haste, he forgot to pay for the fitting. "I didn't even have money on me anyway," Beck said. After landing, he modified the rocket in front of the eager, increasingly impatient crowd.

Just after 2:00 p.m., Beck went into the shed and prepared the rocket for launch. Dressed in his white lab coat* with a black T-shirt underneath, he stood before a couple of laptops and started poking at buttons with Raghu and O'Donnell at his side. "In the tradition of great New Zealand explorers, New Zealand, we are go for space," he said. "Arm igniter. Arm oxygen . . . ignition. Ten, nine, eight, oxygen, seven, six, five, four, three, two, one." As Beck jammed his hand down onto a red button, a glorious whoosh could be heard outside the shed. He jumped out the open door, looked up and saw the rocket flying, and yelled, "You fucking beauty! Yes!" jumping into the air as Raghu cackled. "She's still burning!" Beck yelled. "Twenty-two seconds, and we're home free to space!" While people in the background applauded, Beck reached out to shake Raghu's hand and praised him for a job well done.

Hardly anyone succeeds on things like this on their first try, but the Ātea-1 flew beautifully. It traveled more than sixty-two miles into space and came back down for recovery. No one heard from

---

* "People hated the lab coat, and I didn't particularly like it either," Beck said. "But it was very deliberate. If you're building rockets and trying to do these hard things, you can't look like you don't know what you're doing. Let's be honest, we were on a farm, and it was not a super state-of-the-art facility. We had to try and improve our credibility, especially in New Zealand, where everyone thought space was a bit of a joke."

the lamb sausage again, and that was just fine. "It was this instant feeling of huge relief," Raghu said. "All the blood, sweat, and tests. It worked. We did it. This was a really big deal for little old New Zealand. We had to do it."

Not long after the launch, a boat spotted the first stage of the rocket floating in the ocean, and someone aboard the craft phoned Beck to let him know. Due to some technical issues, Rocket Lab had not received as much data back from the rocket as it wanted during the flight. The booster remains, though, showed that the rocket had burned through all of its fuel and performed great. Back on Great Mercury Island, Fay began opening some of his best wines.

Fay thought about the launch in philosophical terms. New Zealand, he said, was a country with no predators; as a result, some of the birds did not even fly because historically they had had nothing to flee. And although the Māori had a word for space, they didn't have one for a rocket. Beck had now changed the country's relationship with the skies and the heavens. Aotearoa had put something into space for the first time.

CHAPTER THIRTEEN

THE MILITARY IS NOT SO BAD

In Silicon Valley, technology start-ups often try to raise more money after they've hit certain milestones. You scrape by for a while on your initial funding, build a product that proves that you know what you're doing, then encourage rich people to marvel at that product, dream about what could come next, and cut you a fresh batch of much larger checks. Had Rocket Lab been based in California, its first rocket surely would have served these purposes well. Peter Beck had reached space almost on his own and on the tightest of budgets. He'd invented new technology along the way, and he exuded the hungry entrepreneur persona that investors love to see. People should have been lining up to throw money at this engineering wizard.

The reality, though, was that hardly anyone noticed what Beck had accomplished. Rocket Lab's launch had been covered by the local press and did appear briefly on the BBC and a couple other foreign outlets. Beyond that, it failed to make many ripples in the technology or space communities. Beck seemed more like a curiosity than a budding space titan. Weird inventor guy in the middle of nowhere sends hobby rocket to space. Good for him.

The only major players who saw much potential in Beck and Rocket Lab came from people in the US military. If someone had managed to send an object cheaply into space, they wanted to talk to that someone.

DARPA, the batshit crazy arm of the Defense Department, had been looking for ways to fly rockets into space as quickly and inexpensively as possible thanks to the urgings of Pete Worden and others. The agency perused the specifications of the Ātea-1 and came away thinking it might aid the United States' "responsive space" agenda. In particular, DARPA was impressed with the work Rocket Lab had done designing its own propellants and making a light rocket that had almost no metal parts. Beck had pulled off a handful of advances that people in the aerospace industry had long discussed but never demonstrated. Unlike his first trip, when he had showed up unannounced at rocket companies' doors, this time Beck received an invitation from DARPA and other military technical offices to visit the United States and pitch some ideas.

The meetings resulted in a couple of deals for Rocket Lab. The company was first asked to develop a rocket-based system for flying a camera high into the sky at a moment's notice. The year was 2010, and commercial drones that could be chucked up into the air by schoolchildren did not yet exist. The military needed a way to put a high-resolution camera into the sky instantly when a battle broke out and let it snap a constant stream of photos of the action below as it came down via parachute. It appeared that Rocket Lab might be able to design a device for the job.

The project became known as "Instant Eyes" and evolved into a handheld rocket launcher. A soldier could pick up a one-pound device built by Rocket Lab, press a button, and blast a camera and computing gear up to 2,500 feet in under twenty seconds. The device would then go to work snapping high-resolution pictures and feed the data back wirelessly to a phone, tablet, or laptop. "It was essentially a situational awareness tool that could be launched by someone in the field for search and rescue or to help someone who was stuck and needed to see what was around them straightaway," O'Donnell said.

The United States and New Zealand were close allies, but the military still did not quite know what to do about Rocket Lab from a legal standpoint. The US restrictions on sharing aerospace technology made it difficult for Rocket Lab to work on a project like that on its own. Someone with a creative mind decided to pair Rocket Lab with a US-based contractor so that they could take on the project together. Rocket Lab would do the lion's share of building the device, and the contractor would pay for the production and then sell the technology.

DARPA fancied the idea that Rocket Lab could make even better rockets than the Ātea-1 and cut it a check to continue researching small, low-cost machines. The spectacularly named Operationally Responsive Space Office and the Office of Naval Research also found some money for Rocket Lab. Much of the funding was meant to help Rocket Lab explore its propellant advances and went toward another project to make something called viscous liquid monopropellant, or VLM. Those deals were minor miracles for Rocket Lab. DARPA and its military chums almost never gave money directly to a foreign company for that type of work. Funding a dishwasher designer's personal missile program in a far-off land is the kind of thing that could go very wrong and come back to haunt a bureaucrat. Nonetheless, Beck proved intriguing enough to receive about $500,000 from the various deals, with VLM being the main area of interest to the United States.

Rockets typically rely on two classes of propellants: solids and liquids. Solid propellants are just what they sound like, hunks of fuel in a solid form. They make life easy because they're relatively simple to create and, more important, safeish to handle. You build the hunk of fuel and then shove it into a rocket whenever you need it. The major downside behind solids is that once they're lit, there's no going back. After the fuel is ignited, it's going to keep right on burning.

Many of the rockets flying today use liquid propellants, specifically a mix of kerosene and liquid oxygen (LOX). You light the kerosene on fire and then feed it with oxygen to keep the fire burning at the desired rate as the rocket moves up through the atmosphere and into space, where oxygen is in short supply. The downside of dealing with these materials is that they have a tendency to blow things up if not stored and used correctly. Rocket companies have to spend money on special equipment to handle the chemicals, and the propellants can be loaded into the rocket only right before it takes flight. If anything goes wrong ahead of the launch, the kerosene and LOX must be removed from their tanks before anyone goes near the rocket. It's a costly process in terms of both time and money.

The upside of liquids, first and foremost, is that they provide more power, or what aerospace engineers call "oomph," than solids. You can also control how the fuels are being fed into a rocket engine with precision. It's not an all-or-nothing affair, so you can throttle the engines up and down as desired. And if something is going wrong, you can click a button on a computer, close a valve on the rocket, and stop the flow of fuel altogether.

With VLM, DARPA and friends wanted Rocket Lab to make a propellant that would bring the best of both solids and liquids. As the name suggested, Rocket Lab would try to produce a viscous, or thick, liquid fuel that could be pushed into an engine and ignited at a controlled rate. The fuel would start out in a quasi-solid state and would liquify when hit with a shock wave. The properties of VLM made it more stable than pure liquid propellants, but it was dense like a solid fuel. VLM would also combine the fuel and the oxidizer into one product. That was a huge plus because engineers would not have to deal with the mechanics of mixing chemicals under pressure inside a rocket.

The military contracts provided enough money for Beck to start hiring more people. He put out job postings and soon had a handful of young engineers join him in the increasingly cramped IRL office. The deals, however, came with a major cost.

After the Ātea-1 launch, the marketing-savvy Mark Rocket had assumed that companies would be lining up to place their logos on the side of a rocket. All Rocket Lab had to do was crank out more of the small machines and blast away. Human remains would be heading to space aboard energy drink–branded projectiles, money would flow in, and everyone would be happy, including the space-certified dead travelers. None of that, however, came to pass. "The corporate market in New Zealand was a lot slower to support Rocket Lab than I thought," Rocket said. "It was disappointing and not as easy to generate revenue streams as I'd hoped."

Beck felt the brunt of Rocket Lab's commercial woes most directly. In the year between the flight of Ātea-1 and some of the military money trickling in, Beck had been sweating out his future. He wanted to spend all day, every day building rockets but still had to take on contract jobs performing carbon-fiber tests on yachts or retooling some company's gearbox for spare cash. To keep the rocket program advancing, Beck often found himself in a junkyard, crawling around to find pipe fittings or hunks of metal on the cheap. "Everyone in New Zealand thought that what I was doing was nutty," he said. "There were so many nights I lay awake, wondering how I would make payroll." Things got bad enough that Beck took out a second mortgage on his house. "I put our family in pretty serious financial situations," he said. "If it all went bad, it would go really bad. My wife was an incredible engineer in her own right but had agreed to stay home with the kids. She gave up a lot and put up with a lot for me to pursue my dream. But it's either you believe in it, or you don't. That's just what you do."*

Given the circumstances, Beck was more than willing to make a few bits and pieces for DARPA and the US military. He saw no other choice, and beyond that, he was obsessed with making rockets and viewed the contracts as the clearest path to achieving his dreams. Mark Rocket, however, decided that he could not stomach doing business with the bringers of doom. He urged Beck to reject the deals and then walked away from the company when Beck refused.

---

* At one point, Kerryn told Peter, "You can do Rocket Lab, and we will live in a cardboard box, but one day, you have to show me a million dollars in the bank account."

"For me there was a line, and for Peter, the line had moved," Rocket said. "I understand why Peter wanted to go that way. It opened a lot of doors. It put some money into the business. Peter wanted a quick decision, and I didn't want to hold the company back. I thought there could be another pathway, but he was talking to some big shots in America, and it had piqued his interest. At the end of the day, he was the CEO."

Rocket agreed to let Beck buy back his stake in the company. But since Beck did not have any money, the two men struck a deal where Beck had five years to gin up the cash. It was a gracious agreement on Rocket's part. He kept things between him and Beck, so that Rocket Lab would not have any debt on its balance sheet, and he walked away from an investment that could have been worth millions, even billions, of dollars if Rocket Lab turned into a massive success. "There was no yelling that I recall," Rocket said. "We were both passionate about our positions. I think I did pretty well by Peter."*

It could be argued that the military deals were a major distraction to Beck's prime mission of building a small rocket that would make regular journeys into space. With ready access to capital, Silicon Valley start-ups are often afforded the luxury of chasing their main technology dreams for years without needing to make compromises. Rocket Lab, though, had no better options. With the pressure on to please DARPA, Beck tabled work on a planned Ātea-2 rocket and focused first on Instant Eyes.

One of the first people Beck hired for the new project was Samuel Houghton. Armed with a mechanical engineering degree, Houghton had dreamed of working in the aerospace or automotive industries but could not find any New Zealand companies doing interesting work in those fields. He did a stint with Boeing in Australia and then one day received a call from an old professor who mentioned that a

* Spoiler alert: Rocket Lab is now worth billions of dollars. Beck repaid the loan about five years later, after Rocket Lab had raised significant money, in the form of Rocket Lab shares. Whereas Rocket once held 50 percent of the company, he now owns less than 1 percent. "In hindsight, I probably would have done things differently," he told me.

rocket start-up had appeared in New Zealand and was "after some smart engineering cookies."

Houghton had to be a jack-of-all-trades on Instant Eyes, helping out with the design of the launcher and figuring out a place where Rocket Lab could conduct its tests. The company wanted a spot as close as possible to Auckland that had unrestricted airspace. Houghton employed the New Zealand trick of calling a friend who called his dad who called his neighbor, and soon enough Rocket Lab had the run of a sheep and cow farm. "The guy thought it was hilarious," Houghton said. "I just had to convince him that we knew what we were doing and would not launch things into his cows or leave his gates open."

Over a period of about six months, Rocket Lab built and tested many versions of Instant Eyes. The employees would spend the first part of the week designing and assembling a new prototype, another day or two testing it, and then head out to the farm on a Friday. On the test days, they would pack a pickup truck with their handheld missile launcher and try to drive from IRL to the farm without arousing too many suspicions.

Out in the pasture, Beck, Houghton, and a couple of other engineers would fire projectiles up into the air and wait for them to come down via a parachute. They learned about the performance of their hardware and the wind's fickle ways. Much of their day would often be spent hopping over fences as they chased down wayward projectiles that had fallen among gorse bushes and then dug them out of wet, marshy ground. The locals rarely complained other than the odd call from a farmer who was upset because his airspace had been closed off by officials for a couple of hours during the tests, preventing him from flying a plane to check on his crops and animals. "There was an air force bombing range nearby, and they were like 'Would you mind telling us what's going on here with all the rockets?'" Houghton said.

It took Rocket Lab less than a year to produce working Instant Eyes prototypes and perform demonstrations for DARPA. By 2011, the company was already receiving awards for the project and announced that it would start selling the device by 2012. It's likely that a military contractor had not worked so fast and at such a reasonable cost since the 1960s.*

Houghton chalked up the success to the camaraderie of the Rocket Lab team. They were inexperienced but enjoyed one another's company and the spirit of the project. Even in their off-hours, the engineers would conduct experiments to see who could slap a rocket motor onto a toy car and blast it across the parking lot the fastest. Beck almost always won the contests, which only made the other engineers want to try harder to beat him. He also set an example with his work ethic, which carried over to the whole company. As Houghton recalled, "Peter would give you a job to do, just ask you to look into something or to buy a part. Coming from my government contractor background, I'd figure to get to it in the next two or three days.

"One of the things was to find someone local who could sew parachutes. Peter asked me to do that and then checked in with me in a couple of hours to see how it was going, and I hadn't gotten anywhere. Not more than half the day had passed, and it turned out that Peter had jumped on the phone and organized it all himself in thirty minutes. He didn't say anything to me or tell me off. But when he wanted something done, he would get it done. He would also devour hard-core rocket textbooks. There was always a stack of them on the shelves. If there was a problem, he'd dive into the books. It was this foundation of knowledge of how to figure out the next step of what we were doing."

Raghu left Rocket Lab in the early stages of the Instant Eyes push. Some of his fondest memories were running through the IRL offices alongside Beck with one weird contraption or another in hand. The scientists in the building always gave them strange looks and hoped the men knew what they were doing and that they weren't going to create a biohazard or destroy the whole place. It felt

* Rocket Lab demonstrated the technology in Florida, and its US partner hoped to sell Instant Eyes by the thousands. The partner, however, never managed to get the business going, and soon enough, drones that could do the same job were commonplace.

exciting being by Beck's side. Raghu, though, wanted to travel and pick up experience in new fields. The age of small rockets flying tiny satellites to space all the time felt at least a few years away. "It's not like someone knocked on our door and handed us $100 million to go chase SpaceX," he said. "I loved every moment of my time there and had so much fun and so many challenges."*

With Instant Eyes doing well, Rocket Lab turned to the VLM effort and Beck's ever-lofty goals for the project. Not only did Beck want to develop a new class of fuel, but he also wanted to put it inside a Rocket Lab–made version of the US military's AIM-9 Sidewinder missile.

Once again, the six-person Rocket Lab crew sprang into action. They designed, built, and tested over and over again, with a couple of people refining propellants while a couple others concentrated on the rocket body. When Rocket Lab needed to tackle electronics or software, it called O'Donnell for part-time help.

Sometimes testing went well. Sometimes it went in more disconcerting directions. "We worked crazy hours, and I remember being in the basement with Pete one time at around midnight," O'Donnell said. "We were testing the propellant in a rocket and pressurizing it. Something inside the system let go, and the machine started building up this massive amount of pressure. I am trying to control it and stop it, and I turn around and Peter has dived under the desk. He was looking up at me. I thought, should I be under the desk, too?"

Beck bounced back and forth among all of the projects and had an unnerving ability to solve problems that stumped everyone else. DARPA initially gave Rocket Lab money to produce prototypes of the technology and then, liking what it saw, cut Rocket Lab another check to go ahead and fly the thing.

In November 2012, Beck and his crew traveled to Great Mercury Island to demonstrate the VLM-based missile for the US military brass. It had been three years since the first Ātea-1 launch. Rocket

---

* Raghu made his way to the United States and ended up starting a robotics company called Alterra Robotics in Silicon Valley. "There's always that little bit of regret," he said. "That feeling of 'Oh, man, I should have stayed.'"

Lab had accomplished a great deal, but it felt as if a bad test could be the end of the company. Inauspiciously, Beck and O'Donnell nearly blew up their all-or-nothing vehicle during a test run the night before the launch.

Michael Fay played host once again. No media were invited, but officials from the likes of DARPA and Lockheed Martin were put up in Fay's house. The night before the launch, a fight broke out during dinner. Fortified by fine wine, an executive from Lockheed tried to put Beck in his place, saying that Lockheed should take over Rocket Lab's technology after the test.* "It was sort of 'Look here, little child. We are Lockheed Martin. You have some cool ideas, but you need adult supervision,'" said one of the dinner attendees. "It was really ugly. Full-on screaming. Other people told Peter that he'd come so far and didn't need any help at all."

The following day, the test bolstered the case for those who had supported Beck. Rocket Lab had hoped to build a machine that outperformed the US military's state-of-the-art technology—and it did exactly that. The rocket made a near-perfect flight and landed softly in the ocean, where it could be recovered and analyzed.

Al Weston had left the confines of NASA Ames to take a vacation in New Zealand. He'd caught wind of Rocket Lab's activities through friends in the military and on a whim had asked Beck for an invite to the launch. He had worked on weapons programs for the United States and done extensive cutting-edge research at Ames, putting him into a good position to evaluate Rocket Lab and Beck. Weston went to the event with the lowest of expectations and was blown away by what he saw. "I'd thought that there wasn't anyone in New Zealand that could even spell rocket," he said. "The whole idea seemed like horseshit to me, but it turned out it wasn't. Peter was real."

Not long after the demonstration, Beck held an all-hands meeting in Rocket Lab's tiny IRL office. A handful of twentysomethings watched as he unveiled what the company would do next. He had

---

* Lockheed was one of a number of companies that had been buying Rocket Lab's homegrown heat shield material.

spent six years toiling away in the bowels of a research lab and now felt he'd proven enough for investors in the United States to take him seriously. Where other CEO types might have given a rah-rah speech, Beck went with a low-key, matter-of-fact approach. He was going to fly to Silicon Valley, and he was not going to come back without a big bag of cash. Rocket Lab was going to make a real rocket.

CHAPTER FOURTEEN

ENTER ELECTRON

Put yourself into Peter Beck's shoes for a moment.

In 2013, he was in his midthirties and did not fit any of Silicon Valley's typical categories for a start-up founder. He was much older than a hotshot dropout or a brainy kid with a big idea who had just graduated from college. At the same time, Beck did not enjoy any of the perks of being a thirtysomething in the Valley. He had no track record of working at or running a successful technology company. He lacked a well of contacts in the industry. At best, he had done some clever engineering projects for the military on things that few people in the technology world would understand, let alone care about.

Worst of all, Beck hailed from New Zealand. People from New Zealand have modesty bred into them. They're terrible at self-promotion and in fact try to bring down anyone who has achieved success.* Beck was genetically and culturally ill equipped to stand before a room of investors and tell them how wonderful he was and how Rocket Lab would, as the Valley's preferred cliché goes, change the world.

* These truisms apply to all New Zealanders other than rugby players.

The last time Beck had raised money, things had not gone well. He'd given away half of Rocket Lab for a measly $300,000. This time around, he planned to visit a few venture capital offices to ask for much, much more money: $5 million. Unlike the easygoing Mark Rocket, the investors in California were trained to take advantage of Beck and pry as much of Rocket Lab away from him at the worst possible terms. He would also be cold-calling those investors and turning up in their offices on his own.

The Rocket Lab pitch was laughable. Beck would pull up a PowerPoint deck and inform the investors of a coming space revolution. There would soon be tens of thousands of satellites that needed launching into space, and Rocket Lab planned to become the world's most prolific rocket launcher. Yes, countries and billionaires usually try to tackle such lofty, expensive projects. And, yes, rockets always take much longer and cost much more to make than anyone expects. Oh, also, there's no profits to be had in rockets historically. But this time around would be different because Rocket Lab had pluck and grit and would solve the endless stream of technical challenges that had stumped thousands of people for decades. It would make awesome rockets, and it would make awesome money. Trust me, guys. I'm Peter Beck. I have floppy hair, cool charts, and enthusiasm.

Beck would turn up at the meetings like a carnival barker. He had a bag that contained one of the small rocket engines he'd designed and some other parts. He'd also printed up a huge drawing of the rocket he hoped to build and would unfurl it onto the boardroom table, allowing the paper to spread from one end of the table all the way to the other. According to one person, Beck also sometimes dumped a bag of small plastic balls onto the table to represent all the satellites that would soon need launching.

Amazingly, it all worked.

Beck pitched only three venture capital firms over the course of three weeks and came away with millions of dollars from one investor, Khosla Ventures, whose name meant almost as much as the money itself. Khosla was one of the highest-profile investment firms in Silicon Valley. Rocket Lab wasn't getting money from just anyone; it was getting money from people who were generally thought to know what they were doing.

"I remember exactly where I stayed during all of that," Beck said. "It was in a Holiday Inn on the first floor and the third unit down the hall. I picked it because it was an accessible room, and it was the cheapest. I also remember that I missed my daughter's first birthday. I told people that I needed to raise five million dollars for a rocket company in New Zealand and that I was literally going to set their money on fire. They were mostly like 'Okay. That's a big investment round. How about a million?' But I wanted five million and didn't want to raise two hundred thousand here and five hundred thousand there. I wanted one—*one*—investor who was sold on the vision and was going to back the company and back me, frankly, to get the job done."

Part of the reason Beck succeeded in spite of the many odds against him can be traced all the way back to Ames and Pete Worden's crew. As investors performed their due diligence on Rocket Lab, they called Worden for advice. Worden, in turn, passed the investors along to Weston, who had witnessed Beck's work firsthand. Weston vouched for Beck, and his words carried serious weight. Beck also fit into the patterns that venture capitalists cherish: He was obsessed with rockets to the point of being crazed. He could answer any question thrown at him in depth. And he'd already made something real. If you were excited about space and wanted to get into the business, the Kiwi seemed like a reasonable bet.

By the time Rocket Lab's deal with Khosla was finalized in October 2013, other people were starting to entertain the idea of entering the small-rocket industry, too. Richard Branson's Virgin Galactic had spent years building a spaceplane for tourism without much success and decided it could complement that sketchy business with another sketchy business. It put a team together to begin designing a small rocket that would carry a few satellites on each ride. Another company called Firefly Space Systems had pretty much the same thought and started work designing its rocket in early 2014. Those companies would also find serious investors and were legit rivals. Meanwhile, dozens of other less well funded and less legit rivals sprang up, too.

A race to replicate or best the Falcon 1 had begun, and Virgin and Firefly had major advantages over Rocket Lab. Most glaringly, Virgin had recruited a big chunk of the Falcon 1 team to come work for it. Surely they would be able to build a similar machine much faster and with less money than SpaceX had since they were sitting on years of experience. Firefly also had a handful of old SpaceX hands. And both companies were based in the United States, where they could keep tapping into US capital with relative ease. Rocket Lab, meanwhile, could not even bring US engineers to New Zealand and let them work directly on its rocket due to the United States' fear that a foreign nation would gain access to prized military and aerospace secrets.* Virgin and Firefly, by contrast, were unfettered by such regulatory restrictions.

To make matters somewhat easier, Rocket Lab officially changed its headquarters from Auckland to Los Angeles. At least initially, it was a mostly cosmetic move. Beck still lived and worked in Auckland, and so did the entire Rocket Lab engineering team. But by having a small office in the United States, Rocket Lab could attract more American investors with less legal hassle and, more important, make it easier in the future to do deals with the US government and military and NASA. Some people in New Zealand were miffed that Beck would declare Rocket Lab an American company in such a transactional fashion, while others understood that it was the clearest path toward acquiring the resources it would need in the years to come.

"I'm as patriotic a Kiwi as you'll ever find, but goddammit, America gets shit done," Beck said. "There is no other place on Earth where a Kiwi could come into town and walk away with

* The issue is less about the person and more about the way information flows. An American could go and work at Rocket Lab and handle a wide range of tasks. But he or she could not, for example, sit in an engineering meeting in Auckland and tell people details about how an engine or an electronics system works if it was deemed to be protected information. The New Zealand government, by contrast, would allow a Kiwi engineer to fly to the United States and provide intimate details on engineering matters to American Rocket Lab employees.

enough money to start a rocket company. After we signed our term sheet with Khosla, I went straight to the supermarket and bought an American flag."

When he returned to New Zealand, Beck began hiring new recruits. Rocket Lab could finally expand from an R&D shop to a proper company with real offices and desks and a manufacturing floor. Some of the fresh engineers came from New Zealand universities or from industry. Beck also lucked out that a couple of Australian universities had quite good aerospace engineering programs. The students who had gone through those schools longed to work at a real space company but usually had to settle for jobs well afield of their educational backgrounds because no space companies existed in the region. For them, working for Rocket Lab represented the opportunity of a lifetime.

Even though Rocket Lab had been consumed with its military projects, Beck had been sketching out exactly what he wanted to do with his millions for years. He'd filled notebooks and computer design programs with models of the Electron rocket and the Rutherford engines. Beck wanted to create the pinnacle of engineering when it came to small rockets and usher in a wave of technical advances.

First off, the body of the rocket would be made of carbon fiber instead of aluminum. The carbon fiber would make the rocket cost more, but it was lightweight and strong and meant that it could carry a bigger payload. It also played to Rocket Lab's strengths. New Zealand had long been a powerhouse in the America's Cup sailing competition, and the modern craft were all built out of carbon fiber with state-of-the-art technology. That gave Rocket Lab ready access to an inordinately high number of carbon-fiber whizzes who also happened to need work due to the sporadic nature of the America's Cup competition.

Rocket Lab also planned to try something new with one of the most complicated parts of a rocket: the turbopump. The turbopump is a mechanical system with a turbine that spins at incredibly high speeds. It's essentially the only moving part of a rocket and has to serve as an intermediary between combustion gases on the one side and fuels such as liquid oxygen and kerosene on the other. The turbopump's mission is to keep the engine fed with the perfect ratio of fuels and to do so while under immense stress.

Beck's idea was to remove a large chunk of the mechanical components and plumbing and have an electric motor running on electric batteries put the propellants under pressure and pump them into the combustion chamber of the engine. The batteries would add weight to the rocket, but the overall design would be simpler and allow for more precise control of the propellants. The real trick was that no one had ever managed to get an electric turbopump to work on a rocket before. But Beck figured he was the man for the job.

Rocket Lab's other major advance would come through its embrace of 3D printing. Instead of making its engines by hand, Rocket Lab would use machines to build its machines. It would acquire 3D printers that could blast metal powder with a laser, fuse the metal, and build up an engine layer by thin layer. People in the aerospace industry had dabbled with the technique for building engine components here and there. No one, though, had tried to make an entire engine via the 3D printing process. Though experimental, the technology held the promise of allowing Rocket Lab to manufacture new Rutherfords with the push of a button.

From late 2013 through 2014, Rocket Lab went from about ten employees to twenty. Beck had been in survival mode for so long that he did not want to rush wildly ahead with his ambitious ideas until he and his team had proven some of the basic concepts behind them. A young CEO in California likely would have done things differently, hiring people by the dozen and expanding as quickly as possible. Beck, though, was already demonstrating the tendencies that would come to define his leadership. He was frugal and methodical and loathed the idea of needing to raise more money from investors because it would mean giving up more control of his company. He was known to say, "Every dollar that I have to raise today costs me a hundred dollars worth of equity." Sometimes that mindset worked in Rocket Lab's favor; other times it slowed the company down.

Some of the earliest hires were Sandy Tirtey, Naomi Altman, and Lachlan Matchett, each of whom was put in charge of key programs. Tirtey had grown up in Europe and then taken a job at the University of Queensland in Australia before joining Rocket Lab. He was unusual among the first wave of employees because Tirtey had a PhD in the aerospace field and had worked in Queensland on aerospace hardware. Altman and Matchett were more typical hires straight out of university who were thrust right into the ridiculous effort of building a rocket without having the first clue about how one would do such a thing.

Beck tried to set the example for his recruits through his work ethic and general philosophy of engineering. He typically turned up at the office around 7:30 a.m. and left at 8:00 p.m. "Pete outworked everyone," said an early employee. Beck had never been one for idle chitchat or shooting the shit, and any attempt to hold casual conversation with him tended to be uncomfortable. Beck liked to talk about rockets and how to solve problems and little else. He didn't come off as unfriendly, but the idea was that if he was speaking to you, the conversation should serve a purpose.

Since everyone was building a rocket for the first time, Team Rocket Lab had to make things up as they went along. The engineers had books, photos, and online technical documents at their disposal, and they used that information as a basis for how to make a part or design a crucial section of the rocket's body. Beck added a degree of difficulty to all of that in that he wanted Electron to be cheap and easy to manufacture. It was not enough to mimic what had been done before. He wanted his engineers to find new, lower-cost ways to build complex aerospace parts.

When Rocket Lab, for example, needed a vacuum chamber, Beck insisted that his engineers avoid buying an expensive one as every other rocket company would do. Instead, he told someone to purchase an industrial stainless-steel meat grinder and convert the hunk of metal into a vacuum chamber. "It ended up as one of the best vacuum chambers in the world, and we got these things basically for free," said Stefan Brieschenk, a former Rocket Lab employee. "Peter did the same thing with almost every fucking part on the rocket. It gets to the point where people won't believe you when you describe what Rocket Lab tried because there's hardly a single element that is like what a larger company would do."

Beck's past as a toolmaker seemed to give him a sixth sense for knowing in which direction to take an engineering project. Some of his fresh hires wanted to show off their engineering skills and were inclined to create artful, sophisticated parts. Beck, however, would cut them off early and nudge them in a new direction. "No matter what an engineer is explaining to him, Pete understands it and the mechanics of what's involved to make it," Brieschenk said. "He can instantly smell if something will be cheap or expensive. An engineering guy will love the engineering of something for engineering's sake. Pete has stood at the lathe himself. He's welded things himself. There is no piece of machinery he has not used with his own hands."

Beck also seemed to have a built-in understanding of physics and could roll figures and ideas around in his mind with ease. On many occasions, a well-educated hire would challenge Beck and point out the theoretical flaws in his approach to a tough problem. Beck would listen to the criticism, go home to his workshop, and build a prototype to either prove or disprove his assumptions. He often turned out to be right; if nothing else, he was decisive. "Sometimes you get annoyed because this guy has an opinion on everything," Tirtey said. "But, you know, that's his job, and he had a better understanding than most of us around how to make something work practically. We had some very frank discussions and debates, and then, once a decision was made, everyone embraced it. It's better for everyone to look one way even if it's wrong than to be pulling in different directions."

To find people who could keep with Rocket Lab's spirit, Beck developed an exacting hiring process. It was common for him to interview a hundred or two hundred people for a position. Though he considered an applicant's academic record, he paid more attention to objects the person had built. In many cases, aspirant employees would come into the Rocket Lab office and be required to perform an engineering test like analyzing a circuit board or building a pump over the course of several hours. Beck would pick the person who completed the task faster and better than everyone else. "He gives people real physical tasks, and that's the way it should be," Brieschenk said. "You have to demonstrate right there that you can do the job."

Like Musk before him, Beck had a habit of setting hyperbolic timelines for building things. Sandy Tirtey, for example, started at Rocket Lab near the end of 2013 and was shown a plan that called for the entire rocket to be completed and standing on a launchpad by November 2014. "It didn't sound realistic to me," Tirtey said. "But I was the new guy. I didn't want to be the nonbeliever. If these guys thought it was feasible, then let's go for it. Obviously, it was, um, a bit ambitious."

From 2013 to 2015, Rocket Lab experienced the complete and utter drudgery of what it takes to build a rocket. The company started off by making a basic engine and trying to get it to fire for a few seconds. Once that goal was achieved, the engineers went back and tried to make a better engine and get that one to fire for a few seconds. The process was repeated over the course of months and then years and over hundreds and hundreds of tests until finally the machine worked as everyone had hoped so long before. As with all rocket programs, some progress occurred in linear fashion, while much of it did not. Engines exploded for no apparent reason. Brush fires blazed around test stands. People dodged shrapnel and then went back to their computers to try to find out why their engines had almost killed them. In the midst of all that were the occasional massive breakthroughs.

Each part of the rocket—the fuel tanks, the electronics, the software, the outer body—had the same adventures. There would be periods of tremendous progress when it seemed as though a major project was nearing completion. But then a niggling flaw would prove tough to overcome and turn into a drag on the program that would last for months. Throughout its early years, Rocket Lab would hire contractors to build specific parts of the rocket, hoping to save time and lean on the expertise of specialists in their fields. The early employees, however, recalled few, if any, instances where that strategy had worked. The contractors either did not move fast enough or didn't understand the interlocking parts of the rocket well enough to make something useful. Rocket Lab's engineers had to develop the parts themselves and spend time learning and perfecting a new skill.

Even though the company had fallen behind Beck's optimistic schedule, its efforts were impressing outsiders. Potential investors watched as SpaceX launched more and more rockets, and Rocket Lab's rivals such as Virgin Orbit and Firefly were reporting some early successes with their programs. All of that made commercial space feel more real. Beck opted to seize the moment and return to Silicon Valley to find more investors who could supercharge Rocket Lab's business. With Khosla Ventures now stumping on his behalf, Beck found it easier to raise money, and a lot of it at that. He went to Silicon Valley twice in 2015 and came back with around $70 million from venture capital firms. The governments of Australia and New Zealand kicked in additional funds, as did Lockheed Martin.

At that stage, Rocket Lab had moved out of the cramped IRL offices and into a large office-cum-factory in an industrial area near Auckland Airport.* The number of employees at the company had been doubling every year, and soon even that building felt small. Rocket Lab had areas dedicated to activities such as electronics and carbon-fiber construction and the main factory floor where Electron was coming to life, but outside there were rows of shipping containers stacked on top of one another and people working inside them.

In the statements announcing its new funding, Rocket Lab claimed that it would launch the first Electron by December 2015. Even if the rocket was ready for the journey, which it wasn't, that would have been a neat trick—because Rocket Lab had two very major logistical issues before it: it did not have a site from which to launch a rocket, nor did it have any sort of legal right to fling a missile into space.

* It was the same building where I would later meet Beck for the first time in 2016.

At first, Rocket Lab did not envision launching rockets from New Zealand. The United States offered more existing infrastructure and a range of places along the East and West Coasts where the company could set up shop. Locations in Europe, South America, and Asia also had their advantages. In the end, however, the company realized that its logistics would be simplified by keeping things at home and that New Zealand's remote areas and general lack of air and sea traffic could, in fact, make the country the ideal location for a rocket launch site.

In a nod to Rocket Lab's culture, Beck put the then-twenty-two-year-old Shaun D'Mello in charge of finding exactly from where in New Zealand to launch. D'Mello had come from Australia and started at Rocket Lab in mid-2014 as a glorified intern.* He'd bounced from one project to another and built up some cred as a capable human. Since no one else had figured out the logistics behind selecting a rocket launch site, D'Mello seemed like a good enough man for the job.

D'Mello pulled up Google Earth on his computer and began scouring both the North Island and the South Island for unpopulated areas near the coasts. If they could be reached from Auckland by car or a short plane ride, all the better. He placed markers on his map for around two dozen locations and then investigated them further. He built a spreadsheet that included weather patterns, air traffic, the number of fishermen nearby, and, of course, which orbits an Electron could reach from the various spots. D'Mello then took a series of road trips to see the sites in person. In a matter of weeks, he settled on the Māhia Peninsula as a prime place from which to launch a rocket.†

Located on the east coast of New Zealand's North Island, the

---

* During an interview for the job, Beck wanted to probe D'Mello about his hands-on experience and asked the youngster to describe the last thing he'd made. "I had the worst response possible because I'd just made a shoe rack the day before," D'Mello said.

† Rocket Lab briefly considered building the rocket launch site in Christchurch, although the idea met with considerable resistance from locals.

Māhia Peninsula is remote and breathtaking.[*] It's a mix of undulating pastoral lands inhabited mostly by sheep and cattle and steep, grassy cliffs that run alongside unspoiled beaches and turquoise water. About twelve hundred people live in the region, most of them making their living as farmers. If you need to visit a large shop or catch a flight, the closest city of significant size is about an hour and a half away. The only breaks in the peace and tranquility come during the Christmas holidays and summer months, when as many as fifteen thousand tourists can turn up in Māhia to fish, surf, and hike. Many of the visitors take up residence in baches, which are a unique breed of small, rustic Kiwi holiday homes that dot the beachfronts.

Rocket Lab made its first inroads in Māhia thanks once again to the generosity and connections of Michael Fay. When not on his private island, the wealthy businessman spent time on his farm in Māhia. After learning of Beck's interest in the area, he began reaching out to friends to see if there were any properties coming up for sale. Someone mentioned that Onenui Station, a sheep and cattle farm that ran right to the water's edge, had been going through tough times and the owners were looking for new opportunities. Fay rang up the owners and asked if they'd consider getting into the rocket business.

Many of the farms in Māhia used to be owned by Māori families, but over time the land had been sold to large corporate farming operations. Onenui Station, however, was still run by a group of eighteen hundred Māori shareholders.[†] Rocket Lab's mission soon became convincing the landowners and other locals that it would be a good idea to invite weekly rocket launches into their quiet, pristine surroundings.

---

[*] The Māori long had considered Māhia as a safe haven or place of refuge, particularly during wars among different tribes.

[†] Rather than owning specific parcels of land, people own shares in a corporation that manages the land as a whole. This arrangement goes against Māori cultural and spiritual traditions in which people have strong ties to particular pieces of land. This system of land management, however, began in the 1930s in a bid to create larger plots that would be more economically viable as farms.

In September 2015, Beck and Fay first met with George Mackey and his father, both of whom were among the senior management of Onenui Station. "They told us they wanted to launch rockets off our farm," George Mackey said. "We sat there and had a beautiful lunch. Then my dad and I left and hopped in our vehicle and were driving away. We didn't say anything to each other for about ten minutes, and then we looked at each other and said, 'What just happened? What the heck was all that about? Is it real?'"

Onenui Station stretched across ten thousand acres and had sixteen miles of valuable coastline. The considerations for any deal to buy or lease the land would go beyond the immediate financial terms. Building a launch site in Māhia would require major infrastructure projects, including the construction of roads that could support heavy equipment and systems for power, fuels, and computing. Then, if Rocket Lab did manage to start lobbing its rockets up constantly, the locals would have to deal with a host of environmental factors, ranging from noise to the possibility of pollutants fouling fishing areas.

The Māori once owned all of Aotearoa (their name for New Zealand), but their landholdings had been cut to 5 percent of New Zealand's territory after the colonization of the country. For obvious reasons, Beck and his Rocket Lab team were viewed with skepticism. Making a deal would not be as easy as presenting some pleasing financial terms on a piece of paper and promising to be good neighbors. Rocket Lab would need to earn the trust of the Onenui Station shareholders and the residents of Māhia.

Beck made the first round of visits with local leaders. As per Māori custom, he sometimes had to learn and sing songs that introduced him as a visitor to the land. Despite his awkward performances, Beck won over some of the most influential shareholders with his life story, regaling people with tales of his inventions and his trip to the United States in which he had barged into aerospace firms' offices unannounced. People in the audience took to his pluck and wiles. "There's a spirit in him that reminded us of Māui, who is one of our demigods that helped found New Zealand," Mackey said.

"As Māori landowners, we could see Māui's trickster spirit in Peter. He gets things done but in a mischievous way."

Soon, though, the task of winning over the Māhia residents fell to Shane Fleming, an American who had joined Rocket Lab in 2015. He spent about six weeks visiting homes and talking with various groups, hearing their concerns. "We went around to two hundred locals," Fleming said. "I can't remember how many cups of tea and biscuits we had."

The charm offensive worked. After two months of speaking with the residents and another month haggling over the finer points of the business deal, Rocket Lab had itself a private spaceport on a sheep farm.* It agreed to lease the land and to pay a fee to the Onenui Station shareholders for each launch.† The more rockets the company put up, the more the farmers would earn. On December 1, 2015, Rocket Lab had twelve trucks full of gravel lined up to begin creating the foundation of its spaceport.

Mackey took pleasure in the idea that the farm, which offered unfettered views of the Milky Way, would be contributing to the modernization of the world. He liked Beck's stories about the influx of imaging satellites and the coming space internet. "The notion of never being out of coverage would help with search and rescue and emergencies," Mackey said. "It was exciting to be part of that in a small way." Looking longer term, he hoped that the farm would earn enough money from launches to be able to retire some of the land, fill it with native flora and fauna, and perhaps start an ecotourism business.

The site Rocket Lab chose for its launch complex was at the

---

* The official vote among the shareholders was done by a show of hands.

† According to figures provided by Mackey, the fee is about $30,000 per launch. It's a significant amount of money for a farm that struggled to turn a profit over the years. Before Rocket Lab turned up, the shareholders had considered building a prison on the site or switching to growing potatoes. "I think we're now the envy of all the other Māori land blocks," Mackey said. Rocket Lab secured a twenty-one-year lease, but the shareholders must jointly agree to renew it every three years. "Our concern is what happens if the Silicon Valley investors decide to sell Rocket Lab to the Russians," Mackey said. "We may not want Russians on our farm. If there was ever a new owner, we would want them to go through that same process of setting up a relationship with trust and respect for each other's culture."

southernmost end of the peninsula. The company quickly got to work building the infrastructure. It laid down a twenty-six-foot-square launchpad and a few miles of road. It shipped in a premade hangar in which employees could work on the rocket and place satellites into the machine. The company also shipped in its fifty-five-ton steel rocket erector. As the facility began to take shape, it could not have looked more dramatic. There were acres and acres of pasture, then a spaceport, and then a steep cliff that cut right into the ocean.

For a while, the locals were happy with the deal they had struck. Rocket Lab created a couple of scholarships for the local schools. It also hired local contractors when it could and brought new business to the people renting out their baches and to restaurants. The coffee shop in Māhia even changed its name to the Rocket Cafe. Most of all, the Māhia residents were thrilled that Rocket Lab had invested in a high-speed internet system that they could all now use as well.

Over time, however, the relationship between Rocket Lab and the locals soured. Residents who had waited weeks or months for council approvals to do work on their homes and businesses could not believe that Rocket Lab had its every wish granted almost instantly. They came to resent the construction and the demeanor of Rocket Lab's employees. It often felt that Rocket Lab employees catered to influential leaders, while not giving as much respect to the other people in town, and also that they ignored Māori cultural customs. "They have trodden on and burned a few bridges along the way," said Janey Bowen, the owner of the Rocket Cafe. "A lot of it has been from a lack of knowledge on the basic manners of coming to a community this size, a lack of good breeding, and absolute arrogance. When you've been brought up in a Māori community and are part Māori, as I am, there are certain things that you do and don't do. We are not dumb, stupid savages on an isthmus. Don't treat us as such."

In the media and during town meetings, D'Mello and others tried to assuage the situation by expressing their regrets for not being more considerate to the locals, especially in the early days of building the pad. Their efforts won few people over. The grumbles from the townspeople, though, never reached the point of halting Rocket Lab's operations. Over the course of about a year, the company kept expanding its facilities and managed to join SpaceX in the rarefied realm of being a commercial rocket company with a private spaceport.

Actually being able to use the spaceport would require more magic from Beck. New Zealand had never launched a rocket before and had no laws in place to govern what companies could and could not do in space. It was also a very peaceful nation, and the prospect of performing rocket launches and possibly carrying satellites for DARPA or another US military body felt like unusually aggressive actions for a nation of sheep farmers and filmmakers. With its headquarters in the United States, Rocket Lab had to make nice with the US government as well. The United States had a four-decade-long history of trying to thwart the development of anything that looked like a missile in another country.

Since New Zealand is a magical, small place, you can find out exactly how Rocket Lab approached the government by going right to the top and asking former prime minister John Key to brunch. If you do so, Key might turn up in shorts and a T-shirt and explain the country's journey toward becoming a spacefaring nation while casually dishing out hellos to people in the restaurant* who recognize him. At least, that was my experience.

"I remember being completely dismissive of the idea," Key told me. "Really? Rockets from New Zealand? I mean, this is not Cape Canaveral or the Kennedy Space Center. I don't know if you've noticed, but our entire defense budget runs well under one percent of GDP and includes about two frigates, three sort of clapped-out boats, three clapped-out planes, and a couple of tanks. The concept of us being on the cutting edge of space technology didn't really fit with the military capability that this stuff often comes out of."

It was around 2015 that Beck first began floating his hopes and dreams to Key's office. Rocket Lab insisted that it needed a raft of

* Shout-out to the kind people at Ampersand Eatery and their bloody beautiful eggs Benedict.

space laws put into place quickly because it planned to send an Electron up in the imminent future. The first major official to field those requests was Steven Joyce, then the minister of economic development, who had an all-time classic Kiwi exchange with Beck. "He came to me and said he was getting close to being ready to launch," Joyce said.* "I told him that was fantastic, and then he said that we needed a regulatory system. I remember saying 'Oh, well, what's involved with that?' He said, 'You've got to organize some regulations and pass a law and get some stuff done.' And I thought, 'Shit. Okay.' I asked him when he needed all this by, and he told me six months. And I thought, 'Well, this will be a good test.'"

Key and Joyce were trying to run a pro-business government and quickly embraced the idea of New Zealand being at the forefront of such exciting technology. From a standing start, New Zealand would have to write a bunch of new laws *and* create an outer space and arms control treaty with the United States. For its part of the adventure, the New Zealand government assigned a dozen people to work with Rocket Lab on outlining what the space legislation could look like. The group grabbed public documents from NASA and the US Federal Aviation Administration and pared them back, opting for a beg, borrow, steal, and simplify approach. While the process took longer than six months, the government had its new space laws in place by 2016.

The only Rocket Lab request that anyone really balked at revolved around regulations tied to a treaty for operations on the moon. Key noted that although he wanted Rocket Lab to do well, he was not sure that the company would succeed in launching a single rocket, much less putting something onto the lunar surface. Moon negotiations seemed rather optimistic. "I'm thinking 'That's a step too far,'" Key said. Or, as Joyce put it, "We drew the line at the moon. We're a country of five million people, and that was looking pretty presumptuous."

Pushing similar laws through in the United States proved harder.

* My interview with Joyce took place in a café inside a home improvement store.

Rocket Lab was technically headquartered in California and had American investors, including the military contractor Lockheed Martin, but the US government detested the idea of another nation developing missilelike technology outside its control.

As Beck explained, "When you have created a rocket capable of putting a satellite in orbit, you have created an ICBM. You can weasel around the subject, but that's the absolute reality. You have the ability to deliver a thermonuclear weapon, and that comes with enormous responsibility. The technology we were creating was incredibly valuable to people who won't necessarily want to do wonderful things with it. Rightly so, there are huge, huge controls around ensuring that this technology doesn't fall into the wrong hands. The US government had a forty-year policy of denying another country space launch capabilities if they didn't already have it. We had to convince the US government that that was a good thing, a safe thing, a controlled thing, and something that was good for the US."

New Zealand was also the occasional renegade member of the famed Five Eyes intelligence-sharing group with Australia, Canada, the United Kingdom, and the United States. The country had a history of not always doing what the United States wanted it to do by choosing peaceful gestures over military posturing during incidents with other nations. Not everyone in the US State Department believed that a New Zealand–backed company would be obedient.

Key often talked to US President Barack Obama and brought up Rocket Lab and its concerns when convenient. The two countries also had top officials bargaining about conditions that would give the United States some oversight of Rocket Lab's launches from Māhia. Weeks went by without a definitive agreement being reached, though, and it appeared for a time that the United States simply would not allow Rocket Lab to operate out of its Māhia spaceport. It would be politics and not engineering that would halt the Rocket Lab dream.

With the future of his rocket program on the line, Beck flew to Washington, DC, and took charge of the negotiations. "I just camped out in a Holiday Inn and wouldn't leave until some sense was made of it," Beck said. "I became a bureaucrat and had many, many meetings at the highest levels of the State Department." After months of haggling, he got his deal in late 2016. "There was a guy sitting next to me at the New Zealand Embassy after we signed the agreement," Beck said. "He was not happy. He'd spent his entire political career trying to remove a tariff off an apple, and we had just negotiated a bilateral treaty."

Under the terms of the arrangement, New Zealand agreed not to turn Rocket Lab's Electron into a missile and not to let enemies of the United States place nefarious satellites aboard an Electron. The United States would also be allowed to send officials to Rocket Lab's launch site to poke and prod its rocket and to monitor the safety of the launches. In effect, the United States would assign babysitters to watch over Rocket Lab at Māhia. For its part, New Zealand could block certain US payloads if it chose. "The US had to trust New Zealand, but New Zealanders had to trust that we were not going to compromise our point of view," Joyce said. "I doubt New Zealanders will ever want to send any kind of weapon into space."

Beck conceded later that he might have underestimated the amount of legal wrangling required to get into the rocket business. He probably should have started the negotiations with all parties earlier. But Rocket Lab's success in developing Electron as far as it had also made it harder for officials in both New Zealand and the United States to deny Beck's requests when he appeared on their doorsteps. "By the time we got to this point, we had been testing our rocket," he said. "It wasn't like we needed any help from America or any American technology to finish it off. It was done. There was no argument that we couldn't do it without America. It was just an alignment of the goals."

Rocket Lab and Key's government faced some criticism for the deal. In a couple of media reports, critics had earlier expressed dismay that Rocket Lab had become an American company, and now it had formed a tight alliance with the US government. Really, though, hardly anyone in either country paid much attention to what was happening. Neither officials in the United States nor the New Zealand public was yet taking Rocket Lab very seriously.

When I'd first met Beck in 2016, he'd offered the same heavy helpings of optimism as he'd been handing out to everyone else. As he told it, the company would launch its first rocket that year and quickly follow with another launch and then another. The factory in Auckland had numerous rocket bodies in various stages of completion, and what Beck said seemed feasible. According to him, Rocket Lab would be blasting off one Electron a month in 2017 and be well on its way to weekly launches.

From 2013 to that moment, just about everything Beck had predicted while speaking to the first batch of investors in Silicon Valley had come to fruition. Dozens of satellite start-ups had sprung up to mimic Planet Labs, and they all wanted cheap, quick rides to space. Huge companies such as SpaceX, Samsung, and Facebook were talking about sending up tens of thousands of satellites to build their grand space internet constellations. The world would soon have an insatiable need for rockets, and this self-taught guy from New Zealand had somehow put himself in a position to capitalize on an all-out space frenzy.

As history shows, new rockets are always tragically behind schedule. SpaceX thought it could build and launch the Falcon 1 in about eighteen months, and it took the company six years to put a Falcon 1 into orbit. Even with the delays, that was considered a historic pace. Depending on how you counted, Rocket Lab had been working toward Electron either since Beck had founded the company in 2006 or since 2013, when the focus on the machine began in earnest. Rocket Lab would not hit the 2016 launch goal set by Beck, but as 2017 began, the company was finally ready to find out just how good its precious Electron was. In aerospace terms, Rocket Lab was remarkably on schedule. As Rocket Lab sent a team of engineers out to Māhia in May, Beck raised another $75 million, bringing the company's total funding to $150 million. Some very wealthy people and an entire nation were betting that things would go well.

Rocket Lab had contracts in place for paying customers but did not want to risk their payloads on its first rocket. Decades of history almost guaranteed that the first Electron would blow up. The major questions were around when the Electron would explode. The worst-case scenario would see the rocket burst into flames on the launch-pad, taking the entire spaceport infrastructure and possibly a flock of sheep with it. A better outcome would be a flight of sixty seconds or so before some part gave out, so that Rocket Lab could gather data from the performance of its machine and make changes to the other Electrons and Rutherfords in its factory. By some miracle, the rocket might fly for a few minutes and kiss the edge of space.

Naomi Altman, the twentysomething engineer from Australia, had been put in charge of navigating the full range of disaster scenarios. Rocket Lab had asked her to oversee what's known as the flight termination system, which would cut Electron's engines in an instant if it appeared that the rocket was doing dangerous things. Rocket Lab and its American watchdogs would track the rocket's trajectory via sensors and software and kill the machine should they have any fear at all that it was out of control or posed a danger to the public.

Altman had never built a flight termination system before joining Rocket Lab, but she'd spent the past four years reading books about them, designing them, and testing them. It could be argued that she had built the most crucial technology on the whole rocket. People would forgive Rocket Lab for not reaching space on its first try. In fact, they expected the company to fail. They would not, however, forgive Rocket Lab for launching its rocket unsafely. If someone sent a command to stop the rocket and the machine did not stop, Rocket Lab would be exposed as a clumsy, cavalier amateur. It would not matter if the rocket damaged anything or not. People would simply think that Peter Beck and his merry band of youngsters could not be trusted. The company would have to spend years begging for the United States and New Zealand to trust it to try another launch again.

Altman was among the dozens of Rocket Lab engineers who had trekked to Māhia on May 25 for the launch, which had been given the playful name of "It's a Test." Many of them felt that the launch was being rushed—not really in the sense that they were doing things in an unsafe manner, more that they were engineers and would have liked to keep testing and tweaking the rocket forever. Beck had no patience for such thinking. Bad weather had already pushed the launch back a couple of days, and it was time to press the button.

From the morning until the early afternoon, the Rocket Lab team went through hours of procedures leading up to the launch. They pressurized and depressurized valves and tanks, checked the rocket's thousands of sensors, and tested their communication systems. Residents of Māhia began to fill the surrounding hills, hoping to discover the best vantage points for the new show that would be playing in town for years to come. A small team of US minders was given a spot in Rocket Lab's operations center. They monitored Rocket Lab's every move and could call off the launch at any time.

At 4:20 p.m.,* the black Electron lit with fire and roared as it began a fierce battle with gravity. Some people held their breaths, expecting the worst, but the rocket proved all doubters wrong. Electron soared into the sky as the first stage of the rocket burned through all of its fuel, separated, and fell into the ocean. The second-stage engine ignited and kept the flight going for four minutes as Electron traveled 140 miles and well into space. At that point, the rocket had passed most of its major tests and appeared primed to reach orbit, as the data from all of its major systems came back nearly perfect. That was when, out of nowhere, the American team of minders barked out a call for the flight to be terminated.

In the chaos of the moment, no one was quite sure what had happened. The Rocket Lab engineers outside mission control thought everything had been going great and were primed to celebrate their great success. The US safety officers, though, had been having trouble tracking the rocket's position. Data about its location kept arriving in short, unpredictable bursts. When a large chunk of

* A time that would make Elon Musk proud.

time passed without any reliable location data coming back, the officers gave the order to shut Electron down. It began to tumble back from space and toward the Pacific Ocean. "I just walked outside and vomited," said Altman, who was sad to see the rocket go but relieved that her flight termination system had done its job.

In an analysis after the launch, it turned out that the American team had misconfigured their tracking software. The rocket had indeed been flying perfectly and would almost certainly have reached orbit. A software glitch had prevented Rocket Lab from achieving one of the most extraordinary and rare feats by flying successfully to space on the first try. Worse, it hadn't even been a Rocket Lab software glitch that had wrecked the moment. No. It was the babysitters. They were the ones who sucked Peter Beck's soul right out of his body and stepped on it.

Even after the mishap, most of the Rocket Lab engineers were thrilled at what they'd accomplished. The rocket had spent four minutes spewing out data that proved it was a beautiful, well-engineered machine. All those years of effort had been worth it. Bottles of booze were cracked open in Māhia and in Auckland. Of course, a few people could not fully embrace the moment. "I found it extremely brutal because we had done all the difficult steps and there was nothing that could have stopped that rocket," Tirtey said. "Everyone was cheering and shaking my hand, but I was so annoyed that all it took to stop it was some guy flipping a switch. I didn't go to the party that evening."

It took until January 2018 for Rocket Lab to finish building and testing its second Electron and transport it to the launchpad for another try. The company noted that it had made no changes to the machine. It had, however, publicly chided the US contractors and helped them fix their software. During that launch, dubbed "Still Testing," the machine performed as well as a second rocket ever had. It deposited a Dove satellite from Planet Labs in a near-perfect orbit, and it unleashed a secret surprise.

Rocket Lab had tucked something it called "Humanity Star" into the rocket. It was a three-foot-high geodesic sphere built of sixty-five reflective panels. The purpose of the object, such as it had one, was to spin around in space and shoot light back to Earth like a heavenly strobe light. Without asking permission, Rocket Lab had thrust all of us into the midst of a global rave. The Humanity Star would be the brightest object in the night sky until it eventually deorbited months later and burned up in the atmosphere. Beck had hoped that people would view the object as an inspiration. "The whole point of Humanity Star was to try and get people outside and to look up and to realise we are one little planet in a giant universe," he said in an interview with the *Guardian* in February 2018. "Once you understand that, you have a different perspective on the planet and a different perspective on the things that are important to us."

Other people were less enthusiastic about the Humanity Star. In his first brush with the major international press, Beck was derided for chucking space graffiti into orbit and violating the night sky with a gimmick.

Though Beck's gesture had temporarily marred Rocket Lab's grand moment, the company's arrival as a space power could not be denied. Rocket Lab had beaten all its rivals to launching a rocket. It had joined SpaceX as the only other private rocket company in the race to place thousands of satellites into orbit. The small-rocket maker with the least obvious advantages had won the first round of the fight.

"I am not ready to call it a defining moment," Beck told me at the time. "It was a great first milestone. All of that is fantastic, and everyone is superhappy. But now the fun really starts. I won't sit back and relax a bit until we're launching at a superhigh cadence and we really start to have an impact on the planet. For me, the light at the end of the tunnel is slightly closer, and that's it."

CHAPTER FIFTEEN

YOU'VE GOT OUR ATTENTION

In November 2018, Rocket Lab raised another $140 million and became that most mythical of creatures: a space unicorn. Its investors valued the company at well over $1 billion. Peter Beck had been forced to give away a huge part of his stake in Rocket Lab, but he still owned about a quarter of the company. The boy from the Invercargill shed was now worth hundreds of millions of dollars on paper, and he began to play the part.

Beck had used some of the venture capital money to build a palace of a new headquarters. After walking through the front door, you entered a white tunnel decorated with strips of red LED lights from floor to ceiling. At the end of the tunnel, Rocket Lab had written, WE GO TO SPACE TO IMPROVE LIFE ON EARTH in silver letters on a black wall. As you turned left at the inspirational quotation and entered the reception area, everything turned black. Like really black. There were black walls, a black floor, and a black ceiling. A receptionist and security guard sat off to the right and were lifted out of the darkness by a few spotlights beaming down from overhead.

The showstopper at the far end of the large, open room was a glass-encased mission control center. It had three massive screens at the front and a couple rows of desks for the people who would run the launches. A viewing area outside the glass had been marked off with more strips of red LED lights. Peter Beck had more or less built Darth Vader's rocket-launching lair, and he was not coy about it. Music from *Star Wars* pumped out of surround-sound speakers on an endless loop.

Deeper inside the building, Rocket Lab's facilities were less sinister but equally spectacular. The now hundreds of employees had sleek desks to work at and state-of-the-art labs for their electronics and engine experiments. The factory floor had transformed from a cramped, R&D-style shop to an industrial-grade manufacturing cathedral. Black Electron bodies were lined up one after the other in perfect rows, flanked by pristine workbenches. There were special areas to work on carbon fiber, 3D print engines, perform vibration tests, and paint parts. Each area had been marked off with big red swaths of paint along a gray, shiny floor. Two huge flags—US and New Zealand—hung down from the rafters like celebratory banners of the activity taking place below.

The facility punctuated the head start that Rocket Lab had secured over its rivals. Virgin Orbit and Firefly had been joined by a couple of other well-funded American small-rocket makers, Astra and Vector Space Systems, in the race to reach orbit. Those companies all vowed that their first launches were imminent, but little evidence existed that they were anywhere near putting a real rocket onto a real launchpad and matching Electron. Meanwhile, Rocket Lab had its bevy of Electrons ready to go and dozens of contracts from small-satellite makers in need of a boost.

Rocket Lab had, in fact, been adding to the pressure on its competitors with a stream of announcements. In secret, Rocket Lab had developed something called a kick stage for its rocket. The machine's first stage would fly Electron to space, the second stage would start and ferry satellites to orbit, and then the kick stage, with its own small engine, would ignite and place the satellites in superprecise orbits one by one. It was like a valet service for the satellites in which each one would be parked in the ideal spot for its job. Rocket Lab had also sent Shaun D'Mello to the United States to begin building a second launchpad on Wallops Island, Virginia. That would enable Rocket Lab to reach new points in space, increase the frequency of its launches, and possibly fly more sensitive payloads for parts of the US government.

Around the time of its first launch, Rocket Lab had taken another step to improve its prospects by beefing up the Americanness of the company. Its supposed headquarters in Los Angeles had been more for show and paperwork than anything else. In 2017, however, Rocket Lab opened a real office with real people in Huntington Beach.

The new office came with some obvious advantages for Rocket Lab. It could tap into the vast US aerospace talent pool and have salespeople close to many of its satellite-making customers. Beyond that, it would look more legitimately like an American operation, which would be essential if it wanted to win more business with the US government.

The company had already signed a deal with NASA for its planned fourth rocket launch, and that mission came with demands. The United States wanted Rocket Lab to manufacture its Rutherford engines in California instead of New Zealand. Rocket Lab's New Zealand team clearly had the know-how to produce engines in Auckland. The US government, though, needed to go through its usual motions of appearing to protect prized intellectual property and aerospace secrets by moving one of Rocket Lab's key technologies stateside. The implications of the deal were clear: if Rocket Lab wanted to keep selling to Uncle Sam, it would have to help the United States save face and nurture its inner patriot.

One of the key people Rocket Lab hired to build its engine factory in Huntington Beach, California, was Brian Merkel, a mechanical engineer who had spent the previous four years at SpaceX. Before taking the job, Merkel flew out to Auckland to interview with Beck and found him accessible, driven, and more involved in day-to-day engineering than any other aerospace CEO. "Peter was there at the office in a purple jumpsuit painting a shipping container," Merkel said. "He was always very hands on, and it seemed like he did the business things because he had to."

When Merkel started in January 2017, Rocket Lab handed him an empty hundred-thousand-square-foot warehouse with instructions to build the entire factory and have engines done for NASA by August. At the outset, the only other people in the building were an administrative assistant and another young engineer named David Yoon. They were all excited about the challenge ahead. "A giant open warehouse is the most beautiful thing you can see," Merkel said. "It's an open canvas."

Merkel discovered some early differences between SpaceX and Rocket Lab. In Musk Land, people optimized almost every task for speed and were willing to pay a premium to get things done quickly. Rocket Lab, by contrast, had more of a balance between speed and spending and if anything emphasized doing jobs for the lowest cost possible. Before other people could inhabit the new factory, for example, Rocket Lab needed to paint the place and coat the floor with epoxy. Instead of hiring a pricey contractor for the job, Merkel did it himself. "Peter doesn't leave a dollar on the table," said one of the new American workers. "That's for sure."

What impressed Merkel about Rocket Lab was the company's focus on manufacturing rockets quickly, cheaply, and repeatably. It seemed to be a mindset driven into Rocket Lab's engineers by Beck and New Zealand itself.

"They were in New Zealand and didn't have an aerospace industry and didn't have anything beyond fucking theory," Merkel said. "They took every little piece and googled them to find out why SpaceX or Boeing or whoever had made them a certain way. Then they just found supercheap, off-the-shelf components that could do similar things and discovered ways to make them work. I was so impressed with how simple and clever the rocket was. They would take things that you use to fix a flat tire on a bicycle and make it have a purpose on Electron. There were fittings on the rocket that came from race cars made in Australia because that's all they had access to and knew. They were legit engineers and are as good as anyone that I'd ever worked with."

Or as Yoon, Merkel's young colleague in the United States, put it, "They did things that seemed scary—things that no classically trained engineer would do. But then you realized that it worked. It just worked."

The expansion of the US office heightened the technical and legal challenges posed by Rocket Lab's existence. The United States had never before dealt with an entity like Rocket Lab, which had such valuable rocket technology being developed in unison across two nations. A set of US laws called the International Traffic in Arms Regulations (ITAR) forbade US engineers from aiding the New Zealand engineers with technical help on Electron. The laws had been put into place to stop knowledge of how to make rockets from falling into the wrong hands, and they were serious. Engineers in the aerospace industry lived in fear that they might do something as basic as posting a picture of a rocket part online and end up in jail.

Rocket Lab, though, knew full well how to make a rocket already, and the ITAR restrictions created an awkwardness that sometimes bordered on the absurd. The New Zealand engineers could send their engine designs to the United States and tell their US counterparts everything about how they worked and how to make them. But if an American engineer thought up an idea for making the engine better, she could not provide technical advice in return.

"Basically, New Zealand could give us drawings, they could give us information, they could give us any damn thing they wanted," Merkel said. "We just had to be careful about what we said to them. I could not, for example, tell the engine team in New Zealand that they could get better performance if they changed this, this, and this. But since we were making the engines in the US, we could make manufacturing suggestions like 'This would be way easier to build if you used this other material or this other fastener.' If that manufacturing improvement also improved performance, that was just a coincidence.

"For the first six or seven months, it was not always really clear what we could or could not do. With ITAR, you always hear these stories about people being personally prosecuted, and I was like 'Well, it's not worth it to come anywhere near walking a line.'"

The irony during those early months was that the first engines to come out of Huntington Beach would go to a rocket built for NASA. The US restrictions were simply making it harder for an American company to build the best possible product for its own space agency.

Though the American employees took to Beck, they were befuddled by some of his quirks. He would provide minimal guidance when asking them to take on large projects and then stop by for a visit and reverse or undermine many of their decisions. That happened with cosmetic aspects when Beck wanted a different-color carpet or different furniture,* and it happened on engineering projects as well. Beck, for example, demanded that the factory go from nothing to manufacturing ten engines in a few months but refused to okay the purchase of various machines that Merkel and others deemed essential for meeting the deadlines.

Beck had also put a controversial quotation up on the wall at the entrance of the Huntington Beach office. It read, "Make everything you do a work of art, if it looks like crap and does not work, then you have nothing, if it looks fantastic and does not work, at least it looks fantastic." The statement veered far from the usual "To the Final Frontier!" rhetoric beloved by space companies and also appeared to say that Rocket Lab favored style over substance. The Americans could not figure out why Beck would want that to be the first thing every Rocket Lab visitor saw. "Peter's very practical, but he's also very image driven," Yoon said. "That quote was telling about his personality."

The American office had another unexpected knock-on effect

* "He has an eye for design—his own design," said Daniel Gillies, who joined Rocket Lab in 2017 after an earlier stint at SpaceX.

for Rocket Lab: it created serious tension around how much the engineers in various locations were being paid.

New engineers being hired in California were often earning twice as much as their Kiwi and Australian counterparts. As more Americans arrived, word of the discrepancies in pay started to make their way around the company. It dawned on everyone that Rocket Lab had been able to make its first Electron for less than $100 million thanks in large part to the relatively cheap labor in Auckland. Still, the engineers in New Zealand had little bargaining power and few options, as Rocket Lab was the only aerospace game in the country.

At that point, Beck also granted stock options only to the employees who were deemed to be in the top 10 percent of performers at the company. The policy flew in the face of the traditions at California technology start-ups, where employees often took lower salaries up front and put in long hours in exchange for shares in a company and the hope of striking it rich when the company turned into a huge success.

Beck exacerbated the tensions by appearing to sell some of his shares in Rocket Lab, which was still a private company, on secondary markets. After the latest funding round, he had turned up at the office with a new car and built a palatial house. Employees passed around photos of the purchases and grumbled because Beck had discouraged them from selling their shares. It became a bad look for the head of the company to cash in some chips before Rocket Lab had seen its full mission through.

On the whole, however, employees in both New Zealand and the United States found Beck to be an inspirational leader. They were not sure of his motives. Did he just really like making rockets? Did he want to get rich? Did he want to spread human intelligence throughout the universe? But they were sure that he excelled at running the business and making spectacular things happen. Beck could be demanding and forceful along the way, although he rarely raised his voice and never attacked people for sport. People tended to excuse his less flattering moments because he was helping everyone else reach their dreams as well. "This is a man who had been obsessed with building rockets for twenty to twenty-five years," one employee said. "There's nothing that will stand in that man's way. That is what it takes to do something that is so fucking hard."

A prime example of Beck's leadership style could be witnessed at the start of each week, when he held an indoctrination session with all of Rocket Lab's latest hires.* On the day before its third launch in November 2018, Beck allowed me to observe one of those meetings. What follows is a sample of the messages he tried to convey.

> You've come here to lift humanity's core potential. I know that's a very lofty, CEO-y kind of statement. But that's the whole point of what we do. People don't realize how dependent we are on space infrastructure. If we turn off GPS, there's no Uber anymore. There's no Tinder anymore. Humans have become incredibly reliant on space, but it's all hidden. You don't see it, but it's absolutely critical to the way we function.
>
> There's been a huge change taking place in the space industry. The whole industry used to be geared toward flying large spacecraft. But that's kind of old space. Now we have smaller spacecraft like these satellites made by one of our customers, Planet Labs. Inside a small spacecraft are batteries, electronics, a bit of code, and solar panels. Every single one of those technologies has gone through a massive kind of evolution over the last five years.
>
> And the really exciting thing is the companies that are doing things in space aren't the companies that you would think of right away. It's all new. It's the guy who's made a sensor to go into a building to measure something and then all of a sudden realizes that he can put that on orbit and provide that technology to the whole world on a global scale.
>
> This is what really excites me. The things that we think we use space for now are not the most exciting things we will

* Rocket Lab employed about 350 people at the time, with 300 of them working in New Zealand and the other 50 in the United States.

use space for in the future. Those kinds of things are yet to be thought of, and you guys might be the ones that think of them.

We have raised nearly half a billion dollars to date. We've been successful not just in raising capital but in the things we have achieved. Electron and Rocket Lab is the only commercially active small-launch vehicle in the world right now. There's only two private companies in the history of humanity that have ever put spacecraft in orbit, and that's SpaceX with the Falcon 9 and Elon, and us. That's it. There is no other.

The club is very small, and it's really, really hard to get there. The barriers to entry here are just so enormous. We need to achieve about twenty-seven times the speed of sound in velocity to get into orbit. If you're a fraction of a percent out on performance or a fraction of a percent off on mass, you get nothing to orbit. It's just a ten-million-dollar fireworks display. It is unbelievably fucking hard.

It's not just technical, but it's regulatory and infrastructure. We needed to find somewhere where we could achieve the launch frequency, because to me frequency is the most important thing. This launch site at Māhia Peninsula is the only private orbital launch site in the world, and it's licensed to launch every seventy-two hours.

We're all about lifting human potential. We go to space to help people on Earth, which is superimportant. It's also important to me that we build beautiful things. I don't care if it's a spreadsheet or a valve for a rocket—make everything you do beautiful.

You'll notice that every component on that rocket is just beautiful. The reason why I put so much emphasis on that is that if someone has taken the time to make it beautiful, it generally works. And it's not just with components. Take that little bit of extra time to format the spreadsheet and change the fonts, and don't use horrible colors that don't match. Just make it look beautiful. That's really important to me.

> Obviously, we want to be a big company, and we're succeeding at that very well. And I don't like losing, so we'll focus and maintain being the industry leader for sure.
>
> We're doing here what it usually takes a country to do, and we're doing it with a small team at that. If you look at our competitors, they have much, much bigger teams than us. We're just way, way smarter. There's going to be plenty of tough days here because this is a lifestyle. We're trying to have a meaningful impact on the world, so that doesn't come easy. But if you're having a tough day, just go downstairs and touch the rocket. Just stroke it. That's all you need to do, and everything becomes good again.
>
> And just remember, as you're stroking that rocket, it's your DNA, and your DNA is going to space, and that's pretty damn cool.

On November 11, 2018, Rocket Lab began the preparations for its third launch. It called it "It's Business Time," which was both a nod to a song by the New Zealand musical comedy duo Flight of the Conchords and a declaration that Rocket Lab had moved past the testing phase. Yes, it had put some satellites into orbit previously, but those customers had known they were taking a major risk with a still unproven machine. This time around, Rocket Lab and Beck had put their reputations fully on the line and would be flying six satellites on behalf of four customers.

Two of the satellites were made by Spire, a satellite start-up that specialized in tracking ships, planes, and weather changes. Tyvak Nano-Satellite Systems also had a weather satellite on board, while a group of California high school students had built a small satellite for a data-gathering experiment. The last two satellites came from an Australian start-up, Fleet Space Technologies, which planned to use its machines as the basis of a new space-based communications network.

So much about the launch captured the spirit of this latest incarnation of the New Space era. Fleet had yet to put one of its satellites into orbit and had spent the last year waiting to hitch rides as secondary payloads on rockets launched by SpaceX and the Indian government. Those rockets, however, had been delayed, and Fleet was not a big enough customer to warrant a special place on other flights. It had learned that Rocket Lab might have an extra spot on Electron six weeks earlier and rushed its devices to the company. Its CEO, an Italian named Flavia Tata Nardini, had come to watch the launch in the viewing area outside the mission control center. She was thrilled that Fleet would finally be able to start its business, which would enable tiny sensors placed on things such as shipping containers and soil moisture detectors to send data from their remote locations up to space and then back down to Earth-based computers for analysis.

The spot on the rocket had opened up for Fleet because of Rocket Lab's previous miscues. The company had made two earlier attempts with the "It's Business Time" campaign in May and July but had called off the launches after discovering serious technical issues. There were rumors that a major explosion had taken place during one of the launch rehearsals, but Rocket Lab had kept exactly what had gone wrong under wraps. All people knew for sure was that ten months had passed since Rocket Lab had reached orbit and that manufacturing rockets as quickly as Beck had hoped was proving difficult.

The delays added to the uncertainty and tension of the third launch. In a meeting before the big event, a group of Rocket Lab's engineers briefed Beck and the US safety officials on the state of the rocket. The meeting naturally took place around a twenty-five-foot-long boardroom table made of carbon fiber.

Even with just a few hours to go, the machine had some niggling bits and pieces that were of concern, but people were on the case and working to fix them. "Don't be jumpy," Beck said when someone noted that they would likely have to make a last-minute call about one of the troublesome parts. "Let it burn." Rocket Lab needed to satisfy about 4,300 regulatory items in order for the US officials to green-light the flight, while New Zealand had about

40 similar provisions. "The attitude is that if it's good enough for the US Federal Aviation Administration, then it's good enough for us," a Kiwi told me.

After the meeting, people at Rocket Lab's headquarters passed the time by dishing on some of the peculiarities of launching from New Zealand. An automated radio broadcast would play for twenty-four hours straight in the area around Māhia, warning boats and other vessels about the coming event. Most people were happy to clear out of Rocket Lab's way, but crayfish fishermen were limited by quotas and liked to go out when prices for crayfish were high. As needed, Rocket Lab would call the fishermen individually and politely ask them to pause their pursuits for eighteen minutes so that it could put a rocket into space. The farmers who worked at Onenui Station would move away from the pad on launch days, but the sheep were sometimes less cooperative. Rocket Lab's employees told a story of one sheep that had been standing near the edge of a cliff ahead of a launch only to disappear after the rocket took off. "There's no evidence that the sheep jumped, because someone went to have a look," the tale went. "The smoke went up, and the sheep was gone, but it's still not quite clear what happened."

As the hours ticked by and the countdown approached, the small talk stopped. Rocket Lab's engineers and launch leaders took their places in the mission control with Tirtey leading the proceedings. The new headquarters allowed Rocket Lab to have a viewing audience for the first time, and fifty or so people gathered in the Darth Vader watch zone, where they could peer through the glass and see the launch and the progress of the rocket on the giant screens. The observers were a mix of Rocket Lab employees and their families, the satellite customers, and me. The launch was taking place on a Sunday, and not as many people had turned up as had been hoped. Some of the employees had better things to do, apparently. "Such a New Zealand thing," said one of the observers. "I don't get it."

Beck sat in the mission control room, wearing a black T-shirt, black pants, and black shoes. The white lab coat days were done. With ten minutes to go before the launch, he gazed up at the huge screens and clasped his hands near his face. It almost looked as if he were praying.

The horrible thing about the rocket business, of course, is that no matter how well you've done, you're always one explosion away from a crisis. Rocket Lab had shocked the world with its successes to date, but a major mishap on this mission would dissolve its lead over rivals and undermine its credibility. The "It's Business Time" moniker could turn into the punch line of a joke.

Beck had told his team that he personally wanted to perform the countdown for the launch. So at 4:50 p.m., the man from Invercargill with his rolling-Rs accent could be heard over the mission control center speakers: "Tin, nine, eight, suvun, sux, five, fourrr, thrrre, two, one." And off the rocket went.

As the rocket made its way into space, Beck held his bush of curly hair in both hands. Whenever Electron passed a major milestone, he would let go for a moment and deliver a muted fist pump or a pat on the desk. "Keep going!!!!" yelled the customer Tata Nardini from outside the mission control center. "Keep going!!!!" After eight minutes had gone by and the rocket had reached orbit and deployed its satellites, Beck's eyes filled with tears, and he clasped his hands behind his head. Most of all, he tried to breathe.

A few minutes later, he left the mission control center and came out for a chat. He looked drained and overcome with emotion. "It's game on," he said. "This era has been coming and coming. The small-launch race is over. We have proven it can be done." Then he asked me to find out if Elon Musk had seen the launch. "The team would be really pumped," he said.

Much to the aerospace industry's surprise, Rocket Lab launched another rocket a month later, placing a number of satellites into orbit for NASA, just as it had promised. There would be no successful launches from Rocket Lab's competitors that year. Or in 2019. Or in 2020. It was Rocket Lab, SpaceX, and no one else.

In May 2019, I helped arrange a meeting between Beck and Musk. Beck had flown to Rocket Lab's California offices, and Musk had found time in his schedule, even though he'd claimed to have little interest in Rocket Lab. The meeting would forever change the relationship between the giant SpaceX and the underdog Rocket Lab.

SpaceX executives had long pushed Musk to use some of the company's Falcon 9 rockets to fly huge batches of small satellites into orbit. SpaceX had allowed the small devices to hitch rides alongside bigger satellites now and again, but some members of SpaceX thought there could be a decent business in flying tons of the things at once. Pete Worden had made similar requests of Musk in the past to help with machines built by NASA Ames and its partners. During one such meeting between the two men, Musk had gone ballistic at the very suggestion of the idea. "Stop asking about this," Musk had said, according to a person in the room. "This really pisses me off. We are never going to do this."

If SpaceX did decide to fly lots of small satellites in one go, it would pose a major threat to Rocket Lab just as the company was hitting its stride. SpaceX's large Falcon 9 rockets gave it an overall cost and cargo advantage. Someone would need to pay $60 million for a SpaceX launch instead of Rocket Lab's $6 million, but they'd be able to put entire satellite constellations into place with a single rocket instead of buying them month by month.

By that time, Brian Merkel had left his job setting up Rocket Lab's US operations and returned to SpaceX. Ahead of the dinner with Beck, Musk asked one of his vice presidents to debrief with Merkel about Rocket Lab and help ascertain how real of a competitor it was. "I said that I couldn't speak to how successful they would be as a business but that they're great engineers and that their rockets will launch and fly well," Merkel said. "I don't know exactly what was said at that dinner, but people came back afterward and said Elon had come away impressed. I think Pete had conveyed a vision for Rocket Lab that was not that far off from what SpaceX had in general. I think Elon may have used the meeting as a piece of feedback that told him Rocket Lab was attracting a bunch of business and SpaceX should take some of it."

Part of the reason Beck asked to meet with Musk could be

chalked up to ego.* Understandably, Beck wanted people to realize that he and Rocket Lab were in the same ultraelite club as Musk and SpaceX. More to the point, he wanted Musk to recognize him as a peer. Beck retained much of his low-key, humble New Zealand spirit, but he'd always harbored grand ambitions. Rocket Lab's success had started to inflate his self-confidence and given rise to yearnings for adoration. The problem, of course, was that striving to get onto Musk's radar meant that he just might end up on Musk's radar, which is historically a horrible place to be.

In August 2019, SpaceX revealed a new plan to begin regular launches for small-satellite makers. It would free up entire Falcon 9s and let various companies buy space on the rockets. If a company wanted to send five hundred pounds of cargo, about the equivalent of what an Electron would carry, it would cost a bit more than $1 million, or $5 million cheaper than Rocket Lab's charge. SpaceX later put up a record 143 satellites in a single launch via the program.

Musk did not know it at the time, but Beck had a few surprises coming for ol' Elon as well.

To outsiders, Rocket Lab may have still looked like a minor player in the aerospace industry. SpaceX and Musk were like black holes of attention, sucking up any and all available press when it came to New Space. Beck had nowhere near the celebrity shine of Musk in his home country or anywhere else.

People in the aerospace industry, however, marveled at Rocket Lab. Electron was considered to be perhaps the most perfectly engineered small rocket that had ever been made. Whereas SpaceX's first three launches had ended in flames, Rocket Lab's first three had been just about perfect. Only a software mishap outside the company's control had stained an otherwise immaculate track record. No new rocket program had ever started like that. Rocket Lab and Beck had figured things out that had stumped everyone else in the past.

By all accounts, the composition and smarts of Rocket Lab's team had led to that success. The New Zealanders had brought their

* Beck would not tell me much of anything about how the meeting had gone, other than to say that he'd "had a blast" with Musk.

creativity and can-do spirit to the project. The Australians had more experience in industry and knew what it took to advance something from R&D to real manufacturing. The remoteness of the operation had encouraged people to think differently and to simplify. But even all those conditions did not fully explain what Rocket Lab had accomplished. It had built the best small rocket ever for $100 million and almost done it on time. The company and its culture were rarities in an industry that prided itself on being inhabited by the best and brightest people.

Beck would never say it aloud, but he had been the key element in making all of it happen. He had brought a mix of engineering magic, speed, and practicality that seemed lacking among Rocket Lab's rivals, who were busy spending billions of dollars trying to catch up to their competitor.

There were basic stories that spoke to Beck's breed of relentless pragmatism. In the early days of the company, for example, he would take a handful of engineers on field trips to the United States to visit museums and NASA sites. They would end up somewhere like the National Museum of Nuclear Science & History in Albuquerque, New Mexico, analyzing what sort of insulation the old ICBMs had around their tubing. Then they would be on to the next spot and the next, hunting for rocket-making clues hiding in plain sight. "Traveling with Peter sucks," Tirtey said. "You fly all the time, and the whole trip is so condensed. It's not like you see one place per day and stay in a nice hotel at night. You go visit this one place, see it, and then leave and go to the next place and then sleep at the airport. There's never any time to get food, so you just get an ice cream and eat it in the taxi. It's all so intense." Beck lived each day with that same sort of driven, focused spirit.

Another thing that Beck would never mention is that he has a series of folders in a filing cabinet at home that detail all of Rocket Lab's most dire challenges. The information in the folders goes over the technical issues that Rocket Lab has faced and how the company thought its way out of the problems. In almost every case, Beck thought up something in the shower or while alone in his workshop that pushed Rocket Lab past an intractable obstacle. It was the team that ultimately brought Beck's ideas to life and made them work. But without Beck's insights, Rocket Lab very likely would have found itself in the same position as its competitors.

As for what drove Beck, that remained a mystery to those around him and maybe even to Beck himself. One camp of people believed that his prime motivation stemmed from a desire to get rich and be a big shot in both New Zealand and the world. As Rocket Lab did better and better, evidence appeared to bolster this line of thinking. Beck's humbleness would sometimes give way to a desire for more attention. He wanted people to view Rocket Lab with the same near-religious fervor that they gave to SpaceX. Without question, Beck wanted a bigger slice of the attention that was going Musk's way.

After the "It's Business Time" launch, I visited Beck at his vacation home on the South Island. We went panhandling for gold in a river with our sons. My kids were sweet but pretty hopeless when it came to helping the endeavor along. Beck's son hopped into his dad's Jeep, lugged out equipment, and began hunting for gold. When other people drove near us in their trucks, Beck's son would call out things he'd noticed about their engines based on their sounds alone.* After that, we rode jet skis. Beck had purchased the fastest machines available and nearly broke me in two as he raced off with me sitting behind him.†

During our day together, I tried my best to engage in small talk and then to get Beck to tell me what he really wanted out of being a space magnate. Musk's life ambition is to colonize Mars. Did Beck have a secret, equally lofty goal? What was the point of all his work? But I never seemed to make any progress. Beck wanted to discuss the minutiae of the rocket business and the travails of his competitors. He made no mention of wanting to colonize anything or

* My boys would have held their own if a D&D game had suddenly broken out.

† In a nod to my journalistic independence, I refused to hold on to Beck's torso and instead gripped the piddly handles on the side of the jet ski. That meant I was thrown back every time Beck let the jet ski loose and had to use all of my dad strength to stay on. I'm pretty sure Beck was trying to send me a message.

hunting for life in the outer reaches of space. "Don't get me wrong, I think sending a few people to Mars increments the human species," he said. "No argument. I think it's wonderful. But I think you can have a larger impact on a larger group of people by commercializing space and making it accessible. That's how you influence people's lives and improve them.

"I mean, if we're being honest, how does sending a couple of dudes to Mars meaningfully impact your life or my life? We're inspired, and that is an impact. But that doesn't really change the way that I live my life. However, if we put up a ton of weather satellites and give way better weather predictions so that crops can be harvested better or, shit, just so that we can decide whether to go on a hiking trip or not, that has a meaningful effect on my life."

But Rocket Lab and Beck would prove in the years to come that they always had surprises in store. While competitors were trying to keep pace with Rocket Lab, the company had already plotted out its next few moves. Beck had a knack for keeping his real intentions hidden, lest a rival get an inkling for his master plan. Case in point: not long after that jet ski ride, New Zealand discovered that it would need a moon treaty after all.

AD
ASTRA

CHAPTER SIXTEEN

LET'S MAKE A LOT OF FUCKING ROCKETS

Around the middle of 2016, a couple of friends in the aerospace industry sent some odd rumors my way. They claimed that there was an aerospace start-up located near Market Street in San Francisco and that it had figured out how to build tiny rockets that could carry things into space at a moment's notice. Most of these rockets, my friends said, were being manufactured at the behest of the Defense Department. The military wanted to see just how weird it could get when it came to blasting objects into orbit and fancied the idea of being able to chuck up a satellite without anyone noticing.

After a bit of digging, I discovered the name of the start-up—Ventions LLC—and its CEO, Adam London. And, sure enough, its address at 1142 Howard Street placed the company in the part of San Francisco known as SoMa, or South of Market. All of this intrigued me because the internet had precious little information about this London fella, and rocket start-ups are not meant to exist in SoMa, an area made famous during the dot-com boom for giving rise to internet and software companies and coffee shops where venture capitalists liked to display their self-confidence.

I dropped Ventions an email, hoping that the secretive start-up would opt to become less secretive if given an opportunity. It declined my generous offer. "Unfortunately, since most of our work is Department of Defense related, we're unable to talk about it publicly because of release constraints," London wrote back. "Should that change in the future though, we hope we can approach you then?"

That very polite email elevated my interest in Ventions from curious to insatiable. Over the next few months, I peppered people in the aerospace industry with questions about Ventions and what it was up to. Beyond that, I dug into the government contracts Ventions had secured. The story that unfolded before me was one of a small outfit of about a dozen people that had survived for years on an air force contract here and a DARPA contract there. All of the money went to the same goal of building a very small, very low-cost rocket that could put at least one small satellite into space very quickly. In fact, Ventions' rockets were small enough to be placed underneath a plane, which would carry one up into the sky and drop it, at which point the rocket—called Salvo—would ignite and take off into space. (This is very close to the "responsive space" quest that Pete Worden and his DARPA brethren had dreamed about for years.)

For one reason or another, I kept putting off writing about Ventions and what I had learned. Part of me expected London to come around and dish the whole story eventually. And then, in February 2017, a funny thing happened. While I was traveling with Planet Labs' Robbie Schingler in India, Schingler mentioned to me that he'd been spending some time at a secret rocket company in the middle of San Francisco. What's more, his good friend Chris Kemp had just become the company's CEO.

I'd known Kemp for years, dating back to his work running the technology operations of NASA Ames. In particular, I'd followed Kemp's role shepherding a software project called OpenStack, an effort to build a type of cloud computing system within NASA. The organization had struggled to share its data among different centers and scientists and engineers. OpenStack was an attempt to write a layer of software that made it easier to link all of NASA's information databases. The project proved so successful that NASA, at Kemp's urging, opted to make the software open source and let anyone use it. A number of companies were quick to grab the OpenStack code and create their own shared database systems, and OpenStack emerged as one of the most popular open-source projects in the world.

In 2011, Kemp decided to leave NASA and form a start-up based around the OpenStack cloud computing technology. He named the new company Nebula, a nod to the OpenStack code name within NASA, and raised more than $30 million from the most prominent investors in Silicon Valley.* The company built its own computer servers to go into data centers and outfitted them with the OpenStack software. Nebula hoped that companies would buy the server-and-software combo and use the technology to create their own cloud computing systems. That was during the early days of the rise of the cloud computing industry, and Nebula was a competitor of Amazon, which had started to dominate the young market. Instead of renting space on Amazon's computers, a company could create an internal shared database system with Nebula and maintain more control over its information.

Kemp's decision to base a company on taxpayer-funded technology did not sit well with some NASA officials. In addition, Kemp had been one of "Pete's Kids," and Worden had plenty of enemies who were ready to seize on anything that looked controversial and make life difficult for both Worden and his disciples. Before Kemp left NASA and officially incorporated Nebula, he found himself under investigation. As he recalled, "Dozens of companies had been started based on technologies that were invented at NASA. Bloom

* The investors included three billionaires who had cut the first checks for Google: Andy Bechtolsheim, David Cheriton, and Ram Shriram. Those three men were known for having a golden touch when it came to investing.

Energy was one of them that was pretty well known at the time. As I started to get serious about potentially leaving NASA and starting a company, I went to the guys at Bloom very early on to ask for advice. They said, 'Follow the rules.' And that's exactly what I did. I got a lawyer and made sure we did everything cleanly. But before we had even got going, one of my partners who had become a bit exuberant about the whole idea sent out an email celebrating how we had started a company.

"It was not true, but it was forwarded around to a bunch of people.

"Then, one day, I'm sitting at my NASA office early in the morning, and the FBI raids it. I've got about ten guys in black suits and black jackets showing up to seize all of my computers and files. I gladly handed that stuff over, but then they wanted my personal phone. And I'm like 'No, you're not taking my phone.' They said, 'Yes, it's our phone.' I'm like 'No, you're wrong. You're not taking it.' And they put their hands on their guns, and I ask, 'Are you going to shoot me?'

"They keep asking me to give up my phone, and I keep telling them, 'No.' We had a standoff. I said, 'You guys can stand there as long as you like with your hands on your guns. You're not taking my phone. Because it's my phone, so fuck off.' And so they said, 'Well, we're following you home.' That's when I called my wife at the time and said, 'Hey, uh, there's the FBI and inspector general and a whole bunch of people with black cars, and they want to follow me home, and I was wondering if you could give me a ride back to the house. I want to be real thoughtful about what happens between now and when we get home.'

"My wife picks me up in front of the building, and this motorcade of agents follows us. We get to the house, and I clicked the button on the garage door opener, and we pulled the car into the garage, and then I closed the door. That made them very unhappy. They kind of freaked out. They figured I was inside destroying evidence or something. My attitude was that I had done nothing wrong and this is a huge inconvenience for me.

"They came in, and I handed them some more computers. They

interrogated me. You're not supposed to make any comment, and you're supposed to tell them to talk to your lawyer. I didn't. I told them everything. They were asking about the email and all that.

"Anyway, long story short, there was a grand jury investigation, and everybody I knew was interviewed over the course of a year. I wasn't aware of it at the time. It was pretty serious. And it was all because they thought I had started a company that I hadn't started but had talked about.

"Eventually the statute of limitations ran out, and I had a statute of limitations party."

This incident provided a snapshot into Kemp's character. He would try to follow the rules to avoid bureaucratic gotchas. At the same time, he was a fount of ambition and not one to let anyone or anything get in his way. Kemp tended to view restraints and traditional structures as challenges that needed to be overcome. He certainly did not have much time for authority figures or their staid thinking. Many of those traits had been part of Kemp since birth, but his experiences at the Rainbow Mansion and Ames had pumped him full of increased levels of confidence and swagger.

By all rights, Kemp's time running Nebula should have been a blow to his ego. After the NASA kerfuffle passed, he got the company off the ground and managed to attract a lot of attention to the venture. But although Nebula enjoyed some early successes, it never managed to find a big market for its products. In 2013, Kemp relinquished the CEO position, and in 2015, Oracle acquired some of the company's technology and people in a going-out-of-business-style sale. A company very well poised for success had failed.

Kemp, though, shrugged the Nebula debacle off and began searching for his next venture while also freeing his spirit even more. He became an "entrepreneur in residence" at a venture capital firm, which is a fancy way of saying that some investors gave him office space to sit around and think up ideas for a new company. And he increased his involvement with the people who run Burning Man, the annual sex, drug, and art–fueled gathering in the Black Rock Desert that has become enough of a cliché to always be described as a sex, drug, and art–fueled gathering in the Black Rock Desert. A couple of years earlier, Kemp had a revelation at Burning Man that triggered a transformation from data center tech nerd to man of industry and action. Post Nebula, his dealings with the Burning Man* elders were part of a personal commitment to solidify those character changes and embrace the new and improved Chris Kemp.

When not sitting and thinking in his venture capital office, Kemp pursued a consulting gig on behalf of his friends at Planet, Robbie Schingler and Will Marshall. Planet had spent so much time and money placing its satellites on rockets that the company wanted to explore the idea of making its own rockets. The Planet executives asked Kemp to travel around the world and find out what the state of the art in the launch industry looked like, particularly when it came to small rockets. Kemp spent several months doing just that and visiting dozens of companies. Naturally, he made a trip to New Zealand to visit Peter Beck and Rocket Lab. He also ended up knocking on the door of Ventions and meeting London.

London had earned a PhD in aerospace engineering from MIT and looked and acted the part. He was slim, wore glasses, and had a boyish face. He mostly talked when prompted and did so in a soft-spoken, deliberate manner. His word choice was so calculated and economical that it contributed to most people's first impression of the man, which was that Adam London was really fucking smart.

After university, London worked for a few years as a consultant at McKinsey, and then the siren song of rocketry called him back. He started Ventions with a couple of colleagues in 2005 more or less to pursue ideas he'd been experimenting with during school. Unlike many other Silicon Valley CEO types, he did not exhibit much in the way of a drive to get rich and get rich fast. More than anything else, he loved engineering and spending his days solving hardware

* Kemp was sometimes the leader of the Lunar Fueling Station Camp, which gave out water to Burners. "It's the one thing that everyone needs that people forget to bring," he said. He also developed a homemade solar-powered air conditioner for his tent. "It tends to get hot around eleven a.m., but you're still sleeping until three p.m., so this is timed to kick on automatically at eleven."

problems. As a result, Ventions ended up as more a research-and-development firm than a technology business.

The San Francisco office of Ventions looked like a large do-it-yourself workshop. The company had taken over a row building with a roll-up garage door that had a jujitsu studio and a print shop as neighbors. There were a few desks up in a loft where people could sit at their computers, although none of the desks matched because Ventions could not afford real office furniture. A slab of wood on a couple of sawhorses, for example, functioned as a conference table.

Much of the action took place down on the main floor, an open area packed with tools and rocket parts. A handful of workbenches had been set up to run experiments and bend metal. Tinfoil coated bits and pieces of machines, and there was a constant sound of water flowing from a pump that no one had bothered to turn off. The most striking work space had a homemade blast shield made out of a wall of superstrong plastic. It allowed Ventions employees to put objects under high pressure, run to the other side of the shield, and have at least some illusion that they'd be safe if things went wrong. For much of its life, Ventions did not own the entire shop floor because it could not afford the whole rent, so a machinist rented the extra space for his projects. Someone had run a strip of blue tape down the center of the floor to mark where Ventions' territory ended and the machinist's zone began.

Since London wanted to build very small rockets, that was what Ventions tried to do. Day in and day out, a handful of employees tried to create miniaturized versions of engines, turbopumps, and all the associated machinery. "They were trying to see how small you could go and still put up a meaningful payload," said Matt Lehman, an early SpaceX employee who joined Ventions in 2010. "There are certain components that don't like to go small, and so Ventions was trying to figure out how to make them. Small valves. The fuel injectors. The electronics and the guidance. We worked on these tiny thrusters where the thrust chamber was like the size of a Coke bottle. It was almost everything from top to bottom. We would put all the electronics on a restaurant busboy tray. Twenty years earlier the equivalent stuff would have filled a room."

When Ventions needed to perform a major engine test, employees would drive to the Mojave Desert or sometimes to Castle Air Force Base in Atwater, California. They'd spend a few days going through the toil of setting up equipment, watching it not work during the initial tests, and then tweaking and fixing components until fire shot out the end of a metal tube. "We'd sleep outside or in a Quonset hut," Lehman said. "We had one propane blower to try and keep warm. In the winter, we'd all be sleeping with socks on our hands, and you couldn't even boil water because it was so cold. By day three, the productivity started to wane. By day four, you had to get the hell out of there."

Lehman was somewhat used to a lack of luxury. He had obtained a PhD in mechanical engineering from Penn State before taking a job at an aerospace contractor, where he had built replicas of the Scud missile and then tried to blow them up. After that, he had gone to SpaceX when the company had still had fewer than a hundred employees and all of the grit of a start-up. Unlike SpaceX, however, Ventions lacked a sense of urgency, and London certainly did not manage people with the same ferocity as Musk. The company tried to find aerospace technical consulting contracts that would allow it to earn some money and stay afloat a bit longer to keep pursuing London's ideas. "The whole situation was like graduate school on steroids," Lehman said. "We'd go to the Department of Defense website and do keyword searches for contracts we thought we could get. We'd get a government contract for like $100,000 for eight months of work. Then we'd get to the next phase, and it would be more money for eighteen months of work."

On most days, London is a very amiable and kind human, and the camaraderie within Ventions grew from his personality. The team varied in size from three people to about a dozen. Many of the employees were young engineers trying to learn the aerospace craft, and people like Lehman and London were happy to teach them. By 2013, a core group of employees dug in harder on the quest to build and fly a complete rocket, and by 2015, the company was on the cusp of its goal. Through a deal with the US Air Force, Ventions prepared to strap its small rocket to the bottom of an F-15E fighter jet and launch it and a satellite weighing eleven pounds into orbit.

Despite that momentous occasion, Ventions found itself at a crossroads. By that point, companies such as Rocket Lab and Firefly were raising millions of dollars for their rocket start-ups and employing hundreds of people. Ventions still had its tiny team and a proof of concept, and even London wondered whether the time had come to sell off the company's intellectual property and do something else. "People always used to ask us what our plan was, our endgame," Lehman said. "We thought maybe some aerospace start-up would buy our engines. But then you realize no one is going to do that because the engine is the sexy part, and they all want to make their own. To be honest, as time went on and some of these start-ups got funding, I thought we'd missed our opportunity. Then, in late 2015, Adam met this guy through Planet Labs named Chris Kemp."

Kemp and London were both men, but that was where their similarities ended. Kemp marched into Ventions with his optimism and ambition and talked London right out of selling the company. Kemp knew many of Silicon Valley's wealthiest investors and through Nebula had developed a gift for talking them out of their money. Over the course of months of evenings and weekends, he camped out at the Ventions office with London and worked up studies to prove that a small, cheap rocket could be the basis of a real business. Once they managed to create a proposal that had a whiff of believability to it, the men began inviting investors in to hear their pitch. Kemp handled the business aspects; London focused on the technology; and Lehman played the role of personable engineer who gave tours of the facility.

"The whole venture capital idea of risk and the dollar amounts I would hear were so foreign to me," Lehman said. "I was thinking 'Man, this guy Chris is not of my world,' to put it mildly. He had so much enthusiasm and vision to look far out and see what this could be. I think Chris would admit that he didn't know that much about rockets. I had to really think about what he was saying when he'd bring people in and talk to them as far as setting expectations goes. Does this even make sense? Would people really be into this? I mean, if you came to Howard Street, it didn't look like much. It sure didn't look like a rocket facility."

The questions Lehman asked himself were similar to those any sane person would ask. Silicon Valley, though, does not run on sanity. What Ventions needed to peddle in that moment were hopes and dreams, and, boy, did Kemp and London deliver them.

First off, the company would no longer be called Ventions. That was the past. The new company would be so cool that it would have . . . no name at all. If someone needed to refer to it, they were to call it "The Stealth Space Company."

The Stealth Space Company planned to build a small rocket in the vein of the Falcon 1 or Rocket Lab's Electron, only the rocket would be even smaller and made out of the most basic materials available to ensure that it could be manufactured at a low cost. According to the pitch deck math, the company would be able to produce a rocket that could carry 150 pounds of cargo into space for $1 million. It would design, build, and launch its first rocket faster than any other organization in history—possibly within a year to eighteen months—and then perfect mass-producing the rocket. It needed to make a ton of rockets because it planned to launch a rocket every single day. And the story only got better the more Kemp and London talked.

Sometimes the launches would take place from typical launch sites. In such cases, the Stealth Space Company would perfect an automated launch system in which its rocket would arrive by shipping container, be taken to a pad in a truck by a couple of people, and blasted into orbit with the touch of a button. Mission control setups with dozens of people were antiquated. The Stealth Space Company wanted as few people involved as possible. That was in large part because the company's major goal was to launch rockets from automated barges that would pick a rocket up from the factory, ferry it a few miles out to sea, and send it on its way before returning to do the job again.

Add all of that up, and the vision sat somewhere between building both the Ford and FedEx of aerospace. Cheap rockets would come off an assembly line, get stuffed with satellites, and be launched more or less by automated, robotic systems. Customers who wanted a launch on short notice need do nothing more than enter their corporate credit card information on the Stealth Space Company's website and secure their ride.

Whereas Rocket Lab had pursued its work pragmatically and aimed to make the platonically perfect small rocket, Stealth Space was embracing more of a "Let's fucking go" strategy. London had run tons of calculations and simply did not believe that the economics of small rockets worked unless you made them dead simple and achieved the economies of scale that come through mass production.* Just as Planet had revolutionized the design and manufacturing of satellites, Stealth Space would do a similar thing for rockets. It would try to take the rocket science out of manufacturing the machines and turn them into just another product that humans could build quickly and cheaply.

No one on Earth at the time really knew if a single $250,000 rocket carrying just 150 pounds of cargo into orbit would be all that useful, let alone hundreds of them rolling off a production line. A rocket barge going into and out of San Francisco Bay sounded pretty awesome, although surely some timid members of the citizenry would object to daily missile launches so close to their homes. And, well, it had taken Ventions a decade to sort of, kind of finish an experimental rocket, so the odds of this team cobbling together

* The math held true, according to London, even for reusable rockets. To offset the cost of retrieving and refurbishing the rockets and make them economically worthwhile, a single rocket would have to be launched more than twenty times over the course of its lifetime. No company had ever accomplished that. London was inspired by a 1993 paper titled "A Rocket a Day Keeps the High Costs Away," which had proposed an industry shift to a mass production model. You can read the paper here: https://www.fourmilab.ch/documents/rocketaday.html.

a brand-new, mass-producible machine that could make it into orbit in eighteen months felt at the ludicrous end of aspirational. But, like, you know, those details could be addressed later.

The calendar had ticked over to 2016 while Kemp and London cooked up their wild-eyed rocket fantasies, and people were ready to buy into the plan. Investors had seen SpaceX and Rocket Lab, and by God, they wanted to own rockets, too. Kemp, just as he had promised London he would do, raised tens of millions of dollars to take Ventions, the R&D operation, and transform it into the Stealth Space Company, revolutionizer of the heavens.

BY THAT POINT, I'D MET with Kemp and London at the San Francisco office, and we'd come to an arrangement. Kemp wanted Stealth Space to remain a secret all the way up until the moment the company launched its first rocket. If I could abide by the secrecy requirement, he'd let me follow Stealth Space's journey from PowerPoint to orbit. For me, the idea had instant appeal. It would be like going back in time to watch something like SpaceX form and see what it took to make a rocket from scratch. With a bit of luck, Stealth Space would live up to its aspiration of building an orbital rocket faster than anyone had ever done it before, and I'd have a front-row seat to the historic feat.

On the other hand, there were reasons to think I might end up wasting a lot of time following the company. Yes, Kemp had worked at NASA, but on data center stuff. As Lehman noted, Kemp's knowledge of aerospace engineering and what it took to build a rocket did not inspire total confidence. Kemp often described various hardware engineering challenges as if they could be solved with the same techniques he had employed in the software world. I'd heard many people make similar arguments in the past, and things had rarely turned out well for them. Hardware ends up being messier and more time-consuming than software people think.

Kemp's enthusiasm and overall personality also raised concerns. By 2016, he'd taken to wearing all black all the time. Black leather boots. Black jeans. A tight black T-shirt. A black leather jacket. Day after day. He said not having to choose clothes every morning saved time, and he did have a thing for efficiency. Still, it felt forced and overly dramatic. Here's this blond-haired, blue-eyed, forty-year-old guy cosplaying as Space Johnny Cash. More to the point, Kemp's Joel Osteen levels of enthusiasm could be both inspiring and off-putting. He made you want to believe that he could do everything he said, but the little voice in the back of your head also gave off warnings that anyone with that much self-confidence was either deluded or trying to hide something.

To his credit, Kemp did not outright deny the challenges the new company was facing, and he agreed to be more open than any other executive I'd ever encountered. Either Stealth Space would succeed, or it would literally go up in flames. Kemp agreed to let me document either outcome in gory detail. I took the deal.

KEMP'S FIRST MAJOR MOVE PROVIDED encouragement that he might actually know what he was doing. He'd found a new headquarters for Stealth Space in a building located across the bay in the small, sleepy town of Alameda.

Alameda is an island right next to Oakland, and in the 1940s, the US Navy began building a vast air station on the wetlands at its western edge. In the decades that followed, the military outfitted the base with runways and a series of large buildings to test, repair, and house aircraft. In 1997, the base closed. Many of the buildings were full of hazardous materials and left empty. A few of them, however, were revamped as the city tried to turn the old base into a point of interest. Industrial companies set up shop in a couple of warehouses. A couple of restaurants took over offices and opened waterside eateries. Just as its name suggests, the vodka maker Hangar 1 built its distillery and tasting room in one of the former airplane hangars.

Kemp had spent a few weeks touring abandoned military facilities in California. He did not want to have a company headquarters in one place and a rocket test site in a far-off location, as companies such as SpaceX had been forced to do. He thought it might be possible to find a former base that had offices and also things like bunkers in nearby fields where the test engineers could duck down for safety when they put engines through their paces. He also wanted the place to be remote enough that no one would cotton on to the testing taking place.

Eventually, Kemp made his way to the Alameda base and a building at 1690 Orion Street, and it seemed too good to be true. From above, the facility looks U-shaped. It has a large main building at the center flanked by two long structures. It turned out that the navy had used the building to test jet engines in the 1960s. The two long structures were test chamber tunnels. An engine could be placed at one end of the tunnel and turned on, and fire and heat would rush through what amounted to a giant exhaust pipe. At the opening at the other end of the pipe was a metal wall that slanted back and shunted the exhaust up and out of a tower at the top of the building.

Kemp had only heard rumors of all that the facility entailed, and the city refused to let him into the building to inspect it. It had been abandoned for decades, and the city said it would never take the risk of renting such a decrepit, dangerous place to anyone. Naturally, Kemp embraced that as an invitation to break into the building one night. He climbed over a fence, found a way through a door, and snooped around, using his cell phone as a flashlight. "It was in really bad shape," he said. "There was an inch of water on the ground, and it was full of asbestos and garbage. We had to use a lot of imagination to see how we could take this building over and actually use it."

The city had treated the building as a catchall storage facility. There were racks and racks of blueprints. Little league baseball equipment sat in puddles of water. Piles of dot-matrix printers and microwaves were packed into corners. Some refrigerators were bunched together. A fire truck with no wheels capped off the smorgasbord of municipal detritus. The whole place smelled of rot, and mold crept up the walls and onto the ceiling.

No matter how bad it looked, Kemp could not get away from

what the Orion Street building offered. Stealth Space could set up a rocket start-up in a place just twenty minutes away from San Francisco, which would give the company access to the world's top software engineers. Meanwhile, it could build and test engines right at its headquarters and fire them up *inside* the facility, and no one on the outside world would be any the wiser. "I knew we needed a place like this," he said. "I told Adam, 'Imagine building a rocket engine test stand in this soundproofed bunker that's right next to a place where we could put people's desks and a control room.'"

In a testament to Kemp's powers of persuasion, he managed to convince the city to lease this possible death trap to Stealth Space. He went before city officials with a spectacular proposal outlining how he'd bring jobs and a state-of-the-art factory to Alameda. In a bolder move, he had workers clean up and detox the building and paint it, and then he moved employees in while the city was still weighing the deal. Inspectors would turn up and tell him to stop doing things, but Kemp just kept on going and eventually the city sided with inertia. What's more, it even gave Stealth Space a sweetheart deal on the rent.

From January 2017, when it moved in, to April, Stealth Space made a tremendous amount of progress on both its rocket and the building. After walking through the front door, you entered the high bay, the main rocket-manufacturing area. All of the walls had been painted white, and the floor had been stained white and coated in epoxy. It looked fresh and new and as though a surgical operation could be performed there. The ceiling was about forty feet overhead, and a crane hung down from it to help move heavy items around. Right in the middle of the room sat an early version of Stealth Space's rocket, or at least parts of it. The company, now up to about thirty-five people, had built a pair of aluminum fuel tanks and a carbon-fiber fairing, or nose cone. Workstations were set up around the high bay where people prepared wiring and computing systems to place into the rocket.

A small team had taken over one of the tunnels and built a test stand inside it. It was a large steel structure bolted into the ground, and it had all manner of tubes and wires mounted to its body. The chaotic assembly of machinery was designed to replicate the innards of a rocket and feed fuel into an engine that hung out the back of the stand, pointing its tail into the tunnel. Large tanks of liquid oxygen sat near the test stand along with tables piled high with tools, power cords, duct tape, and rolls of aluminum foil. People working in the area had to don goggles and face shields, and frost covered many of the objects in the room due to the LOX's extremely low temperature.

A conference room at the center of the Orion Street building had been turned into an operations center for running the tests. A series of six laptops was placed on a couple of cheap plastic picnic tables in front of a large TV screen, which displayed a video feed produced by a camera near the engine. A homemade control box had a series of switches that controlled the engine operations. That room had also received a coat of white paint, but there was only so much damage that could be covered up. Parts of the walls were missing, and there was a huge, gaping hole over one of the doorways.

Another large work space at the back of the building functioned as a machine shop. There were 3D printers, lathes, mills, and computer-guided metal-cutting machines. The city of Alameda had once told Stealth Space that the building lacked power, but Kemp and his crew had heard some buzzing and humming coming from this room and been beyond thrilled to discover that it in fact had a working electric substation.

In an outside area at the back, Stealth Space had created a place for the employees to hang out and have lunch. It was located in the middle of the U shape formed by the ends of the two tunnels. You could saunter over by the tunnels, peer down, and see deep water reservoirs that had been put in as a safety measure for any very large, unplanned fires. Look up eighty feet or so, and you'd see the tops of the exhaust towers.

Most of the people working at Stealth Space were men in their twenties. A group of eight or so who had come over from Ventions formed a tight clique. At that point, the company was taking its old designs for a very small rocket and turning them into new parts for a small rocket. The Ventions OGs were thrilled that Stealth Space had the money and desire to turbocharge their work and try to turn it into the basis of a real rocket company. On the whole, they harbored skepticism about Kemp and found both humor and disappointment in his flamboyant approach to management. At the same time, they were a pack of rocket junkies, and Kemp had made it possible for them to live their dreams.

Other employees had arrived from defunct rocket start-ups, university programs, car-racing teams, software companies, and other industrial firms. Some of them had loved space as kids and were thrilled at the thought that something they helped create might end up in orbit. Some of them longed to build a real, tangible object that felt as though it carried more meaning than just another app or Silicon Valley bauble. For others, Stealth Space was just a job. They cared about the task and the quality of their efforts but might as well have been welding on an oil rig or tuning a car.

The experiment all of those people had signed on for was to show just how far New Space had come. London had a PhD in aerospace engineering, but there were not many employees with advanced degrees. In fact, Stealth Space had a number of college dropouts on staff, as well as folks that many companies in Silicon Valley would never consider hiring because of their unconventional backgrounds. It had set up shop in a building that probably should have been condemned and was led by a data center geek who had failed in his previous bid as CEO. There were hints that their rocket, if they could actually make it, might be of interest to customers, but no confirmations. When describing the mission, Kemp veered into the hyperbolic, but there was some truth to his sentiment: "We're basically building NASA at one one-millionth the budget and one one-hundredth the time," he said.

SpaceX had needed six years to go from the drawing board to a functional rocket. That had been a record-setting pace. To see how long they had to accomplish the same sort of task, the Stealth Space employees could look up at a countdown clock hanging over the door in the factory. At three in the afternoon of April 17, 2017, it read: 239 days, 22 hours, 59 minutes, 41 seconds.

In other words, Stealth Space was supposed to complete a rocket in roughly eighteen months and have it ready to fly in early December 2017.

CHAPTER SEVENTEEN

CHRIS KEMP ON CHRIS KEMP, SPRING 2017

After I'd spent my first few months with Stealth Space, it became clear to me that I'd stumbled into a special opportunity. Kemp, London, and the rest of the company had not fibbed. They really were going to let me observe their every move and record their every conversation. I would get to experience the struggle of building a rocket and a company right alongside them.

The more time I spent with them, the more I wanted to convey to you, dear reader, how they talked and thought about the space business and how they solved problems. To that end, I'm going to put you next to the engineers both when things go right and when they go horribly wrong, and also simply let you hang out with the engineers and technicians as they chat to get a real feel for them. In a number of the following chapters, you'll be hearing much of the story directly from their mouths. It's a slightly unusual approach, but these were unusual circumstances. You're going to see the religion of space engineering and the religion of Silicon Valley in unvarnished form.

Chris Kemp provided his own unique brand of surprises. Though I pride myself on being a decent writer, I started to wonder if anyone would believe my descriptions of how he talked and thought unless they experienced it for themselves. With that, I present to you Chris Kemp discussing Chris Kemp as only he could.

> **KEMP:** I was born in upstate New York. And when I was a baby, my father, who's a neurobiologist, was given an opportunity to come down to Birmingham, Alabama, and set up a lab. So I spent most of the time growing up in the suburbs of Birmingham. At the time, though, we were just south of Huntsville, Alabama, where they have the Marshall Space Flight Center and the Space Camp, which I attended.
>
> My dad was a professor and taught and did a lot of research. He worked on papers around how neurons communicate with each other. I remember going to his lab as a kid, and he would have all these experiments running around sequencing DNA.
>
> He was a very multifaceted person. He built race cars in the garage and drove them. As a kid, I grew up working on the cars with him. We had about ten cars and a huge garage. He was also a violinist and played in an orchestra. I picked up the violin and still play as a way to unwind. Oh, and he also played tennis quite seriously. Something of a Renaissance man.
>
> My dad was very serious and intense. Whatever he did, he took it very seriously, and I grew up with that focus and intensity modeled for me as a kid.
>
> My mother was a teacher and a pilot. She taught classes at a private school and also flew private planes. But she gave a lot of that up when my parents had kids. She became a dedicated mom for me and my sister.
>
> I was definitely a nerd. The skinny pale kid. I spent a tremendous amount of time as a kid not thinking about school stuff but doing personal projects. In fifth grade, for example, I had more chemistry stuff in my locker than books. Later on with stuff like physics and calculus, I often spent more time writing elaborate

software on my Texas Instruments calculator to solve problems than doing the actual homework. I remember quite distinctly almost failing AP Physics because, rather than solve a problem, I would write software to solve a problem. It was something where you're solving a circuit and want to understand what the current or voltage is at a certain point. You could do some easy calculations to figure it out, but I drew the circuit on my calculator and set up a menu to point to like a resistor and figure out it was at fifty ohms.

I would just write the answers down. And then I would finish the test in ten minutes and they'd be like "Well, you didn't show your work. You cheated." And I'm like "Fuck you, no, I didn't. Not only did I understand the problem, I understood it well enough to write an application. Would you like the source code?" After finishing the test, I'd work on this little Space Invaders game I'd written.

I got on the internet at a superyoung age. Like, I was one of the really early users. This was before CompuServe and AOL. I would connect from my computer to the university computer. I got Kemp .com.

I was super into electronics. My mom loved to go to garage sales, and I would pick up these books on how things worked and how to fix them for like ten bucks. As a thirteen-year-old, I bought a broken TV that had been hit by lightning, and it literally cratered all of the chips. I got all the schematics from Philips and systematically fixed every single one of these boards. Normally you would just chuck it, which is what someone else did, but when you're thirteen, you have time for stuff like that.

I started to realize that computers were valuable and pretty easy to fix and that there was a huge arbitrage opportunity there. My first job was at an Apple store working in the service department. Then, as a teenager, I started buying broken computers and repairing them and selling them. I could easily turn a few hundred dollars into a few thousand dollars. That was the hustle. At the time, I was making tens of thousands of dollars.

This was around 1996, and I was kind of living this drug dealer lifestyle. And I think everybody thought that I was dealing in some illicit drugs. I had a secret room in our house where there was this bookshelf that you could slide to reveal a door. It had a security system with a keypad alarm, and inside was a big-screen TV and a super-high-end audio system and all my computers.

I was also into film production. Our school in Alabama was obviously really into football, and we had one of the better teams in the state. One of the dads had made some money and donated millions of dollars to set up a professional-level TV studio with like a satellite TV truck and cameras. I'd go to the football games and help with broadcasting the games live. Then I ended up spending a lot of my own money on a studio setup at home with a really powerful computer. All in, it was probably like a hundred thousand dollars of stuff. We were middle class, and the house was probably not worth much more than that.

In high school, I got seriously into the video business. People were paying me to shoot video and edit it and put in animations. I did that throughout high school and started working for video production houses. That business brought in tens of thousands of dollars per year, which was a lot for a high school student. It was enough for me to buy a BMW.

I had some friends but was just really focused on the business stuff I was doing. When the time for senior prom came, the school hired me to shoot a video of it. So that's kind of fucked up. I didn't really attend it but got paid to attend it.

When the time came to go to college, I wanted to continue what I was doing but on a bigger level. I wanted internet access, and most colleges didn't have that in the dorm rooms, but the University of Alabama in Huntsville did. Or at least I thought they did. There were Ethernet ports, but they were just connecting to a university server. After I got there, I went to the president of the university and said, "I came here because of this. You misled me. You will fix this." And they did fix it that summer.

During my sophomore year, I started working at SGI [Technologies]. They were sort of the Google of the time, the hottest technology company. They made really powerful graphics computers, and as an employee, I could buy them at a discount and the software was free.

I immediately bought a ten-thousand-dollar computer and got a hundred thousand dollars' worth of software. It was the same software they used to make *Jurassic Park*. I met another guy that was a senior at school, and we decided to start a company. This was 1998, and we basically made the equivalent of Instacart but twenty-five years too early. You had to pick up a CD-ROM at the grocery store, put that in your computer, and then get people to go online for the first time in their lives. But it was the same deal, you would pick out your groceries and then people would pick them up for you and deliver them to your house. It was called OpenShop. I have these adorable videos from the late nineties where I'm on TV talking about how you can buy groceries online.

My first girlfriend at college was a senior. This kicked off a phase of my life where I dated women way older than me. It lasted too long. But they had their shit together.

I also got my pilot's license in college. I put basically every penny I made into learning how to fly. The whole pilot scene blew my mind because the people that showed up were prefiltered as people that believed that they could make a huge impact in the world.

Anyway, we ended up raising ten million dollars for OpenShop from about sixty people in Huntsville. I had to wear a suit and a tie a couple of times. We actually got acquired for twenty-five million or something, and the company that bought us asked me to move to Seattle. I had enough money to buy a house, so I dropped out of college and moved when I was around twenty years old. I said, "Fuck it. I'm going to be a CEO," and never looked back.

I wrote probably the most thoughtful letter I've ever written in my life to explain dropping out to my parents. I don't like to write. I'm much more of a speaker. I was afraid, I think, that if I were to go in and try to say what I wanted to say, I wouldn't be able to get it out because my dad would get very angry at times. I felt like of all the things he might get angry about that this would be at the top of that list. Of course, they said, "Well, you're on your own now." I said, "I know. I got this."

I'd also met Will Marshall while I was in Alabama, and that meeting had a big impact on my life. I was visiting a friend at college and ran into Will randomly. We got to talking and started hanging out a little bit and going on adventures. I think the first thing we did was go caving. There were an incredible number of really cool caves where you could crawl deep down into the earth. Will was like "That sounds good! Let's go!"

We stopped at a Walmart on our way to buy flashlights, and, well, Walmart sells guns, too. Will had just come to the United States from England and thought that was one of the most ridiculous and absurd things in the world. "What? You can buy guns right here?" And so I'm like "This guy is different. This guy is good." He was like a smarter, British version of me. I was impressed with how he saw the world. He was very open and liberal, and that resonated with me because that's what I saw with my parents, who were academics.

He was doing undergraduate studies in England but had come to work on this program at NASA where they were trying to perfect a fusion rocket engine. It involved matter and antimatter being contained by a magnetic field, and the field was being distorted in a way that made a nozzle. Imagine a shed in the middle of a forest with power lines coming into it, and they flipped a switch and this thing goes *bzzzzzzzzzzzzz*, and the whole thing uses as much power as the entire state for about thirty seconds as they try to contain the antimatter.

Needless to say, the project didn't work. If it had, there might be a black hole in Alabama—or at least a bigger one than what's already there—that would kind of suck all of reality into it. Will was an intern on that. It's probably classified, and I shouldn't talk about it.

Later on, Will and I got really into breaking into buildings. We took a trip to England and got a map of every castle that has ever been built. Many of the castles are completely demolished. They're just piles of stones in the middle of forests. We decided to explore as many castles as possible as we traipsed across all of the UK in a week. This was a no-tourist-castles trip. If you had to pay to get in, we were not interested. We had to be able to break in.

We drove up the country and saw Stonehenge. Well, we broke into Stonehenge. We got up to London and found a way to climb over a wall at Buckingham Palace. It turned out the queen was there. We got into some garden. We weren't shot. We weren't escorted out. We joined a group and masqueraded as tourists for a while.

After moving to Seattle, I ended up working for Classmates .com. It was 2000, and things were pretty harrowing. I still believed in the internet, but the internet was cratering. All the investors were fleeing. But Classmates was making money. They were converting people's inherent visceral interest in reconnecting with people from their past into profits. "Here is someone you used to date. Would you like to talk to them? That'll be $24.95 a year." I'm like "I love it! This is great."

The company was printing money, and I had the chance to go in and be the chief technology architect. We went from a few million people to fifty million people when I left. It was huge at the time. It was like the Facebook of the 1990s.

While working at Classmates, I'd also started another company called Escapia on the side. I'd rented a beach house for a vacation and realized that about twenty percent of Americans owned a second home that was sitting idle most of the time. I set up a system to rent out those houses. It's more or less Airbnb but also twenty years too early. It got acquired by HomeAway, which got acquired by Expedia.

The travel thing started out as a hobby but turned into a full-time thing because I was fired at Classmates. I was used to doing my own thing and had encountered politics for the first time. This guy who was the head of engineering was playing games. He ultimately got fired for them, but he got me first.

I had a fun time with that, actually. When I was fired—which

I believe is the only time I've ever been fired—I realized what had happened. I created a website with the guy's name in the address. Then I created websites with all his subordinates' names, and they all had pictures of pawns on them. If you clicked on the pawn, it went to the guy's website, which had a spot where people could anonymously comment on him. It looked exactly like that old website FuckedCompany.com. People did comment and outed all of his treachery, and that's when he was let go. I think he ended up in real estate.

By the way, that was petty. I would never waste time on things like that at this point.

While I was living in Seattle, Will started putting together these events around New Year's Eve. We called them 4D. I'm not sure anyone still knows what the Ds stand for.

One of the first events was at Oxford University. The idea was to bring together a group of people that shared the desire to find a purpose in life that was beyond just making money or beyond what they were trying to accomplish in a particular job. You wanted to try and look at your entire life as a story arc, and we wanted to see if this group of people could work together to accomplish something larger and more meaningful.

I was one of these first ten or fifteen people that Will had collected. Jessy Kate was there, too. She had this young, nerdy Lara Croft Tomb Raider vibe. Adventurous and smart and open. The whole thing blew my mind.

We still do 4D. Every New Year's, we find a site for the event. Ideally, the site will be a place that humanity has built that reflects something larger than life. We went to the Arecibo Observatory once, where the world's largest telescope is in the middle of a jungle, and we took it over. We had another one at the Biosphere 2 complex, which is a set of glass pyramids in the desert that have jungles and oceans and deserts inside of them. It was airtight and completely separate from the outside world. We broke into it, of course, and took that over, too.

These are places where I think we can truly reflect on what

people are capable of doing. Someone in the 1980s raised billions of dollars to build a giant multiacre indoor facility and got a bunch of people to lock themselves inside of it. Okay, well, if somebody can do that, what can we do? And I think that's the inspiration behind 4D.

The core of the event is reflecting on what we've accomplished in the past year and what we intend to accomplish in the next year. We tend to pair up with people that are randomly assigned that we can talk through our goals with and then challenge each other to think bigger. Then we share the conversation with the group. There's an accountability aspect to it. You have to get up in front of everybody and say, "Last year I said that I was going to do this. Well, I didn't." Or, you know, "Last year I said I was going to launch a rocket, and I did," for example.

Will was having one of these events while I was doing the travel company thing, and it was in Washington, DC. And I meet this crazy general named Pete Worden at the New Year's party. He was about twice the age of everyone else, so he stood out.

He's holding a martini glass and telling these stories about "When I was in the Pentagon . . . ," and everything would start with "Well, when I ran Star Wars," and I'm like, well, this is the most interesting person I've ever met. The one thing that really stuck with me was this story about him running a misinformation campaign during one of the wars in the Middle East.

He was thrown under the bus by the Bush administration because he was trying to use information war versus actual war. They were dropping radios and leaflets to try and convince the Iraqis that the war was over versus shooting them. Pete believed in this. But apparently, the press came out and said, "This is Orwellian. You can't do this. It's not right." And Donald Rumsfeld, who was secretary of defense at the time, basically blamed it on Pete, and Pete's picture got plastered on the front page of the *New York Times* and the *Washington Post*. The line that I remember so viscerally was "Donald Rumsfeld fucked me, but they're going to give me a NASA center."

And eventually, Pete did get his center at Ames, and I followed him there. It took me a while to process that I'd been working on stupid little companies that would make travel better or shopping better. Finally, after I arrived at NASA, I got it. I'm like "Wow. I'm the CTO of NASA. I can make very big things happen, and I should. In fact, it's immoral for me not to take this opportunity and do anything but try to have the most impact I can possibly have."

I became the youngest senior executive in the entire federal government. Pretty quickly, I realized that the NASA culture needed changing. It has almost a hundred thousand people, and it's very hard to fire anyone. The question becomes not how do you fire someone but how do you get them out of the way. You need to take people that are thinking about something correctly and put them in a position where they can make a difference. You reorganize the other people and put them on projects that are out of the way and don't matter.

I created an entirely new tech organization and a whole leadership team, which was very controversial. People had expected to have certain roles because of their tenure, but I didn't care about tenure, and I didn't care about loyalty. I cared if they had what it takes to do a certain thing.

The budget we had at the time was seen as barely able to support operations. That was such bullshit. We cut out a bunch of dumb costs and had money left over to invest in new things. That's what allowed us to create the special projects group that did things like Google Moon, Google Mars, and OpenStack.

As for Stealth Space, I love this new challenge. You have to learn and grow and evolve as a person. You have to put yourself in a position where the experience and passion and energy you have is matched with the challenge. Here you have to face challenges and deal with them whether you're ready or not. A lot of successful people get distracted with women and material things and having fun. Those people regret their lives ultimately.

I have never felt better about anything I have done in my life than this thing. Adam is brilliant. We are so perfectly matched and complementary. He has the rocket science covered. I could spend the next five years reading physics books and fluid dynamics books to try and get some grasp on all the things going into a rocket, and I would get halfway to where Adam is.

For my part, I've built four companies from scratch and advised lots of companies. I've raised over a hundred million dollars and helped people raise over five hundred million. I know how to raise capital and build teams. I don't think what we're doing here is fundamentally impossible to do. I think it's possible.

CHAPTER EIGHTEEN

THE GRIND

Stealth Space wanted to remain stealth, and it had the vastness of the former navy base to give it some protection. The restaurants and bars that had opened near the water were about a mile and a half away from the rocket operation, and the cars heading to those spots usually traveled along an outer road. Most of the large warehouse buildings near Stealth Space were crumbling and abandoned, and on a given day, only a couple dozen people were likely to pass the Orion Street building.

That, however, is not to say that the Stealth Space operation was entirely discreet. The company's headquarters were near the southeastern edge of the base. A beat-up fence about a thousand feet from the rocket factory marked where the base ended and regular life in Alameda began. There were houses, preschools, soccer fields, and restaurants right on the other side of a fence less than a half mile from the controlled explosions taking place inside Stealth Space. On the base itself, Stealth Space had a Pottery Barn Outlet and the Pacific Pinball Museum Annex as its nearest neighbors. None of those parties was aware right away that the equivalent of an intercontinental ballistic missile was being built on the cheap in their backyard.

The plan Stealth Space had come up with to make its rocket in record time was to keep things as simple as possible. Whereas Rocket Lab used carbon fiber for the body of its rocket, Stealth Space would use aluminum, which is cheaper and easier to manipulate. Its rocket would be small, only about forty feet in height and three or so feet across, and so would not require a large, complicated engine to fly. Stealth Space would propel its rocket with five modest engines that a couple of people could build by hand. The company's engineers were asked to do things like look for any spot where they could pare back the amount of wiring and electronics inside the rocket. Kemp wanted the bare minimum number of parts necessary to make the suckers fly. All the better for mass-producing the projectiles.

Stealth Space also wanted the ability to blast its rockets off with minimal infrastructure required on the ground. Rocket makers were accustomed to spending weeks, if not months, at a government-run launch site preparing their machine for liftoff, and the operations usually required dozens of people. The plan Kemp and London had come up with was to make a mobile launch apparatus, a motorized scaffold that could shift the rocket from a horizontal to a vertical position and that came with all the needed fuel and electronics connectors built into its body. The launcher would be small enough to fit into a standard-sized shipping container and could be sent via land, sea, or air to any site from which Stealth Space wanted to launch. To go with the launcher, the company planned to build a mission control center that would also fit into a shipping container. That would let the employees use their own familiar gear instead of borrowing a mission control center at a launch site. It would also save Stealth Space from paying the exorbitant fees that launch sites charge for the use of their equipment.

Stealth Space's ideal launch operation would see it pack the launcher, mission control, and rocket into shipping containers and send them off. The machines would be received by a handful of people, who would unpack the containers and transport all the machinery to a launch site. The same people who unpacked the equipment would then be in charge of the launch. In this idealized world, the whole operation would take a few days. Kemp wanted rocket launches to go from cool, rare events to ho-hum, everyday occurrences as Stealth Space emerged as a new-age shipping company.

At that point in the company's history, Stealth Space had not quite worked out exactly where its first launch site would be. Kemp took a trip to Hawaii and visited a spot near an active volcano. He guessed that the locals would not be all that keen on welcoming a rocket company into their midst but thought he might be able to talk them into it if Stealth Space was willing to exist on a perilous hunk of land that no one else wanted to go near. "I'm looking at a lava field," he said. "I think it's the way to go. There are spots where new land is being formed, and no one wants to live on land that has only existed for a couple of years. The good thing about land that didn't previously exist is that it's pretty cheap, too."

The company was also considering a spaceport on Kodiak Island in Alaska. Like Hawaii, it was remote and had less air and boat traffic to contend with during launches than the most used US launchpads in California and Florida. The Kodiak spaceport also had existing infrastructure to work with since the US military had been launching missiles from the site for years. Those options would just be stopgaps anyway; Kemp had already bought a barge with the intention of turning it into the company's pad of choice. "It's a hundred feet by forty feet," he said. "It's beautiful just floating there." (Later, I found the barge docked next to a retired aircraft carrier. It was covered in bird shit and had a rusted-out pilot's cabin topped by a pelican.)

By the summer of 2017, some refinements had been made to the Orion Street building. An asbestos-cleaning crew had gone over every inch of the facility and made it safer for the employees. Kemp had improved the Wi-Fi coverage by climbing up the walls and into the rafters to place a few extra routers there. And a bathroom had been built in the center of the building. It was the only bathroom and had four stalls and a couple of urinals, so women, men, and all other members of the gender spectrum got to take shits next to one another at work.

The company set its priorities for the day and the week during morning meetings that began at 10:00 a.m. Kemp would bring up a screen that listed all the jobs that were in the process of being completed and one by one asked about their status. An hourglass with ten minutes' worth of orange sand sat on the table, and Kemp turned it over with gusto each time the sand ran out to remind people to get to their points. Throughout the proceedings, he maintained a wide-eyed intensity to convey that he was deeply present in the moment and wanted things done faster. London would usually stand near the back and let the information wash over him as his mind ticked through possible solutions to the issues.

The two men were really quite different. After the meeting, for example, Kemp explained that he'd once given himself a thirty-day window to quit his job at Ames, found a new company, get engaged, get a house, and start a family. He'd done all that, had a son, and then gotten divorced a short while later. "The challenge was a success," he said. "Only then it was like 'But I don't love you.'" In the subsequent years, Kemp cycled through numerous girlfriends and appeared to have a thing for skinny blondes. (Around that time he dated a San Francisco 49ers cheerleader who had a chemical engineering degree from MIT.) London, meanwhile, had a wife and children and described family life as if it were another engineering challenge. "My nominal plan is to go home on Tuesday and Thursday at a reasonable hour and have dinner with the kids," he said.

Not too many people at Stealth Space were working reasonable hours, in large part because the rocket engine did not want to cooperate with their work-life balance. Many of the initial months in the Orion Street building consisted of about twenty people working between the test cell and the test control center trying to get their engine to spit out fire for any length of time.

I'd sometimes show up around 7:00 p.m. to watch the testing activities. People in the test cell would begin filling tanks with the liquid oxygen and kerosene, while other folks fiddled around with wires and other elements on the stand. A man named Lucas Hundley, who'd been at Ventions, monitored the action via the video feed from his post as chief of the control center operations. Since people in the test cell often contorted themselves as they wrenched and banged on things, Hundley saw the ass cracks of many of Stealth Space's top engineers. All part of the glory of working on rockets.

It could take anywhere from fifteen minutes to two hours to prepare the engine for a new test, depending on what people thought had gone wrong the last time. For the first few weeks, the engine either didn't do anything or sputtered for a second or three. Ben Brockert, a tall, heavyset young man who approached life with voluminous amounts of cynicism and attitude, was in charge of calling out the order of operations during the test procedures and had gallows humor prepared for most situations. Ahead of one test, he informed the team, "The emergency protocol says that if there's a fire near the engine, we kick on the water and put it out. If there's a fire elsewhere, we have less good plans." Later, in the midst of a long testing session, he rallied the troops, saying "We will keep going until the engine works or we die." When I asked him what he'd been up to one night, he replied, "Mostly working on my poetry."

On a given night, Stealth Space might test the engine five or six times with people staying until 2:00 a.m. The Orion Street building did not have heat for a while, so the employees huddled around little propane space heaters to make the long nights more bearable. London often paced around the building during the runs with a bag of potato chips in hand. When things went wrong, he might blurt out, "Log data!" or "That was far from nominal." When things went more right, he'd deadpan, "Fire has come out of the stand. Hooray." Sometimes Stealth Space would make an engineer go outside the building during the tests to record the sound level in a bid to figure out just how worried they should be about the neighbors phoning in noise complaints.

Matching the typical rhythm of the rocketry business, the mood at each testing session oscillated between excitement and tedium. People really seemed to believe that their fixes would finally work and the engine would burn as intended. On many occasions, Kemp busted out his smartphone to hold a video call with an investor as a test took place, convinced that would be the moment when the investor would see proof that money was being spent wisely. When the engine crapped out, Kemp would smile and talk the failure away. People got hungry and tired, but they kept right on working the problems and usually refused to call it quits until some necessary component such as the LOX ran out.

Eventually, the effort and long hours paid off. The engine burned for three seconds and then thirty seconds and then a couple minutes and then basically as long as Stealth Space kept it fed with fuel. In the background, employees had been building and buying the other key components of the rocket and assembling them on the factory floor. Kemp felt convinced that the whole rocket could be assembled and in near-working shape within the next three months.

In May 2017, Kemp invited Creon Levit, his old friend from Ames and the Rainbow Mansion, to come give a talk to all the Stealth Space employees. Levit spent his days working on satellites at Planet Labs but had a hobby studying aerospace history. His lecture centered on why rockets are so hard to make and fly. "Nothing much has happened in the last fifty years," he said. "We still build rockets more or less the same way. Nothing has gotten cheaper in terms of the cost of kilograms to orbit. How can this be?"

Humans, he explained, had marched across the periodic table long before and found the best chemicals to bash together to unleash the most energy. Unless people wanted to make a rocket powered by nuclear energy, they were stuck combining liquid oxygen with kerosene or hydrogen. "The laws of fluid dynamics make it hard," he said. "The laws of material science make it hard. You have tons of oxidizer and fuel flowing into this thing every second, and you're lighting a match. If you did that in a room, you would call the result an explosion. With a rocket, we call it burning, but it's basically just a constant explosion that is being contained and shaped. Everything is right on the edge of making this feasible."

In the early 1900s, the fundamental math that still governs rockets today was worked out, and it's brutal stuff. The energy produced by the best propellants is barely enough to have an object escape gravity. About 85 percent of the mass of a rocket has to be propellant, compared to a car, which needs just 4 percent of its mass in the form of fuel, or a cargo jet, at 40 percent. That leaves 15 percent of the mass to create a structure to hold the propellant and all the machinery, electronics, and computers necessary to make it do something interesting. When all is said and done, a rocket builder is lucky if 2 percent of the rocket's mass is made up of the stuff you want to ferry into space.

Not too long after a rocket lifts off, it reaches something known as Max Q, the moment when the rocket experiences the maximum amount of dynamic pressure it will face on its way to orbit. You feel dynamic pressure forces when you stick your hand out of a car window while driving and witness them when a hurricane knocks down trees and obliterates houses. Levit noted that the amount of pressure the rocket encounters at Max Q is roughly seventy-five times that produced by a Category 5 hurricane. "Of course, if the rocket goes sideways at all at that point, well, then, the whole project goes sideways," he said.

In most engineering exercises, you would deal with these incredible forces by overengineering your machine. You double whatever needs to be doubled so that the thing will work. But that's not an option with rockets because there's no room for excess anything. To make matters worse, Max Q is hunting for flaws, pushing against every weld and chamber to see if you really made the most of the little leeway you had.

Then, hey, let's say you've made a rocket that works. Well, the actuarial tables used by insurance companies show that the sixth launch of the same rocket usually fails. People become cocky and complacent and miss problems. The next point when things start going wrong is around the fiftieth launch. By that point, mission creep, bureaucracy, and institutional memory have become the perfect enemies. Levit's message to the Stealth Space employees who had suffered plenty to get to the starting line amounted to "Good luck!"

The positive news for Stealth Space, as Levit saw it, was that the company could cope with some explosions. It would not be flying humans, and the satellites it would be flying had come down in price far enough that customers would not mind if they blew up now and again, so long as the rides to space were cheap overall. The other good news was that we could even be talking about such things. "If the earth was a bit larger and its gravity a bit stronger, you are not going to space," he said.

THE CHALLENGES DESCRIBED BY LEVIT are so hard to overcome that we have turned making and launching rockets into a cliché. "Well, it's not rocket science" is what people say to denigrate all other endeavors.* Most often the folks working in aerospace, be they astronauts or engineers, are portrayed as our bravest geniuses who face and then conquer seemingly insurmountable challenges in their day-to-day work. The truth these days is actually much more colorful.

London fit the bill with his résumé full of MIT degrees. He did not ooze charisma or swagger, but he did ooze genius and was also more than capable with his hands. He had an intimate relationship with rockets and talked about them so lovingly that he made them seem mystical. London liked that they "existed separately from the rest of the world" and required an engineer to understand the interplay of different subsystems such as electronics, fluids, and propulsion and get them all working together as a whole. One time he described how hardware can speak to you: "When something seems weird, you need to listen to what the machine is saying and not write it off."

The work he'd been doing at Ventions was at the most extreme end of rocket science. He'd once hoped to make a rocket engine so small that it would need to be manufactured with semiconductor techniques. He relished the thought of shrinking all the electronics and machinery of rockets to their physical minimums. DARPA and Pete Worden were enthralled by his work. London had consulted with Worden for a spell on a project to develop a tiny, six-inch diameter

* Except for brain surgery, of course.

rocket that would fly into space and inspect a rival country's satellite as if the rocket were a covert operative. The US government hoped that London's rockets would be so small that another country would not notice them on radar.*

London might well have continued with his cutting-edge research and lived a quiet life had it not been for Kemp and the call of commercial space. "I saw Rocket Lab doing stuff similar to us at Ventions, but then they accelerated everything dramatically by going out and raising money," he said. "It was either do it or watch other people do it."

Many employees of Stealth Space, however, had not been seeking the higher calling of rocket science. They seemed to have stumbled into the business more than anything else.

Rose Jornales, for example, grew up in Southern California as the daughter of immigrants from the Philippines and Vietnam. Not much more than five feet tall, she'd enlisted in the air force in 2000 and become an electronics repair technician on aircraft flying to and from Afghanistan. After leaving the military in 2006, she'd worked as a waitress, gotten married and then divorced, and eventually found a job performing maintenance on the AirTrain at San Francisco International Airport. In 2017, a friend suggested that Jornales stop by Stealth Space and check it out.

"I didn't know it was a rocket company," Jornales said. "I also didn't know they were going to spring a job interview on me. All of a sudden, I'm talking to four people, and they're asking me all these questions. Oh, man, I don't know if I should say this or not . . . I was drunk when I came in. I'd been day drinking. It was easy. It was a cakewalk. But I don't remember what they asked me."

Jornales took on a job building the harnesses that hold and tame the miles of wiring inside a rocket. Her focus then shifted to the flight termination system that ensures the rocket will blow up in midair if something goes wrong. She was one of only a handful of

* At one point, Planet Labs had considered getting into the rocket launch business and talked with London about the idea. London credited Will Marshall with proposing the notion of launching every day. In the end, however, Planet decided to keep its focus on satellites.

women present in Stealth Space's early days, as it's not uncommon in aerospace to have one woman for every nine men on the factory floor. Jornales radiated positive energy and chased a good time. Her happy-go-lucky demeanor played well among the people turning screws and welding metal. In fact, she bonded best with the most cynical and hardened of the Stealth Space crew. Or at least the ones who also enjoyed a few Coors and maybe something extra after a hard day of aerospacing.

Stealth Space's electrician-cum-jack-of-all-trades was Bill Gies, who'd also stumbled into the mix. His ever-changing hair color rotated among blue, red, orange, green, and a rainbow fusion, and the overall style was natty mullet. He had many earrings and big holes in his lobes from years of wearing plugs. He wore black combat boots. He smoked and had smoke-stained teeth. He appeared unkempt, kind of dirty, and inclined toward a surly disposition. Bill was great.

He'd left home at eighteen to chase a girl and ended up homeless. Gies, however, had dabbled with electronics as a teenager and eventually found jobs here and there as a handyman, fixing elevators, washers and dryers, and arcade games. Right before joining Stealth Space, he had bizarrely worked as an electrician for a secret society. The group was backed by a multimillionaire who had acquired a house in San Francisco and turned it into a sort of escape room with lots of tech twists such as vibrating floors and light and sound illusions.

Gies also heard about Stealth Space from a friend and began showing up to do electrical work for free. "We'd be up until three a.m. or four a.m. in the test cell, and there he was," said one employee. "None of us knew his name." Eventually, he did so much work that Kemp started paying him. He also happened to have skills breaking into things, and if someone had lost a key to a storage container or a gate, he or she would call Gies to jimmy the lock.

Stealth Space employees always followed a series of protocols before a rocket engine test to ensure the safety of the personnel. The steps included clearing people out of the room with the engine (what with its flesh-dissolving heat and all), closing the blast doors of the test cell, and letting everyone in the building know that a controlled explosion would soon take place. Stealth Space also had one protocol that separated it from other rocket companies. It was the "Make sure Bill is not in the ceiling above the test cell" rule.

On one occasion, Gies had crawled into the ventilation systems to repair something, and the rest of the Stealth Space employees had forgotten he was up there as they went about their test. "I started thinking, 'God damn, it's windy all of a sudden,'" Gies said. "When some folks saw me coming down later, they were kind of shocked. Is it okay to tell you this? OSHA* can't sue someone retroactively, right?"

As the months wore on, Gies became a beloved member of the Stealth Space team and applied his broad set of skills to all sorts of endeavors. He joined the company's "top secret department of partying and mischief" and helped spice up otherwise dull Stealth Space affairs. "If I had to come up with a mission statement, it would be to ensure we are never invited back to a hotel that the company has booked for an off-site meeting," Gies said.

Just as aerospace companies tended to have few female workers, they were lacking in Black employees. Kris Smith was an exception, joining Stealth Space early on and proving essential to the day-to-day management of the company's operations. If a building needed fixing or expanding or a new construction project was afoot, Smith made sure it happened.

Of mixed race, Smith grew up in New York, surrounded by drugs and violence during his childhood. He witnessed murders firsthand and watched his friends' parents shoot up heroin in their living rooms. But he was tall and athletic and excelled at basketball, and that helped him avoid conflict. He went to college on a basketball scholarship and then played overseas in places like Syria, Mexico, China, and Spain. While training in San Francisco one off-season, he saw a job offer for a satellite start-up, took the gig, and helped it build its offices and laboratories. The company was Terra

* The Occupational Safety and Health Administration, the US agency in charge of worker safety.

Bella, which Planet Labs eventually acquired. Smith took the money he received from the sale and directed it toward real estate, creating a small Bay Area empire.

"I try to speak to as many young Black and brown kids as I can," Smith said. "They think maybe being a pro athlete is the way to go, but honestly, no one becomes a pro athlete or an entertainer because it's so hard. They don't know that they can become an engineer and make three hundred thousand dollars a year and get some stock options and make a couple million bucks if the company goes public. I didn't know, either, until I came and experienced it. These companies have changed my life for sure."

Then there were the guys who had come from car racing, who were some of the most impressive technicians. After years in the navy, Ben Farrant had traveled the world tuning engines at endurance races such as Le Mans and hopping from one racing team to the next. Even though he had no aerospace experience, he took on the crucial job of building Stealth Space's engines from scratch. "The first thing all of us did before our job interviews was hop on Wikipedia and look up some rocketry stuff," he said. "I Wikipediaed the shit out of it."

Tall, skinny, and bearded, Farrant oscillated between being anal retentive about his hardware and being easygoing toward life. All of the tools at his station were meticulously laid out during engine construction and then meticulously put away each evening. His workday tended to run from about 8:00 a.m. to 6:00 p.m. He bucked the Silicon Valley culture of arriving late and staying late, preferring to work hard with his headphones on, get the job done, and then return home for a beer and some relaxation with his wife. He rarely engaged in day-to-day chitchat with coworkers but was happy to show someone how to do something if they were willing to pay close attention. In some ways, Farrant felt like a throwback. Maybe because he liked to wear a newsboy cap and sneak out for cigarette breaks. Or maybe because he reveled in playing the role of the slow-talking, grizzled veteran who had seen it all and was just there to do a job and make sure other people didn't fuck everything up.

"I tell some of the younger guys that Adam and Chris have been dreaming about this stuff since they were kids," he said. "We are here to realize their dreams. But it's just another machine for me. These other guys are way into the process and want to stay all night. I think it's better to get your stuff done. I like to go home and work on some old cars or just watch TV."

There were, of course, people at Stealth Space who had longed to work in the aerospace industry since their childhoods, but even their paths to the company had been winding and unusual. Several of the hardest-working and most talented engineers had barely made it through college, let alone attended an Ivy League school. They were the sort of people who liked to build and fix things and spent time chasing those passions rather than applying themselves to academic requirements that held little interest. Others had done fine in school but landed in aerospace backwaters for their first jobs, trying to prove their skills at start-ups with very little money or resources.

A trio of engineers—Ian Garcia, Mike Judson, and Ben Brockert—fell into these categories in one way or another. They'd taken different roads but ended up working together at a start-up called Masten Space Systems based in the Mojave Desert. Masten's claim to fame was building a small spacecraft in 2010 called Xombie that could take off and land vertically and was an early precursor of reusable rockets. Garcia and Brockert had been particularly instrumental in Xombie's success, and the vehicle had been good enough to catch the attention of Elon Musk during SpaceX's quest to make its own rockets reusable.

Garcia had grown up in Cuba, where he'd seemingly had no route to working on a US-based rocket. He had, however, scored so highly in international computer science competitions that he'd earned a full scholarship to attend MIT. He'd applied his gift for writing software to the guidance and navigation controls of spacecraft, and after showing those skills off at Masten, he got a job at Ventions and then at Stealth Space as its guidance software chief.

Judson was that rare animal who'd picked working at Ventions and Stealth Space over working at SpaceX. He'd done his time at Masten and other space start-ups and ended up interviewing with Musk and London at the same time in 2014. It's fair to say that no one loved rockets more than Judson, who often spent the night next to the machines just to be close to them and usually outworked everyone else at any company he went to. He recognized that Ventions' facilities were disheartening compared to the grandeur of SpaceX, but he saw something authentic and inspiring in London.

"Ventions had everything that was good about Masten," Judson said. "The passion. The small team. The desire to be quick and break away from what aerospace had been. But it was also real enough. They had done real things and were organized. Adam's technical knowledge was incredible. He's the most intelligent person in the room in most of the rooms you will ever find yourself in. He's a good, elegant engineer in that he can boil down questions into simpler questions. He knows how to make problems easier to solve.

"Ultimately, I picked Ventions over SpaceX precisely so I could get an opportunity to do what we're doing now. It was my chance to go lead something. I chose to be a big fish in a small pond. I joined up to build a small rocket to send things into space."*

As for the resident curmudgeon, Ben Brockert, well, he'd made his way to Stealth Space as a rocket drifter. He'd grown up in a fifty-inhabitant town in Iowa where his parents had sometimes struggled to pay the bills. He'd attended Iowa State on and off for several years. He'd take classes for a semester and then bail out to work odd jobs as a cook, junk mail processor, or machinist until he could afford to go back to school again. In 2007, Brockert dropped out altogether. "I did the math and figured it would take me forever to graduate," he said.†

---

* About Kemp's arrival at Ventions, Judson said, "I liked the energy Chris brought. Stealth Space needed a CEO type. There is a certain amount of showmanship or antisocial personality or whatever you want to call it that is required. From the beginning, Chris had that visionary personality. He has a lot of the opposite traits of Adam, and we could see he would be the one to go raise money. It was like 'If he can do that, we'll follow this guy.'"

† That type of calculation was very easy for Brockert; he'd notched a perfect math score on his SAT.

Having pined to get into aerospace since he was a kid, Brockert bought a $500 very, very used seventeen-passenger van and drove to the Mojave Desert to begin knocking on the doors of space start-ups. It took months to find a gig, living out of his van and trying to fill the days the best he could. "I managed to get a library card and read all the books I had not read," he said. "I also spent a lot of time bombing around the desert and trying not to die."

Through a stroke of luck, a position opened up at Masten, and Brockert soon established himself as an essential part of the Xombie team. He continued reading aerospace books and learned a ton on the job over the course of three years. "Mojave is like a New Space penance," he said. "It sucks, but you put in your two or three years and then you can go do something somewhere else if you're good."

By the time he got to Stealth Space, Brockert had the gruffest of demeanors on the surface. Chris Kemp would, on occasion, refer to him as the Eeyore of the company, and that was fair. He ambled around the test cells and factory floor with a scowl that warned people not to bother him. If someone did opt to converse with Brockert, he would supply their daily dose of sarcasm, cynicism, and pessimism in one serving. The performance was wonderful to watch because Brockert played his part with wit and because his body and attitude unified so well to form the ultimate misanthrope.

Brockert, though, did have a soft side. He was obsessed with rockets and had a deep knowledge of aerospace engineering that stretched across numerous disciplines. If you were young or inexperienced and showed genuine interest in learning, he would go out of his way to help you out and share his wisdom.

At Stealth Space, Brockert worked on a handful of different teams, going wherever the rocket needed the most help. He also emerged as the person who would run mission control and direct launches. He had periodic tantrums that would result in his resigning or being fired. But after a couple of weeks or months passed, he would turn up in Alameda again.

Ultimately, Brockert very much wanted Stealth Space to work, but he did not share the wide-eyed optimism of many others when assessing the company's prospects. "We've made a lot of good progress and gotten a lot done," he said. "But some of the things we talk about around how quickly we're going to build an orbital rocket are totally impossible.

"The whole thing about taking venture capital is that you're trying to raise money for some payoff, and everybody lies about their payoffs either accidentally or on purpose. You have to go to someone with money and say, 'Hey, if you give me some money, in a few short years, I'll give it all back to you plus a lot more because of this business plan that is almost complete bullshit.' None of the engineering schedules are ever remotely accurate, and none of this is as good as it sounds. That said, there are fifty-six companies saying they're going to build small rockets right now. I know most of them. I would say Stealth Space is the most legit in the US."

CHAPTER NINETEEN

PARTY LIKE YOU MEAN IT

By October 2017, Stealth Space could no longer be Stealth Space. The company's job recruiters had found it difficult to present the rocket maker as a legitimate operation given that it had no website or real name. Other employees had run into similar problems when dealing with suppliers. Stealth Space wanted to be taken seriously and to command good prices on tough deadlines, but when its employees placed an order as "Stealth Space," the people on the other end of the phone tended to laugh.

For months, Kemp rejected the pleas to call Stealth Space something else. The name was not just part of keeping the operation secret; it was a statement about its culture. Kemp rightly thought that too many aerospace start-ups sought attention before they were anywhere close to having a functioning product. He had learned a tough lesson at Nebula about overhyping things before they were ready, and he did not want to repeat that mistake. Build rocket. Fly rocket. Then talk about how awesome your rocket and your company are. On a pragmatic level, though, Kemp came to realize that the PR strategy made many things worse, not better. So Stealth Space became Astra Space, or simply Astra.

Astra had fallen behind schedule, but it had also made a lot of progress. The company had what looked like a whole rocket out on its factory floor. It had built several engines, and they often worked. Kemp decided to throw a party to celebrate the new name and those accomplishments. Only people who worked at Astra or invested in it or were close friends would be invited, and the event would take place in the Orion Street building with the rocket on display as the guest of honor. The invite read: "#NoPhotos, #NoSocialMedia! DON'T forget to wear your favorite SPACE SUIT!"

Kemp took the costume assignment seriously. He acquired an all-black jumpsuit bedazzled with leather straps and lots of metal buckles. He also wore what looked like a modified bicycle helmet, only with a couple extra pieces that came down over his face to make a type of mask. The ensemble presented as S&M space cowboy.

At the time, Kemp was in the middle of recruiting Chris Thompson, one of the SpaceX cofounders, to come work at Astra. He was also managing the details of the party very closely. Astra planned to show off its rocket in a plexiglass-enclosed case. Kemp liked that. What he did not like was that the rocket needed five fully complete engines attached to its body, and not all were ready as the party approached. In addition, people were suggesting that a mock-up fairing—the nose cone of the rocket, which holds the satellite payload—be displayed instead of the real one they planned to fly because it was easier, and probably safer, to futz around with the mock-up on short notice. Kemp felt that displaying a faux nose cone would undermine Astra's integrity.

The squabble about the engines and the fairing devolved into a full-on fight between Kemp and his engineers. He accused them of not delivering what they had promised. They accused him of not listening to repeated warnings that the work would not be completed on time. To an outsider, the squabble seemed ludicrous. No one at the party would have any idea about the true state of the rocket, nor would they likely care while being plied with free food and alcohol and dancing the night away.

Kemp, though, had a tendency to fixate on things that he deemed symbolic of Astra's character. On the drive to the October 21 "Dawn of Space" party, he let out his full Kempness, explaining his reasoning behind the blowup and also expounding on some of his other philosophies of life. This is what it sounded like inside his BMW convertible:

> **KEMP:** I do need to warn you about something. I don't actually have a valid driver's license. This car is not registered. And they have canceled my car insurance. So this is a little risky, and I just want to disclose that to you.
>
> We've been running really hard as a company for the past year, and frankly, we don't celebrate a lot. A lot of companies celebrate every time they do something. That's really inefficient. I think if you have a celebratory culture where you're celebrating raising money or building a facility, you're celebrating the wrong things. These are actually all negative things. Raising money is a bad thing. It just means it costs more to do what we're trying to do. A facility is something that it's unfortunate we even need to have.
>
> We celebrate things that really move the program forward, like getting the rocket put together and getting the launch infrastructure put together. It focuses the team on the things that really matter. We have a bunch of these things coming together, and it happens to be the one-year anniversary of the company, so it's a good time to celebrate.
>
> Also, we're a stealth company, so we don't publish videos about what we do or have updates on a website. We don't have PR. It's a distraction, and it's expensive. But that means that our customers, investors, and partners who help us do all this stuff never get to see anything. Now I can bring all these people together and in forty-eight hours knock it all out at once. This party will save me hundreds of hours of email updates and meetings and visits and tours.
>
> Our main competitor, Rocket Lab, raised their fourth round of funding before their first launch. If you have to raise four rounds of funding before shipping your first product, you're not

doing it right. You're giving away a lot of ownership and control of the company to these investors. There are ways investors view businesses at different stages, and there are things they want to see. If you understand this stuff, you will be more successful at doing the right things and derisking things to attract certain kinds of investors. I think a lot of the space companies don't understand this kind of stuff at all.

We don't spend any time on making people think about things that aren't real. For this party, they wanted to put the fairing that we are not planning on flying on the rocket because it's already there and it's less effort. That lacks integrity. Being able to say that we celebrate things that are real is absolutely critical to the culture we are trying to build. So to say "Let's put something out there for people to see that we're not flying" is a complete, fundamental betrayal of these values.

What do you think of my costume? It kind of brings postapocalyptic chic and space suit together in a kind of alien way, but in an optimistic alien way—in black. When I was looking for space suits, there was one that was more traditionally white. And it would have meant wearing a bunch of white shit. And I thought that might have detracted from the overall costume. The white suit was nice, but I just didn't want to break out the white.

Okay, we're going to do this next part illegally and drive the wrong way. Most people are confused when you do illegal things.

You know how I lost my license? Back when Hillary Clinton was running for president, there was this fundraiser, and I was with a friend of mine. We were heading down Interstate 280, and I don't generally speed, well, I guess a little bit, but I don't go excessively fast. But my friend was really stressed that we were going to be late. So I sped up. I pushed the gas, and while this isn't a Tesla or anything, it's like a go-kart with a 250-horsepower engine. If you drop it to fourth when you're doing 80, you can easily be doing 120 in a few seconds.

I did that just to fuck with her, and the moment I gunned it and got up to speed, she is like "There's a police officer."

Generally, my way of dealing with this is that I want to get out of it and get on with my life. If they're going to give me a ticket, they're going to give me a ticket. I don't try to make shit up. I don't try to make excuses.

But my friend tells the officer, "Oh, we're going to a Hillary Clinton fundraiser at Sergey Brin's house." I'm thinking, "No. No. No. Everything about that is wrong." This is a state trooper. Anyway, I proceed to get every ticket there is. Reckless driving. All that. It was bad.

I hired a lawyer, and they actually got me off, which was amazing. I had to do some sort of driving school thing.

But I only check my mail twice a year. I get my mail scanned, and it comes to me in PDF form. Often I forget to read it, and most mail is no good anyway. I just don't like mail.

The flaw in the plan is that a six-month cycle is generally enough to catch most things. But the decision-to-action cycle of the court system was just inside of that window. I was told to do the driving school thing and then didn't do the driving school thing and then I was informed I had to go back to court. I missed all of that. At that point the lawyer had stopped dealing with it, so it stopped getting dealt with.

Eventually I went and looked at my mail, and my insurance had been canceled, and my registration had been canceled. It was this horrible series of things where "You need to do this or we will do that." I don't think there was a warrant for my arrest, but I think that was going to be the next step.

I wanted to deal with this, obviously, so I called the DMV. The California DMV is one of the least productive organizations I've ever seen. You go in at eight a.m., and there is a three-hour line, and then you get in line to get in line, and even then you might not get to talk to them. So I did that. They said, "You need this and that and this, and then you can come back." It was ridiculous. I don't have time for this.

So I figured out all the things I need to do. You have to send them checks, noting this and noting that. I don't have checks. My bank doesn't support checks. It's this new kind of bank that doesn't believe in checks. It seemed great when I first signed up. Fuck checks. Well, the DMV needs checks. It's the only thing that needs checks. So I had to generate a check and send it to the DMV, and the DMV took weeks and weeks and weeks to process it, and then they didn't process it to the right thing.

Long story short, I am now at a point where I believe my license has been reinstated, although it's really hard to confirm because you have to wait hours on the phone just to talk to somebody.

The reason why we can do something like building a rocket is because you don't get distracted, and you can stay in the zone. Things like the DMV take you out of the zone in a really profound way. It immerses you in a part of society that is so broken. You start thinking "How can I fix this?" You start down that path, and it's terrible.

Rockets are easy compared to the DMV.

CHAPTER TWENTY

YOUR FRIENDLY NEIGHBORHOOD FOG MONSTER

Astra had gone from seven people to about seventy over the past year. It was December 2017, and according to the original countdown clock, the company was supposed to have a few people at a launchpad somewhere ready to witness the first flight of its rocket. While that was not going to happen, the company had reached its next major milestone: engineers had assembled the entire rocket and were ready to take it outside and conduct the next battery of tests. For the first time, outsiders would have a chance to catch a glimpse of what Astra had been up to because the next round of tests required the rocket to go from horizontal to upright. Anyone paying attention would see a missile poking out above the Orion Street building fence.

Chris Thompson, the old SpaceX hand, had officially joined Astra as one of its top executives and brought with him experience, a hardened demeanor, and hopefully some much-needed adult supervision.* The once empty high bay was now full of machines and people. A new rhythm had come to Astra. People would arrive early in the morning and coalesce around the rocket body like doctors assessing a patient that had been stabilized but would soon face a difficult procedure. After that, they would head off to their desks in the middle room of the building and busy themselves until midday. Then it was back to the rocket, but this time with serious energy and purpose as they opened the patient up.

The company had gone through several skirmishes with city officials who were concerned about Astra's sometimes cavalier attitude toward rules and regulations. The Orion Street building had been broken into a couple of times, which had led to Kemp surrounding the facility with a razor-wire fence. When an inspector told Kemp he could not do that, he doubled the amount of razor wire. "They said again that we couldn't do this," Kemp said. "I said, 'North Korea.' They said, 'What?' I said, 'North Korea.'" The strategy worked, as the Alameda bureaucrat relented, deciding that he did not want to be the one responsible for letting part of the "axis of evil" peek at the United States' latest rocketry technology.

During one botched engine test, a chunk of the Orion Street roof caught on fire. Someone at Astra sliced into the roof on either side of the burn mark and then refinished the area to make everything look the same, so that inspectors would be none the wiser. The general approach was to stay several steps ahead of the inspectors. City officials would come by and inform Astra that the company needed clearance to work on a particular project. Astra would go ahead and do the project anyway. By the time the officials returned to either deny the work or okay it, they realized that Astra was onto something even grander. That left them exasperated, but there was little they could do. Alameda had bought into Kemp's vision and saw Astra as a major part of the area's revitalization and job growth. You

* The company had hoped to hire Tim Buzza, another key member of SpaceX's early Falcon 1 team, although it never managed to get him.

either let Astra get away with its behavior or shut the whole thing down.

On December 17, the operation began to move the rocket from its stand on the factory floor onto the mobile launcher unit. People would need to hoist the rocket up by the cranes hanging from the high bay ceiling and then slide the mobile launcher underneath the rocket and let the rocket down easy into the launcher's cradle. After that, the whole contraption would have to be wheeled outside the factory into the parking lot and around to the side of the Astra complex, where someone had poured a square of concrete about ten feet from the main building to mimic a launchpad. On top of the concrete sat a five-foot-high, black metal trapezoidal stand, and the goal was to maneuver the rocket from horizontal to vertical, lift it onto the stand, and then begin running experiments by pumping liquids and gases through the rocket's plumbing.

A nervous energy took over the high bay right from the get-go. People stared at the rocket and chattered, wondering aloud how the next who knows how many hours might play out. Chris Thompson recounted war stories from SpaceX's early days. He told a yarn about going on a trip with Elon Musk to Wisconsin and Musk being confused one morning by the best way to cook Pop-Tarts in the hotel breakfast area. "It was like watching the Nature Channel," Thompson said. "He's studying the metal packet. He puts them in horizontal and then rams his finger into the toaster to pull them out. He's screaming 'Fuck! It burns!' in the middle of the lobby." The engineers gathered around Thompson laughed, although regaling the troops with a tale about a very smart man struggling with proper Pop-Tart alignment ahead of the delicate rocket erection procedure felt inauspicious.

Much of the morning was spent peering at the rocket, looking up at the crane hanging from the ceiling, walking around the rocket, touching the rocket, drinking coffee near the rocket, and fiddling with spare parts on workbenches. The entire company had united to build that one object, and no one wanted to destroy it. They had not, however, pregamed a precise plan on how to handle and move their Precious. They were going to wing it, and all the talking and looking and touching of things was an unconscious shared tactic to delay the inevitable. That left a lot of time for people to tell me about the latest happenings at Astra.

By that point, Astra had settled on launching its first rocket from the Pacific Spaceport Complex on Kodiak Island in Alaska. For that to happen, the launch range officials in Alaska would need to believe that the rocket would not kill any Alaskans. One part of the process required Astra to send the officials software simulations of how its rocket might behave. There was an ideal scenario in which the rocket would fly perfectly through its predicted corridor and into space. There were also all the other scenarios in which the rocket would begin to veer off course, at which point one of the officials would hit a button on a computer and disable the rocket. Rockets typically carry an onboard explosive device and are blown up remotely when they start to do bad things. Since Astra's rocket was small and pretty well contained, it did not need a bomb. When the official hit the button, the rocket's valves and pumps would turn off, and the rocket would tumble back to Earth.

A company called Troy 7, based in Huntsville, Alabama, specializes in running these kinds of simulations for rockets and missiles. Its technicians can pull up a computer screen that shows the various possibilities for a rocket's flight, and the animation looks like fireworks bursting in a mushroom pattern. An operator up in Alaska will spend days practicing in front of just such an animation, hitting the kill button anytime the rocket goes out of its designated corridor. To track its flight with serious precision, the rocket has a military-grade GPS unit because most GPS chips are designed to shut down when they reach 59,000 feet or 1,200 miles per hour, which, I'm told, is a built-in safety mechanism that's supposed to stop terrorists and other ne'er-do-wells from being able to aim missiles effectively at targets with off-the-shelf technology.

Though regulators use the simulations and kill switches to try to avoid casualties, they're willing to take the risk of an *occasional* casualty. "There's this one hippie family living on an island near Kodiak," said Kemp. "The calculations have been run that it's okay to launch over them. Technically, they are statistically insignificant."* Part of the reason he wanted to launch from Hawaii or a barge was to avoid the nuisance of having to algorithmically assess the value of human life.

Some of the chatter going on around the rocket was less fanciful and more reflective of the journey. Vita Bruno, Astra's chief financial officer and a consistent voice of sanity, noted with joy that Astra had finally installed individual restrooms in the factory and added central heating. "Last year, we had the propane heaters, but things were starting to catch on fire," she said. All told, Kemp figured it had taken $20 million to get to the point where Astra had restrooms, heating, and a rocket.

Eventually, the Astra employees decided to stop talking about their rocket and get down to moving it. Someone pulled on a chain to open the roll-up door at the left side of the factory. A cool breeze rushed into the building and mixed with the epoxy-scented air wafting through the main work area. A couple of people outside hitched the rocket launcher to a white Ford Silverado pickup truck and backed the launcher to the edge of the opening. A group of people then unhitched the launcher and guided it the rest of the way through the door and into the building. The rocket hung overhead from a blue crane with yellow straps around its body and was slowly lowered down onto the launcher. A few yelps went out—"Slow down!" and "Fuck!"—but things mostly went along smoothly.

The thought had been to bring the Silverado back, reconnect it to the launcher, and drive the rocket out of the factory and around to the side test area. "Whoever is parked by the Pinball Annex needs to move their car!" yelled out one engineer, hoping to clear some extra space for the operation. The strategy, however, did not work so well. The first pull out of the garage went too fast and spooked London, who rushed over and placed his hand on the side of the machine like a protective papa. Then, after the truck managed to pull the rocket out

* It was unclear to me whether or not the family had been apprised of its status.

of the factory, it proved hard to turn with the added eight hundred pounds of rocket sitting on top of the two-thousand-pound launcher. People unhitched the launcher from the truck again and began pushing it by hand across the parking lot. Someone ran up with a bag of sand and began laying sand down in front of the launcher's wheels to cut down on friction.

That worked but only so well. The next step was to get a SkyTrak, which is a drivable machine that can be outfitted with different attachments like a forklift or cherry picker—or, that day, a forklift with a strap on the end. The Astra team attached the strap to one end of the launcher, hoisted that end about three feet into the air, and then dragged the assembly through the parking lot. "It doesn't matter where I go," said Chris Thompson. "There always seems to be a SkyTrak pulling a rocket. We did the exact same thing at SpaceX."

People would drive by now and again and stop their cars to try and process what they were seeing. The rocket did not have its pyramidal fairing on top to give it that rocket-looking shape. It was a large silver cylinder of metal that had duct tape in some places and cardboard in others to hold things in place. At best it appeared to be a mad scientist's experiment. At worst, it appeared to be a bomb, and people were hauling it casually through a parking lot with the aid of sandbags and a SkyTrak lift while they stopped to enjoy the occasional puff on a vape pen or a flock of geese flying overhead.

Even with the machine helping, moving the rocket and launcher took ages. The Astra team averaged about a hundred yards an hour as they passed first through the parking lot and then down the side of the Astra factory—past the razor wire, the porta potties, and the shipping containers—toward the makeshift test pad. "How many rocket scientists does it take to drag a rocket?" joked Kemp, who joined in on the big, slow push.

After much effort, the launcher was finally parked beside the trapezoidal stand. Engineers hooked up wires and pumps to the rocket's body. Someone hit a button on the launcher's control system, and hydraulic pumps kicked in to raise the rocket ever so tenderly.

"Watch your fingers!" yelled out a technician. It took only about a minute for the rocket to go from its side to straight up and—perfecto!—it sat atop the stand just as it was supposed to. About twenty people gathered around their baby to judge its posture.

"Is it straight?" someone asked. "I think it's leaning."

"No, it just looks that way because something is off with the building," someone else replied.

Even though it was well into the late afternoon, Kemp and Thompson wanted the crew to perform some tests ahead of the company Christmas party that evening. Thompson had the cherry picker attachment put onto the SkyTrak, got into it, and began to inspect the higher parts of the rocket to see how well it had survived the journey. In the background, a group of people stacked a couple shipping containers on top of each other to create a wall of sorts behind the rocket to block the houses in the neighborhood from seeing Astra's contraption. But there weren't enough containers to complete the job, and one-third of the rocket remained visible.*

Along with Thompson, Astra had recently hired a number of former SpaceX employees, and many of those people had taken over the testing operations. The old Ventions crew and others were left over at the side, watching and advising. Some measure of tension existed between the two groups, and it was easy to feel bad for the Ventions folks and early hires who had done the vast majority of the work on the rocket.

For this test, the Astra team wanted to pump liquid nitrogen through the rocket's plumbing and fuel tanks. It's less explosive than the liquid oxygen they would eventually use at an actual launch, and it's so cold that it's good at pushing things to their limits and cleaning out parts. Soon enough, the test preparations became a communal exercise with most able bodies contributing. It was the rocket science equivalent of jazz. Cranes were driven, wires were fixed, knobs were turned by whoever had the will and the know-how to perform the task. While the saga of getting the rocket up onto the

---

* "At some point the neighbors will want to know what's going on," Kemp said. "We do basically have an ICBM a hundred yards from their houses."

stand had had its comedic moments, this part of the show provided a reminder that the company had hired dozens of high-functioning, capable individuals who came together well as a team when needed.

By 6:00 p.m., some of the office workers were making their way to the retired aircraft carrier a few blocks away where the Christmas party, which included the employees' family members, had begun. The two dozen people by the rocket did not seem to care. They brought out a pair of floodlights and began their tests. In went the nitrogen, as the workers scattered to a semisafe distance and checked their laptops for readings coming from the rocket's sensors. Liquid nitrogen is so cold that it boils instantly upon touching the outside air, and as the rocket belched the nitrogen out, a two-hundred-yard-long white cloud filled the area around the factory, rose up over the fences, and drifted into the neighborhood. Surely some nearby kid peeked out of her living room window and went running to Mom and Dad to inquire about the encroaching fog monster.

The tests continued until Kemp finally stepped in and told the engineers they had to stop engineering. "It would send the wrong message to everyone to miss this time with their families," he said. At that moment, it occurred to the engineers that the rocket would have to sit outside all night uncovered. Thompson asked if there was a Home Depot nearby and sent a couple of people off to acquire tarps. After they returned, the Astra team lowered the rocket back to its horizontal position, and a few people climbed on top of it and spread out the tarps.

The stragglers ran back into the main Astra building and put on their evening wear. Kemp made one last check by the rocket and found Ben Brockert and Mike Judson still at it. "I appreciate your commitment, but there is an open bar on an aircraft carrier," Kemp said. Brockert and Judson were the junkiest of the rocket junkies, and both of them would have preferred to keep running tests until well past midnight rather than schmooze.

Kemp and I walked toward the aircraft carrier. He'd put on a formal tailcoat suit, and his girlfriend shivered next to him in a tiny pink dress. I asked Kemp what he'd been doing in his spare time.

"I'm working on a foundation to create a permanent human base on the moon that's self-sustainable," he said. "It's one of my hobbies." After that, he hit me with some life advice. "Everyone needs four things," he said. "Your purpose, some other person to share life with, yourself so that you're okay being on your own, and then friends and family. If any leg of the stool comes out, you're wrecked. The people who prioritize work too much sometimes end up fired and are just left with themselves. They're the ones that commit suicide."

And with that it was time to party.

CHAPTER TWENTY-ONE

NOT-SO-STEALTH SPACE

As 2018 rolled in, Chris Kemp had two major goals. One was to expand Astra's factory and start pushing toward its quest of building and launching a rocket every single day. The other was to get Astra's first rocket to Alaska as quickly as possible and launch it.

Astra had not been started from scratch. It had benefited from the years of work Ventions had done designing engines, electronics, and guidance and navigation systems. Although those designs had been for a smaller, different rocket, they had given the new company a head start. But even with that caveat, Astra had done something remarkable. History showed that it took six years to a decade to make real progress with a new vehicle and have much hope of launching it successfully. Astra had built what had the appearance of a viable rocket in a bit more than a year.

Kemp tried to set expectations low for that first rocket. Astra would take it to Alaska, launch it, and hope for the best. The company wanted to usher in a new model of rocket development in which it would build and test at a rapid clip. Its engineers would analyze the data from each launch, make some tweaks, and try again.

Where other rocket makers had strived for perfection, Astra would iterate. Kemp felt that approach would keep the energy at Astra high and help people see progress being made all the time rather than their having to wait years and years to see any action.

Most employees at Astra agreed with the philosophy in theory. Whether they had come from the software industry, the aerospace industry, or somewhere else, they liked the idea of trying a novel approach and proving that things could be done more quickly than others had assumed. As time went on, however, the people with more aerospace experience tended to push back against some of Kemp's demands. They warned him that it made more sense to keep testing and perfecting the rocket in Alameda rather than send it to Alaska in a rush. The bodies and tools were in Alameda. No one knew what the resources in Alaska would be like if something went wrong and needed to be fixed. But Kemp wanted to stick to the approach of getting the rocket ready and shipped and then seeing what happened.

Ever since Astra had moved into the Orion Street building, Kemp had been eyeing a property across the way at 1900 Skyhawk Street as the natural next step in the company's expansion. Building one or two rockets in Orion would be fine for now, but he knew he'd need a much better place to pump rockets out by the hundreds.

The Skyhawk building was a huge warehouse that had been abandoned for decades. Half of the building resembled a forgotten factory floor. It was empty, dirty, and decaying. The other half was shambolic. Vagrants and teenagers had clearly spent years working the property over. They'd bashed in every bit of glass, clubbed the walls, and ripped out the facility's infrastructure. Graffiti spray-painted on the walls said things such as "Drink piss & listen to dubstep," "Cat fucker," and "Don't forget: rate us on Yelp!" But most gruesomely, the first person to check on the Skyhawk building discovered a corpse attached to some machinery—obviously someone who had been trying to steal the last bits of copper in the facility, had hit a live wire, and had been left there to rot.

Kemp saw potential in the horrors. He gave me a tour of the

building, walking alongside Bryson Gentile, one of the former SpaceX hotshots who would be tasked with creating a state-of-the-art factory within it.

KEMP: This is a quarter of a million square feet. It's in really bad shape. They used to test the engines in the other building and then overhaul them in this building. They'd pull them off the planes and go to work on hundreds of engines, retest them, and put them back on airplanes. We call this place Skyhawk—that's primarily because it's on Skyhawk Street—and the other building we call Orion because it's on Orion Street. And these are all really cool names. We figured that whenever we take over a building, we'll just pick the coolest street name that's near it.

We first broke into this building about six months after we took over the Orion building. We realized that in order to build a rocket factory, you need a really big building because rockets are big.

Be careful. There's broken glass everywhere.

GENTILE: There's also birds that like to perch themselves inside here and make a mess.

KEMP: Bryson runs our production team and manufacturing team, and we need to put a production line for rockets somewhere.

GENTILE: I get the epic task of taking this building from this to a full-blown automotive-like rocket factory where we pump out a rocket a day.

There are some old areas here where they used to acid wash engines. They had to strip the paint away with the most caustic solutions you can find. We're not sure what we're going to do about that part. The most valuable section for us is this huge high bay. We have the overhead cranes, and we've got a huge amount of floor space. We can fit the rockets essentially side by side and kind of build them in a Ford assembly-line strategy.

KEMP: Astra has to make a rocket a day, and if we are going to do that affordably, we have to have really inexpensive facilities. This would be the perfect set for a zombie apocalypse movie. There are a few holes that you need to watch out for here.

GENTILE: Person-sized holes.

KEMP: This building hasn't been used in twenty-five years. It took them twenty-five years to clean up the ground under the nastiest rooms. Which is one of the reasons it's available. We had to wait until the navy tested the ground and ensured there wasn't a giant plume of solvents and jet fuel under the building.

This is actually lead paint.

Kemp picked up a piece of lead paint, put it into his mouth, bit it in half, and then spat it out.

Still, you have to ask how does a start-up that had ten people a year ago convince the city to turn over a quarter-of-a-million-square-foot building and ten acres of land? What we did was hire the world's leading architect, Bjarke Ingels, who had just completed a project to imagine a new campus for Google. We didn't have any money to pay him, but we said we were going to reuse an old building and make a rocket factory, and Bjarke Ingels got superexcited and came out here and met with us to figure out how to create a hybrid factory and office building.

The idea is that we're going to build these pods that will circle outwards from Bryson's production line. It'll look kind of like Burning Man where you have all these camps that come out in concentric circles. You can kind of imagine domes surrounded by a big rocket production line. That's what this will look like in about a year.*

GENTILE: I come from a background that's a mix of automotive and aerospace. I spent a number of years at SpaceX and led the manufacturing engineering team and basically built up that assembly line. Rockets are not supercomplicated. If it looks

* Bjarke Ingels actually never built that Burning Man–style office. Astra simply put groups of regular desks off to the side of the factory floor.

complicated, it's because you haven't broken it down into enough chunks yet. The key to building a rocket is breaking it into tiny little chunks.

When you start with a brand-new rocket, you have on the order of a hundred thousand parts, if you include things down to tiny little fasteners. You have to cut that down into half and half again and half again and half again.

Really, what you want to do is optimize the mass of the vehicle. That's what you care about. Get the thing to be light and get it to be easy to produce. If you look back at the auto industry of the seventies, that's probably where the space industry is right now on the metal side of things, with the exception of some advanced welding techniques.

We're going to take the stuff developed years ago for automotive and apply it to rockets. And we're going to marry state-of-the-art automotive technology like robots with the state of the art for rockets.

KEMP: We're going to do this ultrafast iteration on a rocket. We're going to develop a new version of the rocket every six months. We're going to collect so much data by doing so many launches and having so many rockets, and we'll be able to feed all of that information back to the designers and then have better rockets.

We got this building for free for a couple years, and that allows us to think big and think big with the right tools. We want to get to a point where we build a rocket a day here, roll it out to the pier, put it on a barge, and launch it at sea.

I think that I try to see opportunities. Things don't just happen on their own. You tend to need to bring a lot of energy together in order to catalyze something and make something happen that's magic.

Just this morning, I'm talking to one of the people that is very involved in the Burning Man organization. And I'm thinking about this huge facility that we're soon going to have at Skyhawk. We're not going to use a lot of it. Artists need places to keep their sculptures here in the Bay Area. And so we're also going to find a way to build a Burning Man gallery.

Or, like, when we wanted a place to test rockets on the airfield, I said to the city, "We're going to launch rockets here." Which is insane. Obviously, we're not going to launch our rocket twenty minutes by car from downtown San Francisco or five seconds from downtown San Francisco by rocket. But if I change it and say, "Well, how about we just test it for a few seconds? You know, we'll hold it down. We won't launch it." They're like "Okay. All right." But if I had gone in asking to test a rocket, they would have said, "That's crazy. No."

It's the same thing with the art. I'm going to go in and say, "Hey, we've got this architect, and we have this vision for this campus, and we want to support artists." I'm not going to say, "Burning Man." I'm going to say, "artists." We'll say, "We have some of the world's leading sculpture artists that are going to place some of their work here. We're going to create a public space for this, and we'd like the city to allow us to use some of the space that we're not occupying to support this. We'll clean the space up. And by the way, we're going to take those soccer fields out there. We're going to put a road through them, and we're going to put an entrance, and we're going to build all these things."

They're going to be like "A road? Through our soccer fields? What? Children play on those soccer fields. You can't do that. But the artists are okay. The artists are fine." Is it a hustle? Yes. But does it make incredible things happen? Of course.

In February 2018, Astra reached the point where it could no longer keep its rocket partially hidden behind the Orion Street fences. As Kemp said, the company needed to perform a static fire test of the rocket, in which the machine is held down while its engines burn. This type of operation usually takes place in a desert or in middle-of-nowhere Texas because it involves a lot of fire and could involve an explosion. Astra, though, would do it at Nimitz Air Field, which is about one and a half miles away from its headquarters and next to the Hangar 1 vodka distillery, a fitness club, and some restaurants.

You can think of the airfield as a very, very large parking lot that abuts San Francisco Bay on the northeastern edge of Alameda. Though the navy once used it for all sorts of missions, the airfield had been neglected for many years like the rest of the base and used in more recent times mostly for weird one-off projects. Parts of *The Matrix*, for example, were filmed there, and so, too, were experiments done by the MythBusters. I'd been to Nimitz on a couple of previous occasions to take part in experimental runs of self-driving cars, during which I had strapped myself into a robot going sixty miles per hour.

Adam London and Chris Thompson led much of the static fire test. Surprising no one, the dragging of the mobile launcher and the rocket out to the tarmac was as much of an experiment as the test itself and took a number of hours. Everyone realized that the Stealth Space days would well and truly be over from that point on, although Kemp hoped that not too many people would notice the rocket churning out fire.

> **LONDON:** We are going through the final activities for getting the launcher positioned and preparing to lift the rocket up into it for the first time here at Nimitz. We need to get some things bolted down and level. There's a lot of tick-mark lists of stuff that needs to get done.
>
> The whole goal is to go through a static fire where we light the engines while the rocket stays held in place. All possible outcomes will be spectacular. Some of them will be better for the rocket than others.
>
> On the one hand, we could have a very clean test. All five engines could turn on and run for five seconds. It's pretty likely, though, that one or more of the engines may not start. In which case, we would probably turn off the ones that did start. Abort them all.
>
> Then, all the way at the other end of the spectrum, something could maybe happen to a battery, and we could start a battery fire,

> and we could burn the whole rocket down and make a nice big explosion.
>
> We are accepting that one of the possible spectacular outcomes tomorrow is that we don't have to go to Alaska and launch Rocket 1 because there is no longer a Rocket 1. But, uh, I would much rather do things this way, so we'll see what happens.

It took several days for the static fire to work as Astra hoped. And when the engines did ignite for two seconds, they ended up causing damage to the rocket. The fire bounced up off the ground and came back at the machine, as did chunks of asphalt and grime. On top of that, a couple of the engines cut out early and vented their fuel as an automatic safety precaution. The fuel ignited and created more fire all around the rocket body. To mitigate those issues during the following tests, Astra's engineers placed some fire-retardant material around the engines. Though the rocket sort of looked as though it were wearing a diaper, the strategy proved effective, and the company could carry on with its experiments.

Many of the employees, however, were worried that the debris and fire had likely gone up into the rocket and taken a toll on the internal wiring and various components. They were in a rush to get the rocket to Alaska, though, and a debate began as to whether it was best to keep on pushing and hope things would be okay or rip parts of the rocket open to see what might have happened inside. One engineer said, "I think the question is how deep do we look," which is not really what you want to hear when evaluating the health of a large bomb.

A growing contingent of employees thought that shipping the rocket to Alaska was a pointless exercise. The vehicle had been built too quickly and had too many flaws to work. It would be better to keep the rocket in Alameda and treat it as a test unit while work began on a second rocket that would incorporate much of what Astra's engineers had already learned. Kemp and London, though, both wanted to see Rocket 1 fly and argued that going through the launch exercise would be a valuable learning experience in and of itself. And who knew, it might even work.

In February 2018, near the end of the testing process, a local traffic reporter spotted the rocket at Nimitz. He was in a helicopter and asked the pilot to hover near the test site so he could try to figure out what was going on. Shockingly, Astra had been out at Nimitz for days, and none of the locals had paid any attention to its activities. It was the presence of the helicopter that raised their awareness. "I heard helicopters, and when I look behind me, I see a giant truck with a huge missile on it," a staffer at a nearby beer garden told the ABC news affiliate.

As Astra hauled the launcher and rocket back into its factory, an ABC news van pulled up, and out hopped a reporter and cameraman.* Kemp urged his employees to push the rocket into the factory as quickly as possible while he began walking toward the news crew. He informed the reporter that the thing that looked like a rocket was probably not a rocket and that any mention of the nonrocket might jeopardize an important national security operation. "Don't get all lawyerly on me," the reporter replied. Later that evening, the rocket appeared on the local news along with a story on the ABC website with the headline "SKY7 Spots Stealthy Space Startup Testing Its Rocket in Alameda."

Back at the factory, the Astra team created a plan to clean up the rocket as best they could and perform some visual inspections on the charred parts. "It's good and seasoned now," Gentile said as solvents were applied to the machine's body. People laughed about the news crew and Kemp's "these are not the droids you're looking for" schtick. Mostly, though, people feared that the Alameda residents would complain to the city now that many of them knew about Astra. Kemp told his employees that they had ten days to do the best they could with the rocket, then get it crated up and on a boat to Alaska. Mike Judson, one of the original Ventions team, pulled a few people together in a huddle. They placed their hands in the middle like a sports team, and Judson yelled out, "Clean up crew on three! One, two, three!"

* I was standing next to Kemp and could not wait to see how he would deal with the situation.

The afternoon turned into evening and then into night. Most of the people still hanging around cleaning the rocket had worked twenty-one straight days without a break. Their energy amazed me. The test had been meant to give them confidence in their creation. In many ways, though, it had made things worse and caused them to doubt the machine as much as ever. Still, they soldiered on and displayed their ability to face new problems and come up with all kinds of solutions for them. When a couple of younger people finally became tired and griped aloud, Ben Brockert yelled, "If you don't want to be here, go home. But don't sit here and complain. I don't want to be here, either."

"It's going to be an interesting ten days," Judson said.

CHAPTER TWENTY-TWO

NORTHERN EXPOSURE

I was on a small plane heading from Anchorage to Kodiak Island and had settled in for takeoff when the last five passengers boarded and made their way down the aisle. They stood out. The guys were muscle-bound and had clearly spent considerable time in tanning booths. They wore tight clothes and seemed like a variety pack of romance novel characters. There was a young blond guy with frosted tips and an older, more weathered version of him. There was a Black dude with earrings and a brunette with a ponytail. One guy with dark hair seemed to play the role of aspiring rocker, or at least that was where my mind went as it tried to process the men and come up with an explanation for why they were on the plane.

About thirty minutes into the flight, after the drinks cart had passed and the boys had ordered a round of beers, a bearded guy a couple of rows behind me blurted out, "Ohhhhh. You're the strippers!" Passengers all around us giggled and nodded as they collectively shared in a mystery being solved.

For historical reasons that remain unclear, it turned out that every March a strip club promoter in Cincinnati, Ohio, sent a few of his finest specimens to Alaska for a statewide tour. Our lads were on their way to perform at Mecca Bar, one of three bars located next to one another in Kodiak's tiny downtown. Since Kodiak does not have much going on in the way of local entertainment, both women and men would be showing up later in the evening at Mecca to see the strippers at work. Just about all the passengers on the plane looked forward to the annual tradition.

After landing on Kodiak, it became more obvious to me why the locals were so giddy about their stripper delivery. The island is just south of Alaska and has a population of about 14,000 people spread across 3,600 square miles. It's beautiful in all the ways that Alaska is beautiful—and very, very remote. If you're into hunting, fishing, and exploring the great outdoors, Kodiak has plenty to offer. Beyond those activities, you're pretty much out of luck. So you work and drink and then celebrate when some orange-hued twenty-five-year-old comes to town to take off his pants.

Since it's so remote, Kodiak is a handy place to have a rocket launch site, which was why the US government opened the Pacific Spaceport Complex in 1998. The spaceport was placed at the southeastern tip of Kodiak, where rockets could blast off and head over the Pacific Ocean without putting many humans at risk.

The Pacific Spaceport Complex has not been an especially active launch site. Before Astra came to town, about twenty rockets had been flown from there in twenty years. Most of them had been military exercises in which a missile fired from Kodiak was sent out over the ocean to see if it could be intercepted by another missile launched thousands of miles away from Kwajalein Atoll. In 2014, the military tried to launch an experimental weapon, and something went wrong with the rocket, causing it to veer off course. A safety officer hit a button and blew the rocket up just four seconds into its flight, which resulted in a major explosion that destroyed much of the spaceport. Nothing flew again from Kodiak until 2017, when the military conducted a couple of secretive missions.

Many people in Kodiak had expected more from the spaceport. They wanted rockets, and all the people they bring with them, to boost the local economy. Better still if some of the rockets were made by private companies and not just the government.

By 2018, it seemed as though the Pacific Spaceport Complex was finally about to have its moment. Many rocket start-ups had been eyeing the launch site as a primary option for their vehicles, and with good reason. The main US rocket-launching hubs in California and Florida were dominated by the military, NASA, and SpaceX, and their rockets took precedence over those of young, unproven companies. Kodiak provided a similar infrastructure, and the people running it were more willing to go out of their way to help companies such as Astra get going and to tolerate the start-ups' mistakes and delays. During my visit to Kodiak, at least three rocket companies were taking tours of the complex at the same time and trying to ingratiate themselves with the locals.

Throughout early 2018, Astra had been sending small groups of employees to Kodiak to begin its launch campaign. They had to do the basics, like meeting all the right people at the Pacific Spaceport Complex and finding a place for a large contingent of Astra staff to live, and, most important, they would eventually pick up the shipping containers with the rocket, the mission control center, and the other equipment and transport it all to the launch site.

The officials at the Pacific Spaceport Complex had explicitly asked Astra not to send its rocket up to Alaska so soon. They claimed to be upgrading classified areas of the site for top secret military work and did not want Astra people getting in the way. Adam London, however, had made the decision to ship the rocket up north anyway, hoping to force the issue and keep things moving. That turned out to be the right call. The spaceport employees were used to working at government speed and had not actually done things such as pour the concrete for some of the facilities Astra would be using, giving the appearance that they might have been trying to buy time with their "top secret" warnings. When people from Astra began showing up at the facility, it kicked everyone into a higher state of action.

The drive from the city of Kodiak to the Pacific Spaceport

Complex takes about ninety minutes. In March 2018, that meant guiding your car through snowstorms and slush-covered roads with occasional stops added in for cows blocking the way. The scenery was mostly vast grasslands surrounded by mountains and the gray waters of the Gulf of Alaska pounding against the shore. Eventually, the gated entrance to the spaceport appeared and opened up to provide access to 3,700 acres of land and facilities dedicated to putting rockets into orbit.

The complex consisted of seven main buildings connected via a central road. The northern edge had a mission control facility, and a few miles to the south were the main launchpad and a couple of large buildings where customers could work on their rockets and satellites under cover.

Astra barged right into the Pacific Spaceport Complex with its start-up gusto. The company put its rocket into the largest covered structure. It also installed its portable mission control center a few hundred yards from the large mission control center owned by the spaceport. The contrast between the two facilities was amusing. The spaceport officials, engineers, and technicians worked in a nice, bland, rectangular tribute to all things bureaucratic. The Astra employees, meanwhile, worked in pimped-out black shipping containers. The bottom shipping container had a faux wooden floor with faux wooden Ikea-style desks arranged around its outer edges, providing about nine workstations with two monitors at each one. Eight large TV screens placed high on the walls displayed videos of the rocket from various angles and other information about the weather and the rocket's health. Whiteboards covered every other available inch of wall space. Near the container's main door was a mudroom containing essentials such as chocolate chip cookies, left-over Ethernet cables, and a snow shovel.

A metal stairway led up to another shipping container placed on top of the mission control room that served as a relaxation room and viewing center for the launch. There were a pair of white leather couches and a few egg-shaped white leather seats and a white table and a white fridge and white cabinets and walls that were whiteboards. It was as if Steve Jobs had paid a visit to the Alaskan wilderness and built a one-room chamber of whiteness where he could feel safe.

The Astra workers rarely lingered in the relaxation room. They would pop up to grab a cup of coffee or some snacks and take a breather as they looked out the glass door and onto their gorgeous surroundings. Most of the time, they were stuck at their desks in the mission control room, which would become muggy and warm from all the bodies and electronic equipment.

The folks who were not stuck in mission control were working on the rocket inside the large hangar facility. It was more or less a spacious garage with a fifty-foot-high ceiling from which a crane hung overhead. The rocket rested in a horizontal position on its launcher over to one side of the building. Engineers had hooked several cables up to it for diagnostics and power, while tubes stretched from the rocket to fueling stations outdoors. Lacking insulation, the hangar stayed frigid, and people went about their jobs dressed in jackets and gave off cloudy puffs of breath as they exhaled.

As evening approached, the Astra team would make their way to a lodge they'd rented a few miles from the spaceport. An enterprising local had shipped the sixty-room, prefab building to Kodiak on a barge after the spaceport had been built, hoping to profit from people needing to stay for weeks at a time during launch campaigns. Since the spaceport had not seen that much action over the decades, the gamble had not paid off in full.

The Kodiak Narrow Cape Lodge had its charms. It was a two-story, roomy building right at the water's edge. A grand main room on the second floor had large windows that looked out to the water on one side and out to farms and mountains on the other. Now and again, you could spot whales swimming and spouting in the distance. The same room had several large wooden tables where people could gather family style, a Ping-Pong table, a pool table, and a couple of couches in front of a TV. The walls were decorated with a few animal skulls and photographs of people who had stayed at the lodge, hung up in haphazardly spaced clusters.

The rooms were less inspired. They were small, spartan, and

reminiscent of a budget hotel or dormitory. The dining hall was also more institutional than rustic. It had a couple of large tables and a buffet setup to which a pair of cooks delivered three meals a day.

The whole Astra ethos had been to build a rocket superquick and launch it with as few people as possible. Almost all of the Old Space tendencies to test and test and test had been rejected because they were Old Space, and Astra was to be New Space. But that belief system came with costs as the Alaskan adventure continued.

In the race to get the rocket to Alaska, Astra's engineers had not actually had a chance to finish building the rocket. That was bad. In fact, it was very bad, and Kemp had been warned that it would likely result in much suffering and dire consequences. Nonetheless, the show went on.

The rocket had been packed into shipping crates and sent off without passing all of the testing on many of its internal components or the completion of much of its software. A small group of Astra employees had been dealing with the consequences of the incomplete rocket for a couple of weeks. They'd been putting in long days in the rocket hangar fiddling with the machine's insides. Astra employees had taken to gifting the guy at the Kodiak shipping dock cases of wine in the hopes that their part orders would be delivered to the spaceport more quickly. They'd also figured out ways to trick the spaceport officials about the hours they were clocking to get around bothersome health and safety requirements.

Astra kept sending people to Alaska throughout March to deal with the issues that were cropping up. By the middle of the month, there were about two dozen bodies. Adam London and Chris Thompson were there, along with surly Ben Brockert and rocket-worshipping Mike Judson. So, too, were Roger Carlson, a veteran of SpaceX, and Jessy Kate Schingler, who had joined Astra to write software for the rocket.

Raised in Toronto, Jessy Kate had earned an astrophysics degree from Queen's University and then, at Pete Worden's urging, a master's in computer science from the Naval Postgraduate School. At Ames, she'd helped Chris Kemp get his open-source cloud computing effort off the ground and organize the massive annual space raves. Right before starting at Astra, she had focused on building out the network of communal houses throughout the Bay Area and overseas, while her husband, Robbie, worked alongside Will Marshall to build Planet Labs.

Kemp had brought Jessy Kate to the rocket venture in the hope that she could apply her coding skills to the machine. At first, she had specialized in creating systems to pull data from the rocket to check on its overall health and performance. After doing that for a few months, she moved to the avionics team to help with the guidance of the rocket. In Alaska, she found herself under constant pressure to try and code crucial bits of software on the fly.

What had started out as a sprint to hurl a rocket into space morphed into an open-ended science project conducted in the wilderness. It took until March 26, a Monday, for a firm launch plan to come together. Astra's employees would conduct tests through the early part of the week, take Wednesday off so that the team in Alameda could catch up on software development, do more tests on Thursday with Friday as a backup, do a formal dress rehearsal of the launch on Saturday, review everything one more time on Sunday, and launch on Monday, April 2.

The poor shape of the rocket led to much consternation. The people at the Pacific Spaceport Complex had been scarred by the explosion that had taken place in 2014 and did not want a repeat performance. Some of the spaceport officials viewed Astra as an amateur operation. They tried to look over the engineers' shoulders as they went about fixing the rocket and asked many questions that were viewed as unnecessary and annoying. As much as the people at the complex wanted to be part of the New Space movement and the money it could bring, they could not give up some of their traditional approaches. "They want to do the launch sequence the way they've done it for fifty years," said Brockert. "They already think I'm an idiot." At some point during the week leading up to the launch attempt, Brockert openly revolted, tweeting "I am entirely convinced that it is a mistake for any small commercial launch company to launch from any existing orbital range in the US." His opinion was not well received by the spaceport employees or the Defense Department officials who happened upon it.

As the days and weeks dragged on, Astra's budget for the launch expanded in ways that horrified Kemp. The company had to pay for mission control and safety specialists that the spaceport requested to fly to Alaska and hang around until the rocket was ready to fly. Their presence and services cost on the order of tens of thousands of dollars per day. The owner of the lodge was accustomed to putting up a hundred or so people and charging government contractors high rates for their multiweek stays. Since Astra had fewer people, he tried to make up some of the difference by charging even higher rates—someone said $270 per person per night. To bring LOX and helium to Kodiak, the spaceport wanted to charge $50,000, which forced Carlson to call in a favor with an old supplier friend in Texas who could ship the items cheaper but not that much cheaper. And then there were all the flights back and forth to Kodiak to bring up someone with a special skill or fly in a special tool.

Even before the big first launch attempt, a few of the Astra hands were ready to tap out. They'd been in Alaska for about six weeks. On their midweek day off, a few of them went into town to get drunk at Mecca and its neighboring bars. Brockert and a couple other folks had acquired shotguns and went skeet shooting on the beach. (Brockert was the best shot.) Issac Kelly, a tech whiz who had worked with Kemp back in the OpenStack days, got a massage. Carlson went for a swim in the freezing ocean and then flew his drone.

Many of the other employees remained at the lodge to keep working. As on most days, they talked about aerospace nonstop—not really because they were in the midst of trying to launch a rocket but because they were aerospace obsessives. As in, they complained about all the hours they were putting in and whined about their temperamental rocket but could not refrain from discussing approaches to colonizing Mars over dinner or swapping war stories from stints at Blue Origin or SpaceX over beers. It was during some of these conversations that it became clear that even the true believers—the real space addicts—didn't know if their rocket made any rational economic sense. They figured that Astra could turn a significant profit only if it cost $300,000 to make and launch the rocket. They doubted that such a figure could be achieved. But they all agreed to carry on nonetheless.

The days kept going by, and when April 2 arrived, Astra was in no position to launch its rocket. The entire crew had spent the week engaging in what might be described as launch theater. Each morning, eight to ten people would turn up at mission control, a half dozen or so would go to the hangar, and the rest would stay at the lodge. They would begin the day hopeful that things would somehow click into place and perform their duties as if the rocket were a quick fix from blasting off. But as the engine tests, plumbing tests, and communications tests continued, something new would go wrong, and a fresh cycle of troubleshooting would begin.

Carlson, a tall, bald man with a surfer's calm affect, had worked on massive space projects such as the James Webb Space Telescope and SpaceX's Dragon capsule. He took on the role of the experienced, trusted hand guiding the day-to-day operations, and all of the problems filtered up to him and were managed by him. To his credit, he never lost his cool. He'd nod while people explained whatever new disaster the day had brought, digest the information, and then take a deep breath in and relax his shoulders. Carlson seemed to physically absorb the engineers' exasperation before exhaling a plan for fixing the latest issue.

When things really seemed bad, the problems would make their way up to London. He would listen to the issues, go totally silent for an uncomfortably long time, and then offer a possible solution. Almost the entire team shared in their respect for London and his intellect. During his long pauses, I imagined London's mind scouring every inch of the rocket, as he often seemed to enter a nerd fugue state where he tried to become one with the machine.

Brockert also commanded the respect of the people toiling both in mission control and down on the pad. He'd worked at a number of rocket start-ups and had the most experience with the New Space model of launching often, fixing the issues, and returning to the pad as quickly as possible. "I'd do anything for Ben," one engineer said. "If he told me to pull my pants down and hump the rocket, I'd do it because it must be important."

As I observed the Astra crew, it became clear that the lead-up to a rocket launch was not sexy or exhilarating; it looked a lot more like drudgery. Add in the remote location and the peculiarities of lodge life, and the drudgery seemed that much worse.

The lodge was off the grid and ran on generators fed by a massive fuel tank outside. So you would get back after a long day at the range and be haunted by a constant dull buzz. The soft-serve ice cream machine, which had seemed like a blessing early in the campaign, turned into pure evil because it buzzed ever louder and plagued every mealtime conversation. Even the spectacular scenery outside turned on the Astra employees after they went out one evening for a relaxing evening stroll and happened upon a rotting whale carcass being picked apart by eagles. While impressive in scale, the skeleton stank, and the gelatinous goo oozing out of its vertebrae somehow served as a reminder that life had taken an odd turn. Overall, the lodge, with its twitchy innkeeper and its mostly empty rooms, felt too much like the set of *The Shining*.

On April 3, the Astra crew had a group meeting in the all-white lounge at the launch site. It remained frigid outside, but all the bodies in the small space caused the temperature in the room to spike. Carlson informed the team that they had solved an issue with the engine igniters and another with the communications systems that let devices at the spaceport talk to and keep track of the rocket. Now, however, one of the gimbals used to adjust the position of an engine in midflight had started misbehaving. It moved slower than the gimbals on the other four engines, and people feared that could drive the rocket off course, since it could not react to commands to shift its position in time. No one wanted to rip the thing out and fiddle with it, which would eat up more days, so Astra began running thousands of software simulations to try to determine the most likely outcomes of flying a rocket with a gimpy gimbal.

None of the Astra folks had really believed that the flawed rocket would run for the few minutes needed to approach low Earth orbit, but they, especially Kemp, had hoped for something close. The difficulties of the past few weeks had lowered their expectations. During that meeting, people openly talked about thirty-five seconds of flight being a win. That would provide enough data to make adjustments for the next launch and leave people feeling that their work and flirtation with insanity had been worthwhile. Also, if thirty-five seconds passed, it would mean that the rocket had cleared the land and made it out over the water and that no one would have to witness a crash on land and pick up the pieces afterward. One engineer declared that the launch campaign had officially reached the FIFI stage: Fuck it, fly it.

As they continued to discuss operations, Chris Thompson chided the engineers who typically worked down by the rocket hangar. They'd been unresponsive on their radios a couple of times when Astra had performed tests with the rocket vertical in the launch position. "When we're on the pad, someone needs to be on comms," Thompson said. "There's no fucking excuse for that. Pick up your phone. If you hear someone looking for someone, answer. It's not that fucking hard. Thank you." Ian Garcia, who led much of the guidance and navigation work, complained as well because he'd been doing all the gimbal simulations and the calculations took a long time. "Sorry you have to work," said Ben Brockert.

Back at the lodge later that evening, the team pushed together a couple of the large wooden tables for another meeting. The tables filled with beer and a couple bottles of Jameson and Bulleit bourbon. In a rocket-scientist-versus-reality moment, Adam London spent ten minutes battling to open a bottle of wine after the cork broke. Carlson explained that everyone would be doing a launch dress rehearsal the next day that would be quite true to life. A helicopter would do a sweep of the ocean, pretending to make sure there were no boats in the way, and added security guards would turn up at the spaceport gates and pretend to secure things. Early in the morning, the Astra employees would plop down at their stations and begin a fake launch.

During the mock launch, Carlson explained, people in mission control would periodically put their hand into a hat and draw out a piece of paper with an issue on it—helium has been lost, voltage readings don't make sense—and force the team to troubleshoot the problem while on the clock. All of that would be done with an FAA official sitting in the mission control container and monitoring how Astra performed.

"This is not a joke," Carlson said. "We will always have FAA observers in the room from now on, and you have to treat it as seriously as possible. One of the procedures will be what happens if the rocket blows up three seconds after leaving the pad. What do you do? It's not 'Everyone runs away.' We lock down the data, and it becomes part of the asset record."

Brockert jumped in next, as he would be running the communications in the mission control. "If there is something that is a hazard, call it out. Anyone can call out anything at any time. Even with the FAA and the spaceport guys there and their occasional philosophical differences with me, we are going to do things my way for the most part because it's our rocket. As we get near the end of the countdown, you will have one last chance to yell, 'Don't launch the rocket!' It's preferable if you do this in English."

On the day of the dress rehearsal, Milton Keeter gave a quick speech to everyone in mission control to remind them once again of the stakes. Keeter was a gray-haired gentleman who had spent a decade in the rocket business focusing on safety around launches. For much of the past year, he'd been the point man connecting Astra, the FAA, and the Pacific Spaceport Complex, processing voluminous amounts of paperwork to obtain a license to launch the rocket. He was friendly and easygoing most of the time but firm when he had to either push back on his Astra colleagues because they were being cavalier or push back on the bureaucrats for being tight-asses. "We are way past the point of not being serious about what we are doing," he said. "I know this is boring and hard, but it's important. If the FAA sees something they don't like, they could pull our license and we wouldn't be able to launch."

When the rehearsal began, it was reminiscent of a game of Dungeons & Dragons with Keeter playing the role of the Dungeon Master. His hand went into a hat, and out came the disaster scenario—the monster—for people to face. The problems started off easy and progressed to the ultimate catastrophe: Something has gone wrong with the flight computer or the communications, and no one knows the location of the rocket. You try to disable the rocket with the flight termination system, but that doesn't work either. You're flying blind and totally screwed. Seconds pass. You hear an explosion. Your rocket has blown up. You failed. Carlson and Keeter, please report to the spaceport complex and receive your chiding. Everyone else stay right where you are. The spaceport is shutting off the cell tower and internet connections to block your communications with the outside world. Someone will arrive shortly to gather witness statements. If there's a loss of life, no one will be going anywhere for a long time.

Churning through all the practice runs and that last, very dire scenario took about six hours. Afterward, the Astra team dug back in to try to prepare the rocket for a launch attempt the next day. They spent another few hours going through their systems checks. A few issues cropped up, but there didn't seem to be anything that would block the rocket from flying. They would give it a go.

At a meeting back at the lodge that evening, Carlson, Thompson, and a couple of other folks called Chris Kemp for a videoconference. Kemp had been traveling all over the Middle East and elsewhere trying to raise more money for Astra. He made it clear that he was tired of waiting on the rocket. He also stressed that Astra's launch window would soon close. A few of the people at the spaceport were contractors who go from launch site to launch site on behalf of the government, and they were due to leave for a Rocket Lab launch in New Zealand in a couple of days. Once they left, Astra would have to wait weeks to try to launch again. That would hurt not only for the delays it would cause and the costs it would add but also because Rocket Lab would be launching a rival rocket and shoving it in

Astra's face. Peter Beck had exactly that mouthwatering scenario in mind when he had come up with the "It's Business Time" moniker for Rocket Lab's upcoming mission.

"This is the first day we set out and accomplished everything we wanted to do and more," Carlson said. "Had we actually tried to fly today, we would have flown. Some of the changes we could make could help but are scary."

"There's not a thing on that rocket that's not scary," Thompson said.

"The spaceport has already sent a couple of people off for Rocket Lab," Carlson said. "They said they will give us until the weekend if it seems like things are working. They said, 'Look, you guys have been debugging things forever. Sooner or later, you have to stop CPR on the patient.' That was their statement to us, but I think we've made it past that point over the last couple of days."

"Is it perfect? No," Thompson said. "Will it fly? Yes. There are a lot of miracles that will have to take place in a series, but I think overall we are in a good state. I have to commend everyone that is here that has held everything together."

"Everyone is kind of running on empty," Keeter added. "We can't keep running many more days at this pace. We are being judged by the FAA and others as well. There were some concerns from the FAA about Ben and his swearing and being unprofessional."

"Okay, so Saturday sounds like the last realistic shot at doing it," Kemp said. "We have a lot of investor calls on Friday. Four or five groups are interested. People are champing at the bit to come up and participate in the launch. I would love to launch on Friday and tell them, 'Sorry, we launched early,' which is an interesting way of positioning that. It would really strengthen our position significantly to see us get this one off before Rocket Lab does their next one.*

* Earlier in Astra's history, Kemp had dreamed of flying a group of potential investors up to Alaska to sit in the all-white lounge and then to kick off an auction as the launch countdown began. He hoped the excitement and pressure of the launch would create a frenzied environment with bids coming in right until the rocket went off. Given all the delays, it was good for Astra that it never happened.

"I have different communications plans ready depending on the outcome," he continued. "We have a crisis plan if something goes wrong. I will make some comments on our concerns and condolences for anyone affected and our focus on working with the authorities. No one will be thrown under the bus. We will take responsibility as a team.

"It's obviously been challenging and expensive to work at Kodiak. I'm excited for you guys to look at some different launch sites. Let's learn everything we can so that when we do this again in three or four months, we're that much better at it. Good luck, and let's get this thing in the air."

When Thursday, April 5, rolled around, the Astra engineers once again set about priming the rocket for launch and running through their various systems checks. Overnight, they had received updated weather information and come to realize that Saturday would likely be very rainy and windy and not really an option for a launch at all. Everything was riding on Friday.

For weeks, the pattern of operation in Alaska had been the same: People would leave their morning meeting full of optimism. They'd come up with a game plan to fix whatever issue had held them back the day before and were confident that the work could get done and the rocket would be ready to launch. Not too long into the day, however, some other problem would crop up, and the whole day's schedule would go to shit almost immediately. "It reminds me of being on Kwajalein with SpaceX," Carlson told me, "two steps forward and ten steps back."

The whole process had turned into a self-defeating daily ritual in which the employees were mocked by technology and physics. Even though the rocket had one main problem at any given time, the sum of all the issues appeared more as a comedy of errors. Each time the engineers tweaked one thing, it seemed to break something else.

That said, the Astra employees were dedicated and relentless in how they went about solving problems. They still woke up every morning and worked as hard as they could with purpose and belief. Every member of the crew had developed an intimate knowledge of the rocket and all of its nooks and crannies and tendencies. It was like a piece of art they'd studied for years and could recreate from memory. In the face of continuous challenges, the team had come together and proven indefatigable in their pursuit of launching.

AS THE DEADLINE APPROACHED EVER faster, Astra's engineers and technicians turned to their deities and tried to make any available bargains. The situation felt almost claustrophobic. The pressure of a ticking clock was stacked on top of the pressure of a rocket launch stacked on top of Chris Kemp's ambition and Adam London's palpable desire to join his idols as a real rocket man. The rocket, though, ignored the mountain of urgings directed toward it and refused to cooperate. It just would not go on time. Astra had to stand down and watch as the remaining spaceport personnel scooped up their things and headed to New Zealand for a date with Peter Beck.

Several of the Astra engineers also rushed to the airport the moment the launch campaign was paused. They'd had more than enough of being in Alaska and dealing with the stubborn machine. Kemp eventually told everyone to take a break and return home; Astra would regroup and try again in the near future.

Over the next couple of months, Astra did all the things it should have done earlier. Software engineers were given time to finish their coding rather than trying to send updates on the fly each night. A crew of engineers returned to Alaska, and the most troublesome components in the rocket were removed and replaced. By June, a new window had opened up to try to fly the rocket again, and a smaller group of employees flew off to Alaska to help complete the mission.

On July 20, Astra finally got to see what its rocket could do. The machine had been a pain over the past few weeks, per usual, but had started behaving more predictably during recent tests. The Astra employees were now feeling pretty good about their rocket. They weren't totally sure what it would do, but they were reasonably sure it would do something this time.

With fog surrounding the launch site, Astra's team went through their whole lengthy process in the mission control center. They'd been conditioned by now to expect abort commands to pop up on their computers, but the aborts never came. One "Go" followed another and another, until all of a sudden, the rocket was spewing fire out of all of its engines. The immovable object was moving and moving fast.

Some people were too shocked to feel much of anything. It almost didn't make sense that the rocket had really launched. Others went into a trance of concentration as if they were sending beams of encouragement from their minds and bodies straight into the heart of the machine. And for thirty glorious seconds, it seemed as though their yearnings were working. The rocket rose and started to soar.

The cheers were let out first in Alameda, where employees were watching on a video feed, and then followed in bursts in Alaska when the busy staffers could find a moment to celebrate. And then, just like that, the exhilaration ceased. The rocket did not wobble or slowly veer off course, as might happen with a minor issue. Instead it began to plummet and in a worst-case scenario headed right for the launchpad. A few seconds after it began the descent, it plowed into the ground and exploded, sending debris everywhere. Astra had bombed its own launch complex.

The launch sucked for a number of reasons. The rocket's journey had been so short that Astra struggled to obtain much useful data on its performance. Officials at the Pacific Spaceport Complex were also less than impressed. The Astra flight was the first launch by a private company from the site and hopefully the start of a booming business. Instead, they had to deal, once again, with a rocket exploding at their facility. Meanwhile, the Astra engineers who had spent months in Kodiak were forced to go down to the launchpad and pick up pieces of the rocket by hand.

Rather amazingly, Astra managed to keep the launch and the explosion mostly under wraps. People in Kodiak knew that a launch had taken place, but the rocket had never cleared the thick fog. A local reporter covering the event wrote that the result of the test had been "unclear." "Other than the fact that Rocket 1 launched, no one seems to know what happened next," wrote another reporter from the space trade press.

After a day or so went by, some officials were forced to acknowledge the rocket's existence for the first time and to explain some of what had happened. The FAA issued a statement saying that a "mishap" had taken place. The head of the Kodiak launch site told reporters that Astra was "very pleased" with how things had gone and left it at that. Kemp said nothing.

CHAPTER TWENTY-THREE

ROCKET 2

In an ideal world, Astra would have acquired tons of data from Rocket 1 and been able to make a series of engineering adjustments based on the information. But things back in Alameda were far from ideal. The first rocket had so many problems that no one really knew what had caused it to shut down so soon.

Kemp kept telling investors and anyone else who stopped by Astra's factory that the launch had been a success. He did not try to claim that the rocket had flown into orbit or anything close. But he put the rosiest possible spin on the launch and the quality of the rocket. Part of that was simply his nature as the ultimate optimist. The other part of it, obviously, was to keep the employees and observers believing in the Astra cause.

Down in New Zealand, Rocket Lab had been applying pressure on Astra. Rocket Lab's Electron rocket looked magnificent, and it had flown twice and reached orbit successfully on both occasions. Many people observing the aerospace industry wondered if there was enough business for one small-rocket maker, let alone two, three, or four. And whereas Rocket Lab had a perfectly engineered vehicle, Astra had a Frankenmachine that could well have blown up a large chunk of the Pacific Spaceport Complex. Luckily for Astra, Rocket Lab hit a snag in the middle of 2018 and had to delay its launches for a while as it searched for the root cause of a troublesome component. If Astra moved fast enough, maybe it could catch up.

After everyone came back from Alaska, Team Astra regrouped as best as it could. The first rocket had been flawed, but there was no time to overhaul its design in a major way. The engineers did their best to simplify some of the machine's wiring and make it easier to reach parts that tended to cause problems, such as the batteries and guidance mechanisms. The rocket's igniters had often malfunctioned in Alaska, and people worked to try to find the underlying cause. The team of software engineers were thankful to have more time to refine their code and test it on the machine.

Rocket Lab had announced that it would try to launch again in November, and Kemp wanted to beat its rival to the pad. Astra began sending people back to Alaska in September. This time they stayed in rented houses instead of the lodge to save money. Almost right from the start, many of the same problems that had plagued the project on the first go-around reappeared. September slipped into October as a core group of employees tried to make Rocket 2 work.

Each morning, the employees made the drive to the Pacific Spaceport Complex and tucked in for a long day of work. Roger Carlson, the vice president of launch, and Milton Keeter, the head of safety, licensing, and launch operations, were constant companions and often found themselves together in a truck talking rockets, explosions, and the space industry in general.

Other core employees who spent a lot of time in Alaska that year were Bill Gies, the punk rock electrician, and Kevin LeFevers, a young technician who had grown up racing cars on the Bonneville Salt Flats. They often hung out with Issac Kelly, the software and technology systems specialist, who had worked at Kemp's cloud computing start-up Nebula before joining Astra. Two other key members of the team were Chris Hofmann, who had come to Astra by way of SpaceX, where he'd worked on engines at the upper stage of the Falcon 9 rocket, and Matthew Flanagan, an engineer who had done a ton of work on the Astra engine test stands, the mobile launcher, and the rocket itself.

All of those men had lived through the ordeal of Rocket 1 and were learning the idiosyncrasies of Rocket 2. They'd been there from the very first moments that shipping containers started to arrive from California, through to Rocket 1's demise and then starting the process all over again. Their insights proved the best way to understand the experience of life in Alaska for Team Astra and the day-by-day experience of trying to get a rocket off the ground.

KELLY: We got to Alaska the very first time in the middle of February, and the only people there to get the shipping containers are me, Bill, a dude named Matt, and Milton. There were a lot of people that were worried the folks at the spaceport would think we were yahoos. Well, not a lot of people. But definitely Milton was worried.

It was snowing and seventeen degrees, and it was time to do outdoor work because we had not built an indoor space to do work yet. You work for an hour or two hours and then sit inside with a cup of tea and try to get your fingers to be useful again. You're out on this island, and people are telling all these stories of two-hundred-mile-per-hour winds knocking containers over the week before, and we're out there on forklifts trying to get our containers loaded on a truck. It was a lot of very physical construction work.

Bill was taken aback that there were bars where you could smoke. He was like a kid in a candy store at first, and then even he got tired of it.

GIES: It was me, Milton, Issac, and Matt Flanagan. It was a lot of manual labor and plugging things in and running power. There was a moment where Milton was driving a forklift with this massive steel staircase on it. Issac is trying to align the bolt holes. People have crowbars trying to pry things open. The guys back in California had bonded the containers together with this weird black sealant. Unreal amounts of it. Guess they figured they would pour it into any cracks they saw and cross their fingers.

A guy at the launch site kept telling us to weld even the big stuff down. I was like "Why? Some of this stuff weighs two tons." He's like "You'd be surprised what happens when the wind kicks up to seventy miles per hour. It'll kick one of these containers down the street. I've seen it happen. You're gonna want to weld that shit down."

Issac was a savior because he brought a video game. We also had the entirety of *Planet Earth*. Because of the weather and the length of the days up there, we would leave at sunrise and head back before sunset so no one was driving in the dark. It was Milton's call, and I appreciated that. No one knew the roads that well. There would be Alaskan traffic jams with herds of animals on the road.

They have bald eagles up there like we have pigeons. They're majestic until you see one dig a Big Mac out of a dumpster. They are garbage animals.

LEFEVERS: When we first got there, it was just beautiful. We're taking all these pictures and selfies. I'd been to Alaska before but not an island off Alaska. Then we start working, and it's still fun at first.

By the second week, things already started getting a little nutty. We're working twelve- to fourteen-hour days. Sometimes more. Pushing and pushing. The only place we ever see on Kodiak is the launch site. Two months of that became pretty difficult. There were so many times where it just didn't seem like it would happen.

KELLY: There's nothing to do in Kodiak. But it was cool. We got everything set up, and then, after a very long time, I got to push the actual button and launch the rocket.

There was a very satisfying twenty seconds when we let the rocket go. After that point, you only have a couple seconds to decompress from the mission control stress before you're like "Oh, it's coming back down."

An aerial view of NASA Ames in Mountain View, California. (*NASA*)

Pete Worden during his time as director of NASA Ames. (*NASA*)

The outside of the Rainbow Mansion. (*Will Marshall*)

Planet's founding team in the garage at the Rainbow Mansion. Will Marshall seated in center. Robbie Schingler seated on the left in front of Chris Boshuizen. (*Planet*)

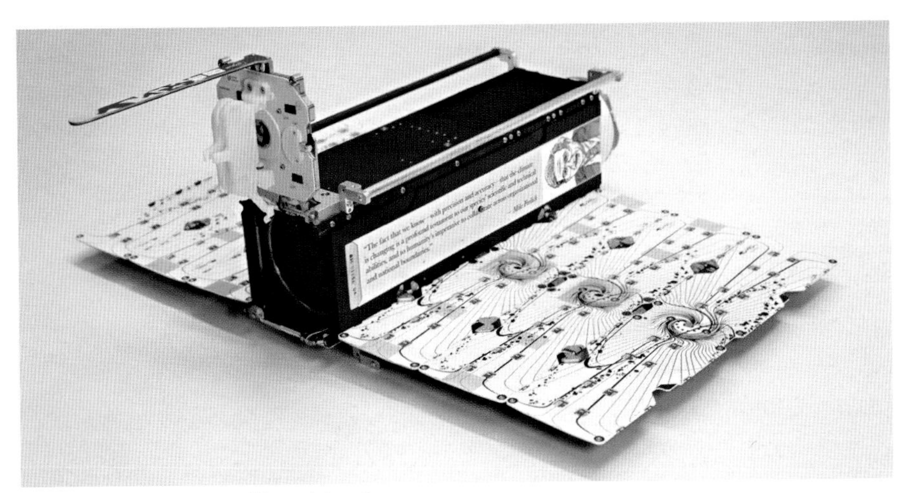

One of Planet's Dove satellites. (*Planet*)

A pair of Dove satellites begin their journey into orbit. (*Planet*)

Planet's Dove satellites spot patches of rainforest being illegally cut down in the Amazon. (*Planet*)

In the months before the war in Ukraine, Russia begins building up its military supplies along the border. (*Planet*)

A Planet image of one of the Chinese missile silos—aka bouncy houses of death—spotted by Decker Eveleth. (*Planet*)

Peter Beck on his rocket bike. (*Rocket Lab*)

Peter Beck during his rocket pilgrimage in the US. (*Rocket Lab*)

Rocket Lab's Electron rocket awaiting launch. (*Kieran Fanning*)

Rocket Lab's launch complex on the Māhia Peninsula in New Zealand. (*Rocket Lab*)

Rocket Lab's production line in New Zealand. (*Rocket Lab*)

Rocket Lab's Darth Vader–inspired Mission Control center in Auckland. (*Rocket Lab*)

Rocket Lab's CEO, Peter Beck, at the company's launch site in New Zealand. (*Rocket Lab*)

Chris Kemp and Adam London at a party held inside Astra's Orion building. *(Ashlee Vance)*

Ben Brockert working in one of Astra's engine test cells. *(Ashlee Vance)*

Chris Kemp and crew push their rocket toward its testing area at Astra's Alameda headquarters. (*Ashlee Vance*)

The sad state of Astra's Skyhawk facility when it was first acquired by the company. (*Ashlee Vance*)

Chris Kemp looking proud and confident as Skyhawk's revamp gets underway. (*Ashlee Vance*)

Jessy Kate Schingler coding a rocket in real time on Kodiak Island. (*Ashlee Vance*)

Team Astra contemplates life and rockets at their lodge in Alaska. (*Ashlee Vance*)

An Astra rocket waiting to launch at the Pacific Spaceport Complex in Kodiak, Alaska. (*Astra*)

Astra's technicians try to repair their rocket on the fly in Alaska. (*Ashlee Vance*)

Chris Hofmann, Astra's launch director, getting ready to Send It. (*Ashlee Vance*)

Astra's revamped Skyhawk facility. (*Jason Henry*)

The Alameda neighborhood near Astra's rocket factory. (*Jason Henry*)

Max Polyakov at his home office in Menlo Park, California. (*Ashlee Vance*)

A rocket in the parking lot of the Dnipro aerospace museum. (*Ashlee Vance*)

The rocket engine test facility in the forest of Dnipro. (*Ashlee Vance*)

Max having some fun while checking out the state of Firefly's rocket at Vandenberg Space Force Base. (*Ashlee Vance*)

Tom Markusic at Firefly's rocket factory in Texas. (*Kelsey McClellan*)

The first launch of Firefly's Alpha rocket from California. (*Firefly*)

A rocket test stand in the background on Max's Firefly Farm in Texas. (*Ashlee Vance*)

(*For more on the new space landscape, see the illustrations that appear after Chapter 27.*)

We couldn't see it, but we could see the telemetry numbers. It was going up, and Ben Brockert was calling out the acceleration numbers, and then it stopped, and then it started accelerating again, which meant it was going down. Ben was the first to notice.

When it hit, you could hear it and feel the impact from mission control. It was only about a mile away. There were two booms. It crossed the sound barrier. And then it hit the ground. Ben was like "We have impacted the earth. Well, it's your show now, Milton." I laughed.

LEFEVERS: I think we had like a total of six attempts or something. Every time, we'd clear out and then have to go back in and fix things. When it eventually went, we were all pretty shocked. We were so used to someone saying "All right, let's get back down there and get into it and turn everything off." This time something was a little bit different.

It takes off. And as soon as it released, I did not care anymore. I was just excited that it made it off the pad. And at that point it was just like "All right, I'm good." All I wanted to do is clear the launchpad. That was my goal from the very beginning with that rocket.

I was outside with my buddies, and we were laughing and crying like babies. We were having a good time, and then you hear this kind of whistling. The fog was pretty heavy, so it was really hard to make out the sound. We didn't see it. We were just listening because "Where the fuck is it? Are we all going to die or what? What's the deal?"

And then it hits the ground, and you hear a boom, and my buddy is like "Oh, it blew up!" I said, "Fuck, yeah, it did." Good riddance, get rid of that thing, you know? Time for number two.

KELLY: The next day there was sort of a camaraderie from the spaceport guys. They'd had a massive explosion before and spent ages picking up pieces. The old-timers were like "Yeah, this sucks for you."

It was really weird to see the people back at Alameda with all the jubilation and shouting and champagne and shit. Because in Alaska we were working like twenty minutes later. And it was like six of us. It was a very weird, lonely experience to launch the rocket and then pick up the pieces. There was certainly no euphoria. It didn't feel like we had achieved nearly as much as we wanted to.

LEFEVERS: We go back down, and it's in a million pieces. We had to find all the batteries, which we did. And find the helium tanks. We found one and some of the second one. We had to turn off all the propellant tanks and everything, which was pretty sketchy. I went first since I didn't have a family or anything, and I was trying to act supernonchalant about it. But, uh, it was a little scary. There's hydrogen spraying out of a tank near the pad. It looks like a war zone, and there's all this hissing.

It was nice to see it all in parts and not see it in one piece. We didn't want to take it back home in a shipping container. I'd rather see it in a million pieces than have to live with it in the work yard or some shit like that next door, and seeing it every day and being like "God, you stupid thing, I hate you."

When I first went out there, I thought one of the guys from the range was pissed. But they couldn't really say anything. It was like "Oh, way to go, guys." We were excited, so they didn't want to really rain on our parade.

KELLY: It took us so damned long to get to the launch. I was wrecked. There were many times where I thought about resigning. The reality is that the rocket was not finished. We didn't ship a rocket up there that could have launched. We spent twelve weeks in Alaska across three launch campaigns. It was supposed to be a three-hour tour.

LEFEVERS: Overall, it was really nice. It was really good. There's some cool stuff that you can do in life, but launching a rocket, I think, is probably one of the coolest things that you can do. Especially with a small team and group of people, and be there with them and celebrate.

You know, I took back a souvenir from Alaska after that launch. There was that whale carcass by the lodge, and we put one of its vertebrae in a shipping container and sent it back to California. So I get the thing, and it really stinks.

I take it home and put in the bathtub to try and clean it. I spray it with some bleach because I'm a genius, right? And, uh, it turns out the marrow section of the vertebra is like a sponge. This bone that weighed about five or six pounds ends up weighing forty pounds.

I left it in the bathtub for a few hours, and it still smells like shit. So I put it in like twenty trash bags because it's dripping water everywhere. And then it's summertime, and I figure I will leave it in my car and let the water evaporate. I poked a hole in the bag and left it in the car. I come back, and it's terrible. Rotting blubber. It's a stench like leaving fat on your kitchen counter until it gets moldy and then mixing that with trash. Since I'm stupid, it only occurred to me then to leave it outside. I think I didn't want anyone to steal it. I didn't want anyone to take my fucking whale bone.

I hid it around a corner at the factory. And it dried out. And now it's in my apartment. It's underneath my coffee table.

CARLSON: Now we're on to Rocket 2. We've been up here in Alaska a lot this year. Milton and I have been in Kodiak a third of this year.

KEETER: I think I'm 115 to 120 days now.

CARLSON: It's mid-October. We're two weeks into this stint. Instead of putting the rocket on a cargo ship, we flew it up here on a C-140 cargo plane. That got it here in two or three days instead of eight to ten days. We did a full dress rehearsal of the launch the other day.

We found a liquid oxygen leak and spent the last day and a half working through that leak, finding it and fixing it. And we're waiting for our launch license from the FAA to come through.

With Rocket 1, the goal was simply to fly it and get data, and we did that. The goal on Rocket 2 is to go for the full duration and have the second stage separate. We want to get through Max Q, which is the toughest part of the flight and will occur after about sixty-five seconds.

You're going up, and you start out going pretty slowly, and as you're going up, you're also going into thinner and thinner atmosphere. But there comes a point when you are flying very, very fast and there's still some atmosphere left, so there is a point somewhere in the flight where you've got the maximum dynamic pressure, the maximum buffeting, the maximum turbulence. It's the roughest part of the ride, and that's the part of the ride where if you haven't built the rocket right, it could break in half. If you make it through there, you know you've probably built the rocket well and have good control systems that are able to control its flight.

The most critical question with all of this will always be public safety. Don't hurt somebody. Beyond that, the next critical question is if you're worth the trouble, you're worth the risk. It's proving to people that you've solved all the easy problems and then you've got something worth flying and worth taking a risk on. You're not just some guys in a garage that don't write things down, and you're just going out with a rocket for the fun of it.

KEETER: There is engineering stuff and legal stuff to all this. A lot of the guys that are here on-site are the inspectors. They look at it more from a lawyer perspective. You documented something, and then you did what you said you were going to do. The other side is the licensing side, which is a group that is looking at all the technical details where they actually assessed the design, assessed the flight safety, trajectories, [and] hazard areas and make sure they're good.

In the end, I have to do enough analysis of the rocket's design to be able to show it's not going to cause a risk to the public. You typically control the risk through the flight termination system. You create a safe area that you can fly that doesn't involve the public or at least keeps the public risk to the required acceptable level. The beauty about Kodiak is that we're not around a lot of population. There is something we do called an expected casualty analysis. It's very low in this area, so we get a lot more flexibility from the FAA.

CARLSON: We had to close out that investigation of our first crash to be able to fly again. It's a requirement. It didn't really slow us up significantly, but it was extra work. They termed it a mishap, and that's the lowest-level investigation there is—a mishap investigation. The rocket came down on controlled land within an area that we had fenced off and away from the public. It was easy to clean up, and there was nothing environmentally sensitive. In a sense, it caused very little damage.

If you want to make a rocket that is incredibly reliable, then it ends up becoming very expensive. That's the right rocket to put a person on or to put on some crown jewel space telescope that takes twenty years to build and costs billions and billions of dollars. But if you're putting up a satellite that costs a few thousand dollars or perhaps you're putting up a few satellites that are part of many in a constellation, then they're replaceable and you don't need all the reliability.

We want to be able to fly a very affordable rocket. I don't mean to say we intend to make rockets that we're going to drop all over the ocean, but we are trying to make something where we accept a little more risk and keep things more affordable. We go and buy a commercial part that's out there, very affordable and is working in things all over the world, instead of buying a fifty-million-dollar part that they make one of at a time and is handcrafted and is tested to death in rocket programs. We're trying to accept a little more risk, and it's really important to us to see that the authorities like the rocket ranges and the FAA are willing to accept that and work with us. And the good thing is we really saw that from the FAA and personally I didn't really expect that.

KEETER: On launch day, obviously, I want to be prepared for the worst. I have to be, because I have to be able to react to the worst scenarios that could happen. I usually don't sleep well the night before, so mentally I just prepare for the worst. I think the most important piece is for me to at least appear in control and calm in a crisis-type scenario because those guys—you know, the young people—they don't know how to handle some of those "Oh, it's going to blow up," catastrophic-type situations.

HOFMANN: I'll be the launch conductor for Rocket 2. There's not really a training class or anything for this kind of thing.

The launch countdown procedure itself is twenty-two pages of small-text font of things to do. It's the last five or ten minutes where the real heart-pounding part is. You're trying to make sure that you're definitely on your time so the zero is zero. But that's the exciting point of "We're loaded, we're ready to go, everybody's green and ready to send it."

It's hard. Like, you're confident in what you're doing. You have to be because you're leading it, you're running the procedure, you're calling to everybody, you're keeping everybody in check and moving forward, but you can't ever get cocky with it or feel like "I've got this." There's that healthy level of nervousness that you should always have and I've always had with the rockets. You've got to be on that adrenaline edge where you're like "I'm ready and I'm looking at everything to make sure I don't miss something and I don't skip something important that'll bite me later."

CARLSON: On a launch day, I try to be very relaxed. I'm not stressed and wound up, and I don't know what my blood pressure is, but I feel relatively relaxed compared to the day before. The state of mind I want to be in on a launch day is that I've done everything I can. You have to put aside all those thoughts running through your head that keep you up at night and just run the countdown procedure and, you know, really be in the moment and just do that one thing. Everything else has gone away.

You've done all you can. Now you just have to take what comes when you read the next line of the procedure. The stress in the control room is really, really high. Everybody handles it differently.

You do everything you can to train people up. You do everything you can to let people know it'll be okay and we'll get through the day safely. Some people don't belong in the control room. You have to learn who has that executive decision ability and who is good at pressure.

A lot of what we're doing, to me, is just hard engineering, not science fiction. The science fiction for us is going to be setting up a rocket assembly line, which hasn't been done before. It's been done by the military. It's been done by places that build aircraft. It hasn't been done properly for rockets yet.

EVERY JOB HAS ITS DRUDGERY, tensions, and moments of glory. The rocket business, though, amplifies these experiences, particularly during the launch campaigns.

As Astra's launch campaigns dragged on, people struggled to deal with the mix of frustration and excitement that would intermingle each day. It felt like the misbehaving rocket would never leave the pad. Still, each employee had to prepare mentally as if it would. Day after day, people needed to concentrate as hard as they could and problem solve under pressure. On the occasions when the countdown got close to zero, the employees also felt the surge of adrenaline that accompanies any launch attempt and then had to release all of that energy and come back and do their jobs the next day.

By the time October 27 rolled around, the crew in Alaska had already tried and failed to launch Rocket 2 several times. They were exasperated. As was often the case, Kemp wanted the rocket to fly more than anyone else did and wanted to put on a good show for Astra's investors and potential customers. He had devised a system whereby the people closest to Astra could watch a secret webcast of each attempt. Kemp did the play-by-play announcing throughout the webcast. Even though the rocket failed to launch again and again, Kemp tried to leave the viewers enthused. He would dish out fresh sets of technical reasons for why the rocket could not launch each time and came across as competent and in control.

That day, Astra had two hours left to try and launch the rocket before its window would once again close and force the company to wait weeks for another opening on the pad. The igniters on the machine were once again giving the engineers trouble. For a while, they considered running tests on the igniters but then decided there was little to be gained from such an operation at such a late hour. Better to go through the regular motions, light the rocket again, and hope that it happens to feel like going into space this time.

The internal debate to get to that point took about ten minutes, with Chris Thompson, who was monitoring the situation from Astra's offices in Alameda, discussing the options with the mission control team in Alaska. After everyone finally decided to move ahead with the launch, Kemp really, really wanted to tell the webcast viewers so that they would hang around and witness Astra's magical moment. But he didn't want to tell the viewers that the attempt would happen until all of the Astra employees and the people at the spaceport knew that going forward with the launch was the plan.

As Kemp tried to deal with the webcast and Thompson tried to deal with a couple of last details, things between the two men turned heated. They began yelling at each other in the middle of Astra's office, and Thompson threatened to bail on the operation and leave the company. Kemp's desire to push things forward often collided with Thompson's pragmatism and gruff demeanor to create added pressure on launch days.

THOMPSON: If you want to keep pushing, I'm done.

KEMP: I'm not pushing. It just wasn't clear to me that the intent [to launch] was communicated to Kodiak after that conversation.

THOMPSON: It has been.

KEMP: Okay, good. That's all I needed to know. All I wanted to do was not get ahead of you and communicate something to the public before Kodiak heard it from you.

THOMPSON: I don't care about the public. I care about making sure our people up there in Kodiak are safe.

KEMP: That's all I care about, too.

THOMPSON: I don't want to stand here and have this argument. Let us do what we need to do.

KEMP: That's all I'm trying to do, Chris. I am just trying to make sure your teams are all on the same page before I communicate anything externally. So we're good?

THOMPSON: Yeah.

KEMP: Great. That's all I needed. Excellent.

Kemp jumped back onto the launch webcast and updated everyone in his most cheerful, optimistic Kemp voice.

> All right. We have an update from mission control. After reviewing data from the engines and the rocket, it looks like the igniters all performed.
>
> Over the next ten to fifteen minutes, the teams in Alaska are working to get the rocket fueled up again and put the rocket back in a state where we can step back in at T minus eight minutes. So stay tuned. We have about another hour and a half in the launch window before the end of the day today.
>
> In summary, we will be making one more launch attempt today.

The rocket did not launch that day.

On November 11, Rocket Lab returned to its pad in New Zealand and launched its third rocket. Technical delays had pushed back the "It's Business Time" campaign, but now Peter Beck could gloat: Rocket Lab was signaling to everyone in the industry that the company had reached a new milestone and intended to launch one rocket after another for paying customers. Since aerospace engineers always share gossip with one another, word of Astra's struggles in Alaska had reached Beck. He considered Astra's rocket to be largely a joke and took pleasure in his company's success as Astra wasted time and money in Alaska waiting for another chance to launch.

On November 29, Astra did its best to try to match Rocket Lab. This is what it sounded like over the course of fifteen minutes inside Astra's mission control center.

HOFMANN: Forty, thirty-three, thirty-two, thirty-one, thirty, twenty, fifteen, ten, nine, eight, seven, six, five, four, three, two, one, zero.

The rocket engines light. The machine soars into the sky. Many high fives are exchanged at Astra's office in Alameda. A few cheers go up in Kodiak as well.

KEMP: YEAAAAAAAAAAHHHHHHHHHHHHHH!!!!!!!!! FUCK, YEAH!!

HAHAHAHAHA. All right. Oh, my God. Oh, my God. It's so beautiful!

Twenty seconds pass.

HOFMANN: Loss of engine. Loss of engine five. Lost all engines.

Chatter in the mission control. The rocket comes crashing down very close to where it took off. Someone decides to pull the webcast before people see video of the crash and wreckage.

KEMP: We had how many seconds of flight? You killed the webcast? Did everybody get kicked off when you ended it?

Oh, shit. Fuck. Well, that's bad.

People in the office are looking at video of the wreckage. They see it's right by the launchpad. "It's outside the fence. Oh, geeeeeez."

MISSION CONTROL: Let's safe the GSE.

Kemp looks at the video and notices something.

KEMP: What's . . . on fire? There's a fire over there. Shit.

MISSION CONTROL: Ground, you can close CXV201.

RADIO COMMS: Closing CXV201.

KEMP: Just say "Test completed" on the webcast. We can't just kill it like that. What you did is, you just killed it for hundreds of people. We can't close out like that.

MISSION CONTROL: Ground, you can close the water valve.

RADIO COMMS: Closing WV201.

KEMP: Okay, now I am getting feedback from *everybody.* That was a bad call. For the people that are still watching, what can you do?

MISSION CONTROL: Do me a favor. Open LOX valve 107.

KEMP: Can you get back to that screen? Can you go back to something that's safe? And can I continue broadcasting?

Kemp gets back onto the microphone.

And that concludes today's webcast. We had a flight that did not achieve full duration. We are in the process of studying the telemetry data, and we'll be providing an email update to everyone. We did have a more successful flight than our first one, and we are looking forward to focusing on Rocket 3.

Kemp steps into the hall for a minute to talk to the other Astra employees who watched the launch.

Congratulations, Adam. We don't have to bring it home. Make Rocket 3 fly longer, okay?

Kemp goes back into his office and sits down at a computer. He starts talking aloud to himself.

> The second flight was beautiful. That was fantastic. Fantastic. Okay, let me prioritize my responses here. Investors are texting. I need to call board members.

Kemp makes a call.

> Hey, I just want to give you guys an update. We didn't quite hit sixty seconds, but it was a really beautiful flight, and it got a lot of great data. Nothing was damaged in the flight like last time. We are still trying to figure out exactly where it ended up. But we got about thirty seconds of flight is the rough number.
>
> **BOARD MEMBER:** Okay, that's really, really exciting.
>
> **KEMP:** It's a better outcome than it not launching, and we were twenty-four hours from that. It was really the last opportunity to launch this rocket. I'm pleased that we launched it. Obviously, I'd hoped it would fly a bit longer, but we got a ton of data to study, and it will definitely make Rocket 3 better.
>
> **BOARD MEMBER:** Wonderful, wonderful. Well, I am so happy for you.
>
> **KEMP:** I really appreciate your support. We'll have a complete download once we get all the data and the video together, and we will be sharing that in the next few days. . . . Yeah, the night launches are always spectacular. I'll pass along your support to the team. Thank you. Bye.

Kemp hangs up. An engineer comes into the room to talk to him and says, "Well, we got about a hundred meters higher than last time."

> **KEMP:** Well, that's consistent. I'll call Sam.
>
> Hey, Sam, did you have a chance to watch that? Yeah, we got it.

It didn't quite make the sixty seconds, but it was a beautiful flight, and we collected a ton of data, and the team is ecstatic. We didn't hurt anything.

I think it was an engine. We lost an engine just before thirty seconds. We are studying the data now. We made it further than the last flight but not as far as we wanted to. But another successful flight. A lot of people got to learn how to operate the vehicle, and we did the entire operation in four hours from rolling it out to launching. It will all make Rocket 3 better, which is what we're doing here.

Yeah, I was getting done with Rocket 2. I am glad it went out with a beautiful night launch. Night launches are great. We will get better at doing these things the first time on Rocket 3.

Kemp hangs up.

Everybody is saying "What happened? The broadcast stopped."

Kemp gets a doorbell alert on his phone.

Why is someone at my front door? It's probably a package or something.

An employee comes into the room. "Anyone have witness statements for me? We need to do them right now. You say exactly what you saw. It doesn't have to be an essay. Just what you were doing, what you saw, and what happened."

KEMP: This is because of the anomaly?

"Yeah."

After he had a few minutes to think and assess the situation, Kemp turned to me and explained his philosophy on talking about and coping with failed launches. "We've told our investors that they should plan for launches not working," he said. "We should be encouraged to succeed, but it shouldn't make it impossible if we fail. I think that it's all how you frame things. If you frame a launch failure as a big problem and an unexpected disaster, then it will become one. If you frame a launch as a fantastic success, then it will be."

CHAPTER TWENTY-FOUR

IT'S A JOB

A small trailer park community had formed outside Astra's original Orion Street building in Alameda. At any given time, three to four RVs were parked about thirty feet away from the rocket engine test facility. The people living in the RVs often had families in other states, and they were trying to avoid the high cost of apartments and houses in the Bay Area. Astra let the employees park their RVs on the company property at no charge, and in exchange, it got a free team of security guards who were on the grounds 24/7 and could hear if anyone tried to break into the facility.

It was not the most scenic of RV camps. The trailers were lined up next to one another on a patch of gravel, surrounded by razor-wire fence, shipping containers, tanks full of various gases and liquids, tool sheds, and piles of miscellaneous hardware. Often, one of Astra's rockets was stood upright about fifteen feet from the trailers, where it would be put through tests. Billowy clouds of gas would wash over the trailers along with waves of thunderous noise.

The RV folks had a distinct vibe. They were gearheads who mostly worked on building the engine test stands and the mobile rocket launcher and dealing with other assorted hardware issues.

They tended to be cynical about the space business, and they tended to drink beer in the evening, which facilitated their cynical chatter.

Two of the RV park residents were Les Martin and Matthew Flanagan.

Martin hailed from Texas. He'd had his first kid at sixteen and at eighteen had joined the marines, where he had been part of the infantry and specialized in antitank weapons. After four and a half years, he enrolled at Texas State Technical College in Waco and learned electronics. Martin spent the next few years working in the semiconductor industry until a bust cycle in the chip business forced him to look for a new job.

In 2008, a friend of Martin's saw a billboard advertising jobs for a company called SpaceX in McGregor, Texas. A rocket company in Texas? It sounded ridiculous, but Martin figured he'd give it a try and stay there for six months or so until he could find something more likely to succeed. At SpaceX, he had helped with the electronics for the company's rocket testing systems, and he'd turned out to be good at it. He'd spent three years there and then gotten a job with Virgin Galactic in Mojave, California, and then a job with Firefly Space Systems back in Austin, Texas. After that, he had joined Astra. Such is the story of how a marine from Texas ended up as a rocket-testing guru.

Flanagan had grown up in Virginia and earned degrees in mechanical and civil engineering from Montana State University. He'd done some hands-on engineering work after school before landing a job at Firefly Space Systems in Austin. After working there for a year, he went to a start-up that was trying to build a version of Elon Musk's Hyperloop high-speed transport system. Then he wound up at Astra, too.

From time to time, Martin would fall out with Chris Kemp or with the idea of working at Astra and retreat to Texas to spend a few weeks—or sometimes a few months—with his family. That opened up opportunities for Flanagan to take over a lot of the testing and launch infrastructure projects. Flanagan was an easygoing, hard-working guy who traveled back and forth between Alameda and

Alaska and endured the many delays the company faced with its first vehicles.

On December 9, 2018, I went to hang out with Martin and Flanagan at their RV compound. We watched the Dallas Cowboys, beloved by Martin, go up against the Philadelphia Eagles. Flanagan had recently returned from Alaska after Astra's second launch attempt. Unlike some of the hard-core rocket enthusiasts at Astra, these two men were not overly romantic about rockets. They mostly saw Astra as a job.

MARTIN: Dude, I don't watch other companies' launches. I couldn't care less. I worry about my own rocket, and that's it. I don't have time to worry about everyone else's. It's not my hobby. It's not my first love. I cared about SpaceX's rockets right until I liquidated my stock options, and then I quit caring about SpaceX's rockets. I mean, if I was a foot taller, I could be playing for the Cowboys. But . . . turned out I didn't have a lot of control over that.

FLANAGAN: When I left school, I wanted to get into aerospace. And now I just want to get out. Nah, it's all right. I mean, it's cool.

MARTIN: The problem with aerospace is all the jobs are on the West Coast, you know? And I want to live in Texas. Matt wants to live where there are no other people. His house is in Montana. Mine's in Round Rock, Texas. The only way we really make this work is by living in RVs. If you want to live in Alameda, you have to pay fifteen hundred dollars a month plus bills, and that's with a roommate. That's my mortgage back in Texas.

The RV is like three hundred a month, and we ride the company Wi-Fi. They don't go into our search history, which really helps. We've got a water hookup. All we have to worry about is emptying our tanks. I avoid using the bathroom in here unless I have to because the black tanks fill up, and that's annoying.

You know, our CEO, Chris, borrowed my trailer to go down to Burning Man. I stayed at his apartment in San Francisco.

FLANAGAN: He poked a hole in it.

MARTIN: Yeah, he did poke a hole in my trailer.

FLANAGAN: It's, uh—it's different living out here. Because, like, your yard is this gravel fucking pile of shit that nobody wants to claim.

MARTIN: It's hard walking outside and seeing a LOX tank first thing in the morning, instead of, like, your kids under your oak tree, you know?

FLANAGAN: One of the nice things is that I can take this with me when I go back home. Can take the kids camping. So it works out.

MARTIN: My kids aren't into the camping scene. I raised spoiled little rich white kids. They're digging the suburbs. I've got to go to karate. That's their big outdoor adventure.

FLANAGAN: I try to go home every two weeks. But I've spent three months out of this past year in Alaska. So that always throws a monkey wrench in the plans. I try to get home, like, on either end of the Alaska visits, and sometimes that happens, sometimes it doesn't.

MARTIN: You know, this is a funny business. Like, take a company that's as successful as SpaceX is. If I'm just doing some simple math in my head, I can't figure out how they make money. Then, if you look at something like Virgin, I don't know how much money has been dumped into that company, and they still haven't done a damn thing. It was an amazing company to work for. I've never had better benefits or a better culture.

But you get frustrated. You know, you're working, working, working, and there's nothing in the sky. That's disheartening. And I mean, even in the back of your head, like, yeah, we've got all this money, but eventually we have to put something into space or I won't keep doing this. It seems like a new rocket company pops up almost every six months or something.

FLANAGAN: Especially lately.

MARTIN: I can't even remember all the stuff I wanted to be when I was growing up, but it—I mean . . . uh, aerospace engineer was not in that group, um, at all.

FLANAGAN: Paleontologist. That's what you told me.

MARTIN: I did. Paleontologist. That was one I wanted to be for a long time. I think I wanted to be a lawyer for a while, or a doctor, just cause I was from a small rural town and, you know, those guys made money. But yeah, growing up my first love was dinosaurs.

But, you know, I'm from a poor rural area. I've always worked. I never had time to think about what I loved. I've been reading a lot of these leadership books, and they're like, oh, you just have to find what you love and then everything else is going to fall into place. I'm like . . . awesome, how the hell do I do that? And how the hell do I do that at forty-two, about to be forty-three, years old, with five kids, and a mortgage? Yeah. That ship doesn't sail, dawg.

I'll just be the ninety percent of Americans that aren't doing what they love. You think the guy that runs Discount fucking Tire is passionate about tires? No. But he runs Discount Tire good, so I mean, who cares? They're making money.

I do enjoy building shit. I do like doing that. It's the getting shit done fast to kind of set that example and that pace, that's what I get off on. I enjoy that.

FLANAGAN: The cool thing about space is that there are a lot of hard problems to figure out. It's one of those areas where people are still trying out a lot of different things. There's hard engineering problems across the board. I wouldn't say I would like to be confined to aerospace the rest of my life, but it's an exercise in engineering passion.

MARTIN: I mean, bottling Coca-Cola is way more complicated than what we do. Have you ever seen Coca-Cola get bottled? There's some hard-core engineering in it.

FLANAGAN: They always get it right.

MARTIN: They always get it right. They don't ever fail. You know what, when this thing says twelve ounces, guess what? It's twelve ounces on the dot, every time. And that's amazing. Millions of bottles a year. Millions of bottles.

FLANAGAN: It's also, like, one of the most boring things I can think of.

MARTIN: Yeah, look, working on rockets is better than working a real job. I wish it wasn't in the Bay Area, though. Let me preface this by saying that I am not some ultraconservative guy. I vote mostly for Democrats. But I don't want my kids around homelessness, weed, and what have you. I am a Christian. There are lots of things here that are not okay in my house.

FLANAGAN: Let's talk about the launch that just happened. It was, like, one of the most unbelievable things that happened that day. I was definitely not expecting it to happen. You know, you do the same thing over and over, like, again and again, and the same thing happens, so when it actually does something different you're like "Oh, that's weird."

MARTIN: That's weird. Heh.

FLANAGAN: But especially, like, the engine hadn't even fired. Every problem that we'd had so far was, like, some avionics thing and the stupid gyros. And then, that day we were swapping out an engine controller, and then right when they were about done with that, they were like "How long do you think it would take to swap out that gyro again?"

And so we did that. I had asked Thompson earlier in the day, "So what's the cutoff? When are we going to say, all right, we're going to have to push it until tomorrow or whatever?" He was like "Eh, we'd better be into count by, like, one or one thirty." And it was like coming up on two o'clock, two thirty, I think, they finally got it all buttoned up, and they were like "All right, here we go."

MARTIN: I thought it was crazy, man, because, well, I was hearing about the problems the night before, and I was like "Well, that sounds pretty fucking serious." Heh. And Chris Kemp was sending out his email: "We're going to try again tomorrow." I was like "Not likely." And then, of course, that was the day it actually goes.

FLANAGAN: I mean, it didn't get to orbit, so technically, it didn't work. But it left the pad. That's one of the things that sold me on Astra. It was Kemp being set on getting there as quickly as possible, and I don't care if it lives through the sky. It was all "We're going to get there, and then the next one will be better, and the next one will be better." There was some indecision on this one. There's lots of regulations that you have to jump over and stuff.

But the willingness to just be like "Fuck it, let's try it and see what happens and we'll do better next time" instead of sitting around and, you know, optimizing, perfecting, going over and over until you make sure that it's going to be the best thing in the world—and then you're likely to fail on something trivial and mundane anyway. So you might as well get there as quickly as possible and have all the headaches up front than try to power through. And try not to burn through too many people along the way.

MARTIN: You know, it took someone like Elon to get all this going. Someone who had his own money. None of us would exist if there wasn't an Elon. For all the good and bad that comes with him, that's why we exist. That's how we get money. We are all riding his wave. That's just the facts of it.

But what makes it hard is just the physics of space. If you build an engine that's going to put out that much thrust, I mean, college kids could do that, you know? But once you get it light enough to actually go where it needs to go, that's where the margins are. Your safety factors and your margins are way low.

FLANAGAN: Making fire and thrust is not difficult. Especially compared to, like, an internal combustion engine. Obviously, a lot of moving parts there. But getting something to space is difficult. Making thrust and then optimizing enough and selecting the right materials and parts so that you're light enough to get there.

MARTIN: And the challenge is to get there before you run out of money.

FLANAGAN: Right.

MARTIN: Because it doesn't take long. I mean, you raise a little money, you can run through it quick. All the money that is being dumped into this is absurd. There's some charlatans in this business for sure. And what's it all for? I don't see the need for all of it. It doesn't make sense. These VCs made their money in software or whatever. And, you know, they just love space. I think a lot of it is it's just cool for them to invest their money in this.

Even with us, the whole goal is for us to launch daily. I'm either going to be dead in the ground before that happens, or I'll be walking down the street with money falling out of my pockets and won't care that they're launching daily.

FLANAGAN: It would amaze me if there was demand for daily space delivery. That seems crazy to me.

CHAPTER TWENTY-FIVE

THE RESET BUTTON

Decades of history show that newly developed rockets often blow up during their early flights. The aerospace industry almost prides itself on the failures because rocket science is supposed to be hard. The machines and the people who build them would not enjoy the same mystique if rockets always worked.

Still, no matter how many times someone at a rocket company tells themselves that they're expecting an explosion, some part of them believes that they've created the exception to the rule. They're the ones who will make the rocket fly on the first try. They were smarter and worked harder. The rocket knows this. The fates know this. This rocket shall be willed into orbit.

It's that shred of belief coupled with the spectacle of an explosion that makes a launch gone wrong such a deflating experience. You allowed yourself to consider the possibility of success and then had to witness just how wrong you were in such a clear, binary fashion. The rocket didn't really almost make it. It blew the fuck up. The physical manifestation of your self-deception is cascading down from the sky in bits and pieces, and everyone can see exactly how wrong you were to believe in yourself.

When governments ran all of the rocket programs, the failures carried with them blows to national pride. But everyone knew that the United States or the Soviets would keep on trying because flying the rockets was an imperative. A commercial rocket maker, however, has a different set of pressures. Investors want to see results. Employees want to believe that they're working at the right company. Explosions raise the question "How long can we keep doing this before the money runs out?"

If the explosions rattled Kemp, he never let on to me. It's true that the rockets had not done what they were supposed to, but Astra did learn from the missions while not harming any people or property in a major way. "Nothing has changed," he said. "Our rockets will be better as a result of these launches. They've made for a company that understands how to operate. I don't look at these rockets as they succeeded or failed. They all launched, and they've made us more efficient."

According to Kemp, all the first rocket needed to do was leave the pad, and it did that. Astra expected the next rocket to reach Max Q and have its second stage fire up, go into space, and open its fairing to simulate the release of a satellite. While the rocket never reached Max Q, some of the important events did happen. "That's the conversation I had with investors, and that's how we shared the outcome with the board," Kemp said. "They were great with it because it's what they signed up for. The court of public opinion is not in session because we don't have a Twitter feed. So I think the strategy is working. If we were very public about what we were doing, people might evaluate what we were doing through another lens, but it's not the correct lens."

Behind the scenes, Astra had started to make major changes to its rocket as it went into 2019. The company had been out to prove Adam London's thesis that small, cheap rockets would be revolutionary. Now, however, its rockets would start getting bigger. Astra's engineers decided that the engines would need to produce twice as much thrust as before and that the rockets would be wider and longer. The company would also do away with its expensive carbon-fiber fairing used to protect satellites during launch and go instead for metal, and it would simplify its mobile launcher. Chris Thompson, who had been so instrumental in the development of the Falcon 1 rocket at SpaceX, would oversee the next generation of Astra's technology.

Rocket Lab had flown its Electron rocket successfully in November and December 2018 and again in March 2019. Kemp bluntly acknowledged that Rocket Lab's success had forced Astra to make a bigger rocket more in line with what its rival offered and make it quickly. Astra had raised about $20 million in 2016 and then another $75 million in 2018. Kemp now pressed Astra's board for additional funds and asked to increase the company's head count from 115 to 140. The board agreed to his requests.

Even though its only two rockets had exploded, Astra had managed to sell flights on its future missions to paying customers. For the bigger rocket, the price had increased from $1 million per launch to $2.5 million per launch for a 220-pound payload. "Rocket Lab claims they can do 440 pounds for $5.6 million," Kemp said. "We have estimated that the bill of materials on our rocket is about five times lower than theirs. We could be wrong. Maybe it's three times lower. Maybe it's seven times lower. Our rocket may be twenty to thirty percent lower performing, but it will be five hundred percent cheaper to produce."

Kemp continued to push the idea that Astra would outflank Rocket Lab through simplicity. Rocket Lab used too many exotic components and engineering techniques ever to be able to mass-produce its rocket. Astra, by contrast, would rely on metal and robots to make its machines by the hundreds. "Our production line will look like Tesla," Kemp said. "There will be robots that are positioning parts, welding, riveting, drilling. This is going to look like a modern car factory." Beyond that, Astra had hired one of Google's top executives to build an automated software system that would unite all of the company's operations from the test stands to the rocket to the launcher.*

Chris Kemp and Peter Beck never feuded in public, but the men

* That executive, Mike Jazayeri, would leave in January 2020 after eighteen months on the job.

did not really care for each other. Kemp had visited Rocket Lab and Beck while scouting launch companies on behalf of Planet Labs and had been given the royal treatment. Beck had flown Kemp by helicopter to Rocket Lab's launchpad on the Māhia Peninsula. He had also revealed a great deal about Rocket Lab's technology and future plans, hoping to win a number of launch contracts from Planet. After the trip, Kemp did tell Planet that they should use Rocket Lab, and the companies had formed a partnership. Once Kemp started Astra, however, Beck viewed the visit to New Zealand in a new light, seeing Kemp almost as a spy on an intelligence-gathering mission.

In his chats with me, Kemp complimented Rocket Lab's engineering in a backhanded way. Beck had made a nearly perfect machine, he said, but that was Rocket Lab's undoing. It was too costly for the job. Kemp also characterized Beck as an amateur when it came to fundraising and playing the Silicon Valley game. Beck had been desperate for money, and the venture capitalists had used his desperation to force him to give up a large stake in his company on unfavorable terms. Beyond all that, Rocket Lab took too long to develop and fly its first rocket. "We raised money and then did a launch," Kemp said. "Investors love that. When we raised money again, we were able to secure a high valuation for the company. It took Rocket Lab five years to get the same level of funding as us."

For his part, Beck viewed Astra as an almost comical operation. He read reports about the failed launches and mined people for bits of competitive data. From what Beck could tell, Astra was wasting its time and not approaching rocket making with the rigor needed to succeed. He also considered Kemp disingenuous and borderline reckless. "I can't build shit," he said. "If you want a vehicle that is not going to just lob you up there with the best of luck, come and fly on my rocket. If someone wants to build a rocket that throws things into orbit that is superrough and superinaccurate and thinks that's what the market needs, then let them win. But I'm not built to build shit."

As Rocket Lab raced forward with its launches, Astra turned its attention both to making its larger rocket and to building the huge new factory inside the once derelict Skyhawk Street building. By February 2019, Astra had given the 250,000-square-foot building its first makeover, removing all of the dead bodies and decades of debris and carting in the Burning Man sculptures crafted by Kemp's friends. In the months that followed, workers painted the floors and walls white, hauled in all sorts of welding and metal-cutting machines, created pods where people could work on their computers, and developed a real manufacturing line where several rockets could be built at the same time. For the first time in Astra's history, its people had room to move around and the feeling that they were in a building that made some sort of sense.

Kemp's ability to convince the city to let people work inside Skyhawk was at least as impressive as anything his engineers were doing on the rocket. While using the building to fix jet engines, the navy had dumped tons of paint thinner and other compounds into the groundwater underneath the building. A multibillion-dollar cleanup effort had helped the situation, but people feared that chemicals were still being released into the air. To put the city and his workers' minds at ease, Kemp coated the floor with a special epoxy designed to trap any nasty compounds.

"Out of an abundance of care, I also bought a gas chromatography laboratory and started operating it myself," he said. "It looks like that Geiger counter thing from *Ghostbusters*. It's like thirty thousand dollars, and I sample the air every seven minutes right where my desk and my team will be. I have not detected a single molecule of trichloroethylene, benzene, or a few other things people were concerned about. This will be the safest, cleanest air in America. I love regulatory hurdles because they keep other people out of buildings that I want to rent because no one else will succeed at this sort of stuff."

Just as he'd done with Orion, Kemp began moving people and equipment into Skyhawk before the city had approved such actions. "At one point, I told them I wanted to occupy the building by April first, and they told me that they would take me to jail," he said. "And so I'm like 'Okay, good. Now we're talking.' After I threatened to move into the building anyway, they gathered their requirements. They see us building things out the entire time. It's basically the unstoppable force of Astra meeting the immovable bureaucracy of the government."

When people told Kemp that it would take twenty-six weeks to get a power substation for Skyhawk, he found a way to make it happen in a week or two. If Astra was told that it could not install a certain machine or make a particular change to the building, the company would perform the work in the dead of night, and the city officials were usually none the wiser or were unable to undo the situation. All of the maneuvering was worth the effort. The city charged Astra 57 cents per square foot for the factory, which was about one-sixth the going rate for a finished building in Alameda. Best of all, Astra negotiated a rent credit from the city that let it more or less operate in Skyhawk for free, as long as they brought the building up to code.

Once the baseline infrastructure for Skyhawk was put into place, Astra set to work turning the building into a state-of-the-art factory. For the first time in the company's history, it had a proper lobby at the Skyhawk entrance with a couple of seats, some magazines to read, and an old Ventions rocket displayed on a stand. Inside the factory, Astra put up a real mission control center with glass panes that allowed onlookers to view the operations. The desks of most of the employees were positioned just off to the side of the mission control and organized in clusters: the executive team, the propulsion team, the avionics team, and so on. Deeper into the factory were workbenches dedicated to particular tasks such as building engines or antennas. At one station, Astra had spread all the wiring and computers that go into its rocket across a couple of tables. That allowed engineers to replicate the rocket's innards and run quick tests on software updates or new components. About half of the factory space went to the actual manufacturing of rockets and included the areas where people used the largest tools and assembled the rockets' fuel tanks and nose cones.

It turned out that building a new rocket and a new factory at the same time slowed Astra down. For the first couple of years, the company had been on a mad dash to try to make and fly a rocket in record time. The failures, though, had given rise to more caution. Months and months ticked by as Astra bulked up the engines and body of Rocket 3. The company had committed to doing far more testing this go-round, as its people no longer fancied the idea of having to carry out life support on the rocket in Alaska. It had also decided to do major tests of the rocket where it lit the engines out at Castle Air Force Base, just as Ventions did, instead of the nearby airfield.* All of those procedures required time and money.

Rocket Lab launched more of its Electrons in May, June, August, October, and December 2019. The company also revealed its secret program to make its rockets reusable. That meant that Kemp's previous economic calculations would soon need major reworking, as Rocket Lab's cost per machine was about to plummet. None of that pleased Kemp.

By the end of 2019, Astra had completed the design of its third rocket and felt confident that the machine would serve it well for quite some time. Kemp took full advantage of the Skyhawk complex and began building not just one Rocket 3 but a number of them before the company even knew if the vehicle would work. Astra was eating through its venture capital money at an alarming rate, and raising more money on favorable terms—or at all—became harder with each successful Rocket Lab launch.

Kemp also had to face the reality that Astra could no longer keep itself secret. The company had signed up to participate in a contest put on by DARPA called the "Launch Challenge." DARPA was offering $12 million to see which rocket maker could launch two rockets from two different locations with just a few days between the launches. Making things more difficult, the contestants would not know ahead of time where they would launch from or what their payloads would be. Dozens of companies had applied to take part in the event, and DARPA whittled down the contestants to

* There were too many prying eyes around the airfield and not enough infrastructure to do lots of tests.

Astra, Virgin Orbit, and Vector Space Systems. DARPA planned for the contest to take place in early 2020 and to accompany it with a grand public relations campaign.

IN LATE JANUARY 2020, THERE was a tense all-hands meeting in Skyhawk. More than a year had passed since Astra's last launch attempt, and money was getting tight. Its executive team had decided to impress the importance of controlling costs on the employees by rolling a series of seventy-five-inch screens onto the work floor. The screens were customized with information relating to each team's major projects. The engine team, for example, had the name of its engine—Delphin—displayed in big block letters atop its screen along with the names of the people on the team, including Ben Farrant and Kevin LeFevers. A countdown clock on the left-hand side of the screen showed how much time remained to accomplish a major task, a countdown clock in the middle was tied to the next major test that depended on the team's parts, and a countdown clock on the far right showed the time until the next launch.

Underneath the row of ticking clocks was a chart showing the team's budget for the month and how much of it it had consumed. In this case, the engine team could spend $40,000 for the month and had spent $34,160 so far. Underneath that was a list of recent purchases, including thermal coating spray, shielded power cables, a wall-mount circular connector, and a low-profile wash-down limit switch with a plunger actuator. An image to the left of those figures showed an angry cartoon bunny with its arms crossed, which obviously meant that the engine team had work to do. Launch Operations had a screen. First Stage Avionics had a screen. Everyone had a screen.

Though Kemp remained ever confident that Astra could raise money as needed, the rank and file were less sure. Many people figured that Astra had enough money left to launch only a couple more machines, and if they blew up as their predecessors had done, the company would be toast. Kemp bet that the rocket would work and felt that it would be better to try to raise money after a success, when he was in a position of strength, than to go begging to keep the company alive after more explosions. Adding to the ominous vibes, Vector Space Systems, Astra's rival in the DARPA Launch Challenge, had filed for bankruptcy in December without even trying to fly its first rocket.

As the money dwindled, London was forced to return to his skills as a McKinsey consultant. He began managing the day-to-day finances of Astra and looking for ways to buy snacks and tool cabinets more cheaply. Privately, London told friends that he thought Astra might not be able to raise more money and would soon meet a fate similar to that of Vector.

The all-hands meeting took place in the lunch area just outside the mission control center. The employees, previously used to catered meals, picked up their lower-budget, prepackaged lunches and sat down to listen to Kemp and the other executives. Many of the employees had been working for weeks, if not months, on end without a break. They were frazzled. Still, some recent tests had gone well, and it seemed as though one more big push might get Rocket 3 to Alaska.

KEMP: Right. Let's kick things off. You guys have done an incredible amount of work the last couple of weeks. I've never been more excited about this company.

As you guys know, Adam and I spent a lot of time huddled up at our cubes while you were at your test cells, just trying to figure out how we can buy the company as much time as possible to get the rocket up to Kodiak.

Last quarter, we spent $5.5 million per month on average. We looked through this about halfway through the quarter, and it was a lot more than we forecast. I've been talking to you guys about this a lot.

What we're trying to solve for is having enough cash to get

through the first half of the year without making any big changes in terms of the program we want to execute on and the people we want to have to execute it with.

We've been looking at our forecast and wondering "How did we get it so wrong in the fourth quarter?" Frankly, it was a combination of poor planning and just the work that you need to do to get the rocket ready to launch. No one made a mistake. We just had a really bad plan. As we figured out how to make the plan better, we needed your input. We now have a way better plan.

I am personally writing up in the error log everything I have personally learned from this experience. You can read it if you wish. I encourage every member of the executive team to really reflect on this.

Now we have to be very focused on spending money and time on the right things. If your team needs more money to get your job done, we don't want you to find out by not being able to buy something. We want to anticipate that and reallocate money before you even realize it's a problem. What success looks like to me is that you have the money you need. You don't have excess money. We thread the needle over the next six months. I'm going to go raise as much money as I can to give us more buffer so that our plan closes.

If you think the plan doesn't work for what you need to do, don't complain about things. Share your concerns with your manager. Let's communicate and be transparent and hold each other accountable. We're here to help you do the things you need to do to get a successful launch attempt off because, when that happens, we have a really, really successful business here.

I'll let Adam make a few comments as well.

LONDON: We didn't do a good job of rolling all this out and communicating this stuff well initially. I'm sorry. We're going to make it better.

So far in January, things are actually looking pretty good relative to what our updated plan is. Through this morning, we have spent approximately $2.5 million, and we'll spend another $1 million by the end of the month on payroll. That includes a onetime $500,000 payment to Alaska for our launch site that has to go out at the beginning of the year. We've really only spent $2 million thus far, which is good, but we need to be careful. We need to make sure that you're paying the suppliers as necessary to keep things moving and being careful about how we use our cash. Remember that, in general, we are always going to do the right thing for the business. When you believe that we are not doing that, I want to hear about it directly. Please find me.

I believe we have enough cabinets and tool chests and tables and other such things to get us through the launch. If you don't think that's the case, let me hear about it. And, as you experienced today, we are making some changes to lunches and snacks. We're going to do this new eating club twice a week for lunch. On average, five dollars per person, per meal, is less expensive than what the other food thing was. So it makes a difference. We are going from two meat choices to one meat choice. I think the salad bar is great, so we will continue to do that three days a week. Dinner will continue. We switched to Costco for snacks.

Overall, these things, which I think are fairly minor, will save us a third of what we were spending on food. Last year, we spent about $1.5 million on food. This will get us closer to $1 million.

Our commitment to all of you is that we will do everything in our power to give you the three chances you deserve to get to orbit. I think we are on a path to do that. I recognize that it's stressful. I know we are all working incredibly hard. But we are on a path, and I think this is going to work.

I think that we all need to recognize that there is a limit to the amount of work people can do. And if you are approaching that limit, you need to let people know because it doesn't make sense. It will take more time to get tired and make a mistake and do the wrong thing than to take a break. We have a very large team. We have the opportunity to make sure that team members get a day or two break when required.

For the people who have worked three weeks straight, figure out how you get a day or two off. We absolutely need to get to a point where this becomes sustainable. At the same time, we have a deadline of shipping the launcher on the fifth and shipping the rocket on the eleventh. We need to help out the people who through no fault of their own are facing the brunt of all this.

KEMP: Let's continue to push hard to get this rocket out the door. We're doing everything in our power both from a financial perspective but also, as your families and as our investors and customers are watching, to lower expectations as much as possible on this launch. Maybe even the next launch. So that we have every opportunity possible that we can to exceed expectations.

If Rocket 3 doesn't work, we have a small team of people to go figure out why, fix it, and launch as quickly as possible. You can't raise billions of dollars and spend infinite time on anything in this world. These deadlines are important because they focus us. If we can continue to do great work and simply tell people about what we have done, this will be unlike any space company ever.

We have a way to thread this needle and buy this company three launch attempts without doing anything crazy. Without changing control of the company. Without giving control to investors. Without laying people off. We don't need to make any massive changes to make this work as long as we all work together. Let's just get this rocket shipped. And we got this thing.

On March 2, the DARPA Launch Challenge began in earnest. The government body sent a film crew to Astra's headquarters to document the launch of Rocket 3. It was the first time anyone outside Astra's core employees, investors, and family had been inside the building for a launch attempt, and this one would be broadcast to the whole world via the Web. Given the circumstances, Astra decided to throw a full-fledged launch party and invited about a hundred guests to hang out in its lobby and watch the proceedings on big-screen TVs.

Rocket 3 had made its way to Alaska about a month earlier and gone through the usual trials and tribulations. In the background, Vector had gone bankrupt and Virgin Orbit had pulled out of the launch contest altogether. That left Astra as the only contestant. If it could launch a rocket successfully, it would earn $2 million from DARPA. If it could do it again by March 18, it would net another $10 million.

Against the spirit of the competition, DARPA had controversially stacked everything in Astra's favor. Originally, the idea had been that DARPA would spring a launch site on the competitors on short notice, testing their ability to ship a rocket and all the necessary launch infrastructure to a far-flung location on the fly. The second launch would take place in a totally new location. In this case, DARPA had given Astra home-field advantage by telling it ahead of time that the launch would take place in Alaska. It had also told Astra that it could do the second launch from Alaska as well. It seemed clear that the various parties involved wanted to try and save face and make the contest with a lone competitor somewhat of a success.

DARPA had originally given Astra a large launch window that stretched across several days, but technical delays, coupled with a blizzard, had chewed through the extra time. Now Astra had to launch on March 2 if it wanted to win the $2 million and have a chance at the larger payday. Otherwise, DARPA would end the competition altogether.

Astra's employees could not wait to get the damned thing off the ground. Many of them had been dealing with the rocket under immense pressure for four months without a break. The people in the lobby at Orion who were noshing on hors d'oeuvres and knocking back drinks knew almost nothing about Team Astra's recent struggles or the battles with the past launches. Most of them expected everything to go off without a hitch.

And at first it looked like the day might be special. Astra started its countdown about an hour before launch, and the clock kept right on ticking down toward zero. Unlike past launches, which had been marred by so many holds and lengthy trouble-fixing procedures, this one progressed smoothly. By that point, I'd become cynical toward launches taking place at their anointed time on their anointed day, and I remember being a little shocked at how well things were going. As the minutes kept going by, my cynicism disappeared completely and my body filled with adrenaline. I wanted to believe.

But then, as I stood next to one of Astra's hopeful investors, the countdown clock held at fifty-two seconds. One of the guidance systems in the rocket was sending out data that did not make sense. The team needed some time to analyze the information and see if they could proceed.

Over the course of the next hour, the Astra engineers reported back that the sensor must have been malfunctioning. The data claimed that the rocket had either tipped over or been moved a dramatic amount, but there it was still standing tall and proud on the pad. Fixing the sensor would likely require that Astra stand down for the day, which in turn would lead to its missing out on winning the challenge. There was a good chance, though, that the rocket would fly fine without the sensor, as other systems would kick in and make up for the wonky part. Per usual, Thompson and Kemp squabbled over the right course of action. The rocket was Thompson's baby, and he did not want to blow it up. Kemp wanted to fly and win the contest. In the end, Astra called off the launch, and all the DARPA people and guests left feeling deflated.

After the launch window closed, Astra had to wait a couple of weeks before it could try again, which gave it time to fix the faulty part and do some additional checks. On March 24, Astra prepared to launch again, only this time the company returned to its secretive ways and told no one about the event, which was probably for the best.

On the day before the launch, Astra's engineers went through their usual series of tests. As the day dragged on, they started to run out of the helium used to prime parts of the rockets. Someone decided to borrow some of the helium used by the rocket's second stage and transfer it to the first stage, which was in the middle of being tested. The helium, though, had been cooled more than normal over the course of the day, and as it went into the first stage, a plastic valve could not cope with the low temperature and froze in an open position. Helium flooded into the tank and built up too much pressure for the metal tank to handle. The rocket exploded on the pad.

Locals in Alaska noticed the explosion, and it made the news in Kodiak. Officials at the spaceport would say only that an anomaly had taken place. Kemp told an inquiring press that the rocket had been damaged during a test without providing any other details.

Astra's people were devastated, as the explosion marked the worst of outcomes. The rocket had been undone in part because of laziness and in part because someone had okayed the use of a cheap plastic valve instead of a stainless steel valve. Astra would not be able to pull any data from Rocket 3 because it had never flown for a single second. And now the company had to wait for a new rocket to be finished before it could try yet again.

CHAPTER TWENTY-SIX

CASH ON FIRE

Chris Kemp wanted to launch another rocket, but the world had different ideas.

The employees at Astra were humbled and somewhat humiliated by their last launch attempt. Not only had the rocket blown up during a routine test, but the explosion had taken the mobile launcher with it. Rocket 3 had been the responsibility of Chris Thompson, the man who had come with all the SpaceX accolades, and people wondered how he'd allowed the crew in Alaska to try and do something so risky as pushing freezing helium from one tank into another. The rumors being traded on the Astra factory floor, however, were that Thompson had gone to the bathroom and been absent when the helium decision was made. He'd literally been taking a shit as the rocket exploded.

To get back to Alaska, Astra needed to rebuild its mobile launcher from scratch and go through its battery of test procedures with Rocket 3.1. Only now the covid pandemic had taken hold, and doing those already difficult tasks became that much harder. The shipment of components slowed. Many partners were banned from operating their factories. Liquid oxygen rose grimly in price and then became hard to come by as hospitals began to purchase oxygen in huge volumes to try to keep their patients alive.

Even though it had yet to prove to be of much use to anyone, Astra was deemed crucial to national security by the Pentagon, as were many other space-related companies. That designation gave it an exemption from the government to avoid the shutdowns affecting many businesses and keep operating. Still, only about 15 percent of Astra's employees would turn up at the factory, as the company tried to limit their exposure to covid. Astra also laid off about 20 percent of its staff in a bid to save money. The company continued to need a few million dollars each month to sustain its operations, and raising more money was nearly impossible in those early months of the pandemic. No one knew what would happen to the global economy, and fearful investors were hanging on to their cash.

Given all those issues, it took Astra until September 2020 to build another rocket, test it, and ship it to Alaska. On September 12, a small group of masked employees descended upon the Skyhawk building to try and send their latest machine to space.

Chris Hofmann, the launch director, took center stage in mission control. His orange mohawk made him easy to spot, and the giant bottles of ibuprofen and antacids by his computer underscored the stress of the job. "We need it to go," he said, noting that Astra had now been through the process about two dozen times among the launches, explosions, and aborts on the pad.

The small number of people in Skyhawk made the launch feel more relaxed than normal. The only employees allowed outside mission control were the core group of people who needed to be on hand to troubleshoot problems or analyze data afterward. Bryson Gentile made a beer run earlier in the day just in case there was anything to celebrate. Really, though, a mood of cynicism had started to permeate the complex. People complained that they'd made a steady stream of improvements to the rocket, but it still kicked up new errors day after day. The night before, a helium leak had stopped a launch attempt, and it had been caused by a part that had never been a problem before and hadn't been touched in weeks.

Matt Lehman, one of the Ventions veterans, shared a quiet moment with me and declared that his next job would not be in aerospace. "I've been with Adam all these years and have to see this through," he said. "But really, I sometimes wonder what the point of all this even is. We already have GPS systems in space. We have surveillance systems. We can make a space internet, but I already have all the cat videos I will ever need."

Kemp looked even fitter than usual. He'd dropped several pounds as the result of a fasting diet during which he'd consumed nothing but water for seven days. Even with covid running rampant, he'd recently been to a gathering with Elon Musk and hoped the encounter would bring good luck. "We just have to launch this rocket, and things will get better," he said. At that moment, someone walked out of the mission control, grabbed a bottle of whisky off a desk, poured a couple shots into a trash can, and went back to his station.

As Friday afternoon turned into evening, Astra reached the point where it was ready to let the rocket go. Once again the countdown began. Once again the rocket took off and found a new way to fail. Right as it launched, it appeared to be trying to go in the opposite direction of its intended flight path. Not long after it started heading skyward, officials at the spaceport hit the kill switch and terminated the machine, preventing it from flying back over Kodiak Island.

After the launch, the company tweeted, "Successful lift off and fly out, but the flight ended during the first stage burn. It does look like we got a good amount of nominal flight time. More updates to come!" In a virtual press conference a short while later, Kemp declared it to be "a beautiful launch." "We are incredibly proud of what we have accomplished," he said. "We have designed this launch system from the ground up to be mass-produced, and we think we can produce it at scale in California."

Inside the Skyhawk building, the employees analyzing the launch in private put a different spin on the event. "Something really gross must have happened," Lehman said. "It was going the wrong way." The rocket had started wobbling after a few seconds with the engines seeming to fight what they were being told to do. Pretty much everyone in Skyhawk agreed that a software error must have commanded the rocket to fly backward or maybe that a directional sensor had been installed upside down. It could have been a horribly basic mistake such as someone typing the wrong number into a guidance system. Whatever the cause, it was a disaster.

London was dismayed but took heart that the explosion had occurred in a safe place. "A nice fireball consumed almost all of the fuel," he said. "It's quite a benign thing environmentally." He thought that the guidance systems on the rocket would have eventually figured out what they were supposed to do and that the rocket would have completed its mission if it hadn't been blown up by the safety officials.

Watching from afar, Peter Beck hoped that the latest gaffe might stop Kemp from slagging off Rocket Lab's machines as being overengineered and wondered if it might spell the end of Astra altogether. "He must be out of money by now surely?" Beck asked me.

In a previous era, that latest explosion might have led to Astra's demise. Before Musk and SpaceX, a handful of millionaires had tried their hand at the rocketry business, and all of them had given up as the years ticked by and the ventures ate away at their fortunes. The reward for success had never been clear anyway. The US government and NASA were protective of their space program and loomed as very well funded competitors of all the private space ventures. Plus, there was not much profit in launching rockets. People simply did it because they wanted to and it seemed like a fun way to spend some excess capital. Musk had been the only rich person to see the enterprise through to the end, and if his fourth rocket had not flown successfully, he probably would have left rocketry, too.

Though Astra's bank account was dwindling to dangerous levels, Kemp managed to keep the whole thing going because this was a new era in which venture capitalists had descended on space and Kemp knew how to speak their language. He'd pitched all of Astra's rockets as "beta" projects. They were simply the physical equivalents of software applications that a start-up might test on the Web or on a smartphone. Sometimes they worked, sometimes they didn't. The important thing was to keep moving forward until something clicked.

Of course, outsiders in the aerospace industry watching Astra's travails had more sober opinions. Astra's initial idea to build a rocket superquick, fly it, and learn from it had appeal. Most people, however, assumed that the company would have been more successful by this point. As far as rockets went, Astra's machine was small and relatively simple. If it was going to keep blowing up, it should at least do so after a few minutes of flight, allowing Astra to secure the vital data it needed to avoid future explosions. Something seemed amiss with Astra's team, the way it had approached rocket building, or both.

Astra, though, had no time for deep introspection or to overhaul the way it went about things. The company simply worked to figure out the basic fuckup that had derailed the last rocket, implemented the fix on an existing rocket awaiting launch on its factory floor, and shipped the new machine to Alaska.

On December 15, Chris Hofmann performed yet another countdown and let Rocket 3.2 loose. And this time, by God, it flew—and flew for a long time. The rocket did everything it was supposed to as the first stage fell off, the engine on the second stage ignited, and the rocket kept making its way into orbit. Chris Kemp and Chris Thompson, so long at odds with each other, embraced inside Skyhawk's mission control with their masks doing little to muffle their exuberant cries. Cameras aboard the rocket showed the earth behind it and only the black of space ahead of it.

After a few seconds of ecstasy, the Astra team refocused on the launch and realized that things had gone well but not perfectly. The rocket had run out of fuel just as it had approached its goal. It had made it into space and was probably the fastest privately funded vehicle ever to do so, but it had not quite reached orbit. The launch had been a strange mix of success and failure wrapped together, and no one knew exactly how to feel about it—except, of course, for Kemp.

"It's been a tough year," he said. "The most important thing was ending the year with a win. I think a lot of people have put their blood, sweat, and tears into this. We had the money to do it again. We had another rocket in there. But it would have been hard to get everyone to come back and do it again."

From there, Kemp moved right into backhanded compliments for rivals such as Rocket Lab that had needed many years to reach space for the first time. "Frankly, I commend them," he said. "Keeping a team motivated and inspired to persevere for eight or ten years would be an incredible challenge."

Come January 2021, Kemp was really feeling it. "The rocket works as is," he said. "We are handing it off to production. It's just like 'Go!' We will fly our first payload this summer, and then we will immediately start flying every month."

Astra planned to build a second spaceport to give it more opportunities to launch. Like SpaceX before it, the company had been eyeing Kwajalein as its next best bet.

"It's a long way to get out there," Kemp said. "Hopefully it will all get easier when we have a plane. We now have a spaceport team. It is like picking out where a Starbucks goes. There are all these factors. How much does it cost? Is there an airport nearby where you can land a C-130? What does the regulatory stuff look like? We will be trying to get approval to kick off several more. Our plan is to build out nationally initially and then get up to a global footprint of spaceports. They are just concrete pads with fences around them."

All of that would cost a lot of money, but Kemp said the recent launch had solved that problem. Right after the rocket had gone up, he started firing off text messages to investors, and cash started arriving. "A ton of it," he said.

The quasi success also brought on a change in the way Kemp talked about Astra. He still rolled out the mass-production-of-rockets pitch but had added a new twist. Once again borrowing from the software world, he talked about making Astra into the basis of a platform—and not just any platform but one focused on improving life here on Earth.

"I can see better where all this is going now," he said. "This isn't a story about settling Mars. What Elon has going is so Elon. I'm trying to make sure Astra has its own very, very unique story to tell. And it really is all about Mother Earth. It's about improving life here on Earth."

Astra had always been thought of and presented itself as the FedEx of space. It would send new satellites into orbit daily and as cheaply as possible. Only now Kemp had hit on the exact language with which he wanted to describe the mission. The billionaires were about sending things and people to far-off places and abandoning the earth. Astra would embrace Earth.

"We are going to enable a whole new frontier and a whole new group of pioneers to go build things in space," he said. "Planet Labs took a decade and a billion dollars and thirty generations of Doves to make their system work. Well, there are four hundred space companies now that have been formed since Planet Labs. They've raised tens of billions of dollars. It's a crazy amount of money. They are all focused on Earth. They are connecting things on Earth. They are observing things on Earth. They all have this common narrative around improving life on Earth.* The biggest play is to make all these ideas possible—to make it so someone with an idea in a dorm room can send something to space. Let's get the cost out of all this stuff, so we can make it big."

Kemp was clearly testing some new material on me. He felt as if the launch had kicked off the next phase of Astra's existence and the future would require a grander message than merely mass-producing rockets. He was chasing something and preparing for the company's evolution. "Some big things are happening," he said. "And some really big people are getting behind them."

---

* Astra went so far as to trademark the phrase "Improve Life on Earth from Space."

CHAPTER TWENTY-SEVEN

IT MAKES SENSE. RIGHT?

Chris Kemp called me on a Sunday afternoon at the end of January 2021 with big news: on February 2, Astra would go public and begin trading its stock on the Nasdaq exchange. In normal times, the idea of Astra going public would have made no sense. But these were not normal times.

Leading into the covid pandemic, the ever-artful geniuses on Wall Street had fallen in love with a financial instrument designed to let investors buy into companies that had little in the way of profit but lots in the way of hype and future promise. The instrument was called a special purpose acquisition company, or SPAC.

To create a SPAC, some wealthy people would come together, raise money, and vow to use the money to acquire a company in the future. That is to say, the SPAC itself didn't make anything at all; it was just a pool of cash, usually hundreds of millions of dollars. The people who had put the pool of cash together would go out into the world and look for a company to buy and merge it into the SPAC, creating a single entity. One crazy part of this is that a SPAC could be publicly traded before it even acquired a company. Investors could see who had put together the pool of cash, and if they liked the fundraisers' backgrounds or thought the fundraisers seemed smart or smelled great, they could buy shares in the SPAC and hope that the pool of money would eventually find something interesting to acquire.

Another crazy part of all this was that SPACs had historically been looked at with a side-eye. They'd been around for decades and were considered among the shadiest of financial instruments. In the past, desperate people had used SPACs to buy up questionable companies, put some lipstick on them, and then pitch them to unwitting investors as something special. Quite often, the process had led to those investors losing all their money.

Around 2019, however, some folks in Silicon Valley and on Wall Street figured that they could put a new spin on SPACs and change their reputation. They argued that there were lots of unprofitable tech companies that would one day turn into huge empires. Investors should have the chance to buy into the companies early and profit as they grew, and the companies should have a chance to raise tons of money so that they could grow faster.

In normal times, a company tries to spend at least a couple of years turning a profit and boosting its growth rate before hitting the public markets. This is, in part, because US regulations prohibit public companies from speculating about their future financial performance. Investors are expected to judge a company on its past financial metrics and make their own guesses about where things are heading. The better a company has done over the last two, three, four years, the more likely investors will be to buy into its story when it goes public.

But because from time to time Wall Street decides to do away with things that make sense, SPACs and the companies they acquired were allowed to more or less say whatever they wanted about their future performance and push investors toward a state of maximum titillation. Will your drug cure cancer in five years? We kinda think so, yeah. Will you solve global warming? We feel pretty good about saying "Yes." Will your revenues double, triple, and then quintuple? Sure. I mean, why wouldn't they?

SPACs, in effect, let all kinds of investors act like venture capitalists. Investors could place risky bets and hope that a company would really hit it big. It's not the most stable of ways to run a public market, and there's plenty of reason to think that many investors do not really understand the risk they are buying into, but whatever. Rich people wanted to make more money, and the volatile, often delirious state of the financial markets during the pandemic made the great SPAC revival seem appropriate for the time.

Astra's suitor turned out to be a SPAC called Holicity. It had been set up a year earlier by the billionaire and telecommunications industry legend Craig McCaw. He'd raised $300 million from some buddies including Bill Gates and used it, along with another $200 million ponied up by the investment firm BlackRock, to back Astra and take it public.

Under the arrangement, Astra would first merge with Holicity and then, after clearing some minor regulatory hurdles over a couple of months, start trading on the Nasdaq under the ticker symbol ASTR. But since Holicity was already a public company, investors could basically begin buying shares in Astra right away while various parties completed the paperwork needed to close the deal. When Kemp announced the deal on February 2, the share price of Holicity jumped from $10.34 to $15.00. By the end of the week, the shares were trading at $19.37. Over the span of a few days, Astra had gone from being nearly bankrupt to being worth $2 billion to $4 billion.

In our conversation ahead of the announcement, Kemp took advantage of everything SPACs allowed by talking up Astra's bright future. Potential investors were going to be informed that Astra's rockets had just missed orbit, but the company had figured out what the issues were. Astra would launch again by midyear; after that was a success, it would begin launching every month through the latter half of 2021. After that, it would endeavor to turn out a rocket a day from its factory and launch a rocket a day from pads that it intended to build all around the world. On the new and improved messaging front, he said that Astra would also get into the satellite business by building many of the most common satellite parts right into its rockets. "This will allow customers to focus on their camera or their software or the unique radio they want to fly," he said. "We are trying to build a platform that has launch as its foundation."

Clearly feeling the moment, Kemp also found time to bag on Rocket Lab. "I don't know what Rocket Lab is doing," he told me. "They are tinkering with their high-performance Ferrari of rockets that they are launching at a low rate. They got to orbit a few years ago and are still only doing a few launches per year. As a public company, we now have the full capability to get to daily space operations."

Upon learning about Astra's fundraising and move to go public, Peter Beck at Rocket Lab nearly stroked out. He wrote me saying that it was one thing to work over venture capitalists who know the risks they're taking on but another thing altogether to work over mums and dads trying to save for their retirements. "Call me old fashioned but is integrity really lost?" he asked.

Beck had long doubted Astra's technical chops and did not care one bit for Kemp's salesmanship. Why, he wondered, would anyone think there's a market for a rocket that can carry only a hundred pounds or so of cargo to orbit? More to the point, why would anyone trust Kemp? "Their numbers don't even add up," Beck wrote. "I think finally Kemp will learn his lesson."

To me, one of the most incredible parts of the whole announcement was the presence of Craig McCaw and Bill Gates as backers of Astra. In the 1990s, the two men had funded a satellite start-up called Teledesic to build a communications and internet system in low Earth orbit using hundreds of low-cost satellites for a mere $9 billion in capital. Like Iridium and other similar ventures before it, Teledesic did not go well and mostly ceased operations by 2002 before accomplishing much of anything.

For some reason, those guys had decided to get back into the space game in 2021, but instead of chasing lucrative satellite services, they would venture into the low-profit world of rocket launches.

"In the end, the whole system we were pursuing at Teledesic

was just wrong," McCaw told me. "It didn't make sense to have men in lab coats building satellites, and the more defense and Old Space people you hired, the more they gummed up your company. If you tried to use a rocket that the big space and defense companies didn't control, they would say, 'If you don't use our rocket, we'll destroy your company.' Even just the insurance costs back then were as much as about thirty Astra launches are today."

He seemed to view the investment in Astra as an act of revenge or at least redemption. "We are coming back to the item of greatest frustration where the most change has yet to occur, which is the rocket launches," he said. "The defense contractors fought things like SpaceX tooth and nail, but Elon Musk obviously rammed a new philosophy of doing business right down their throats. The old guys will now become the asterisks of space. They will be put aside because commercial space projects will do so much better."

When the pandemic first engulfed the world, it looked as though the entire New Space gang, excluding SpaceX, Planet Labs, and Rocket Lab, might disappear. The rocket-making start-ups had all been struggling. The satellite start-ups were plowing through cash while they awaited rides to space. Things were getting desperate, and it looked as though they would only get worse as the global economy shut down and investment capital disappeared overnight. No investor in his or her right mind would dump hundreds of millions or billions of dollars into the riskiest of all ventures as the apocalypse took hold.

As we all know, however, the darndest thing happened: governments began printing money and handing it out, stock markets the world over rose, and obscene amounts of cash started flowing toward technology companies. Rocket and satellite makers were suddenly wonderfully positioned to capitalize on the investment euphoria with their idealistic business plans that soared above the drudgery of Planet Earth. Not long after Astra did its SPAC, Rocket Lab did one, and so, too, did Planet, along with more than a dozen other space companies. None of the companies was profitable, but each one was suddenly worth billions of dollars as if by magic.

Astra started using its newfound riches right away. It began by hiring a slew of executives from big-name companies in Silicon Valley. Most bizarrely, it named Benjamin Lyon as chief engineer. Lyon came from Apple, which, as of this writing, does not make rockets. His experience revolved around developing track pads for laptops, iPhones, and allegedly Apple's struggling and secretive self-driving car program. Kemp argued that Lyon brought fresh perspective and the knowledge of what it took to make industrial-grade products to Astra. In other conversations, people told me that Lyon had been hired to make Astra's investors feel good. He'd built up a strong reputation in the Valley, and the board thought he might check some of Kemp's more brazen impulses. Even though he'd never come close to building a rocket before, Lyon essentially became the rocket chief, pushing Thompson and London to the side.

Since the last rocket had run out of fuel before reaching orbit, Astra had also decided to spend money making a new, bigger rocket. It increased the length of the machine by five feet and made it a bit fatter so it could hold more LOX and kerosene. The company once again changed its price per launch based on the new design. It looked to charge about $3.5 million per launch now to carry just 110 pounds of cargo into orbit. On its website, it touted its ability to fly more than 1,100 pounds of cargo to space, but rather mysteriously, there were no details about how it could do so with its tiny machine, and no one ever bothered to press the company on the matter.

On July 1, 2021, all of the SPAC paperwork went through, and Astra officially began trading on the Nasdaq under its ASTR ticker. It became the first pure rocket company ever to go public, marking a major moment in the commercial space industry. Kemp and many Astra employees flew to New York to ring the Nasdaq's opening bell.

During a short speech at the exchange, Kemp, dressed in his trademark all black, tried to inform potential investors about Astra's journey. "Four years ago, Adam London and I quietly founded Astra with the bold mission to improve life on Earth from space," he said. "Our vision for a healthier and more connected planet has inspired the most talented team in the world to build a rocket and to launch it faster than any company in history at a fraction of the cost. We now have the resources to execute our hundred-year plan. . . . Let's go make space for everyone." He also vowed to send off Astra's first commercial mission that summer and to fly daily missions by 2025.

After the speech, Astra's team gathered around Kemp to celebrate on the Nasdaq bell-ringing stage. Lyon, the new recruit from Apple, was right up front alongside Kemp and London, and so, too, were some of the new executives Astra had just hired. Chris Hofmann and a handful of the Astra veterans were there, although mostly off to the side. What most people watching the event did not know was that many of the people who had built up Astra over the last four years had departed in recent weeks. Les Martin had left. Matt Lehman had left. Ben Farrant had left. Roger Carlson, Ben Brockert, Milton Keeter, Bill Gies, Issac Kelly, Rose Jornales, and Matt Flanagan were all gone, too. Some had been let go. Some of them had burned out and chosen new careers. A number of them had not liked the arrival of the new management and felt that Astra's investors had pushed the company in a bad direction. And whether they were still at the company or not, Astra's current and former employees were wondering if they should hold on to the company's stock or sell it while outside investors were pushing shares higher as they frothed excitedly at the idea of being in the rocket business.

In late August, the revamped Astra team had its first chance to prove that it was moving the company in the right direction. Astra would not be performing a true commercial mission as Kemp had promised because no customer wanted to put a real payload onto the company's rocket just yet. The US Space Force, however, agreed to back the launch as a test of Astra's abilities and put some sensors aboard the rocket to see what life might be like for a satellite hitching a ride inside. Per usual, a few people went to Alaska to prepare Rocket 3.3, while the rest of the Astra employees gathered in Skyhawk to do their jobs in the mission control center (or just to watch).

Unlike times past, Astra managed to launch that machine quickly and in one of the most spectacular fashions anyone in the rocket industry had ever seen. Right at liftoff, one of the five engines failed and triggered a small explosion. The lack of thrust caused the rocket to list to one side. Rather than topple over, however, the rocket kept moving slowly sideways while a few feet off the ground. It drifted away from the pad and toward the metal fence surrounding the launch area. The rocket looked almost like someone who had had far too much to drink, gotten up out of his chair, and then started stumbling in one direction. Comically, the rocket kept hovering above the ground and then went right through an opening about twelve feet wide in the fence, as if someone had known all that would happen and had left a gate open for it. After drifting for an uncomfortably long time, the rocket did what no one watching it expected and actually started to fly up into the sky. It had burned off enough fuel for the four remaining engines to do their jobs and push the machine toward space. Astra employees and those watching a video stream of the launch had their jaws open as the rocket made its way through Max Q, reached supersonic speed, and flew for two and half minutes before it began to veer off course and its existence was terminated by the spaceport officials.*

For a couple of days, the rocket was a minor internet celebrity. Neither grizzled aerospace veterans nor the space geeks who spent hours watching launches online had ever witnessed a rocket fail so bizarrely or have such a redemptive moment. Once again, people were confused about how the machine made them feel. It was mind-boggling that an engine had given way the moment the launch had begun. Yet someone had to be doing something right for the rocket's software to have held firm, figured the situation out, and corrected itself.

The closest Astra observers, however, knew that the launch was in some ways the worst yet. Rocket programs tend to make steady progress: The first rocket blows up. The second one flies for a while and then blows up. The third one almost makes it, and then there's a success and another success and another success. Astra, by contrast, had kissed the edge of orbit the previous December and then

* https://www.youtube.com/watch?v=kfjO7VCyjPM. Enjoy.

gone yakety sax eight months later. Worst still, the most tested and analyzed piece of equipment on the machine was the one that had felled it. Astra did not appear to be learning key lessons. "Space may be hard, but like this rocket, we are not giving up," Kemp tweeted.

In the past, private rocket companies had been able to do most of their experimentation behind closed doors. They had been under no obligation to let people watch them blow things up and usually started showing footage only once their rockets were working. That had previously been true for Astra, but now, as a public company, it had taken the bold step of making the video stream of its launch attempt live. The company that had relied on secrecy for so long was now out in the open and willing to be judged—and the judgments came quickly.

When the markets opened on the Monday following the launch, Astra's share price fell by 25 percent. The internet's finest commentators rushed to Reddit and other forums to express their dismay with Astra's performance. Many of them had invested in the young, promising rocket company and were dismayed to find out that rockets sometimes flew sideways and then up and then exploded. Where most companies have a few months or years to show whether their products were a success, Astra existed in a far more binary realm. You either made it into space and succeeded, or you failed as your product literally burst into flames.

In another first for the rocket industry, Astra immediately faced lawsuits as a result of the launch. Courageous lawyers from multiple firms offered to help investigate whether Astra and its top executives had oversold their abilities and violated securities laws by having their rocket blow up.

Those lawsuits, though pathetic and cynical, underscored the comedy of Astra's being a public company. So few people on the planet had any idea what it took to build a rocket or whether or not rocket making could be a viable business. The ones who knew the most, in fact, generally thought getting into rockets was a very poor idea. Astra, though, had found itself out on a stage being judged by a public that thought themselves space savvy, that dreamed of participating in space in some way, or that simply wanted to capitalize on the fluctuations between idealistic hype and gloom.

Going public was a modern Faustian bargain for Astra and for Kemp. Under no other circumstances would an organization like this one have been given a seventh attempt to fly a rocket into orbit. Yet because it had raised so much money, Astra could, in theory, keep building and blowing up rockets for years to come. It seemed, at least to me, completely irrational that things had turned out that way. Hundreds of millions of dollars had flowed toward a product that didn't work and had questionable value even if it flew just fine.

Looked at another way, though, it was perfect.

Chris Kemp embodied the Silicon Valley spirit. Within his fit, well-fasted frame existed limitless supplies of optimism, energy, competitive fire, and intelligence. Those qualities merged with a need to push laws, authority, and just about everything else to their limits. He had spent his whole life training to play the game, and he played it well. Plenty of people disliked him and thought him to be unscrupulous. Others found him endearing and, more than anything, useful. In many ways, Kemp was purpose built to lead a company in the bonkers New Space era.

One of the Astra employees who had been let go perhaps put it best. He was not a Kemp fan, but he was a realist. He had seen many of the people with rocket-building experience come to Astra and struggle to build a working rocket. He had seen Adam London, perhaps the smartest person at the company, being pushed away from the machines he loved to deal with budget issues and lunch options. He had seen people arriving from big-name companies and receiving titles and stock options as they tried to cash in on the next big thing, be it a rocket or an advertising algorithm. "The one person who was transparent through all of this was Kemp," he said. "He's the only person in the building who did exactly what he said he would do, which is get a bunch of money to let everyone else try and make their dreams come true."

COMMERCIAL SPACE IN PICTURES

LAUNCH VEHICLES

The latest generation of rockets, some reusable, can deliver stuff to space cheaper than ever before.

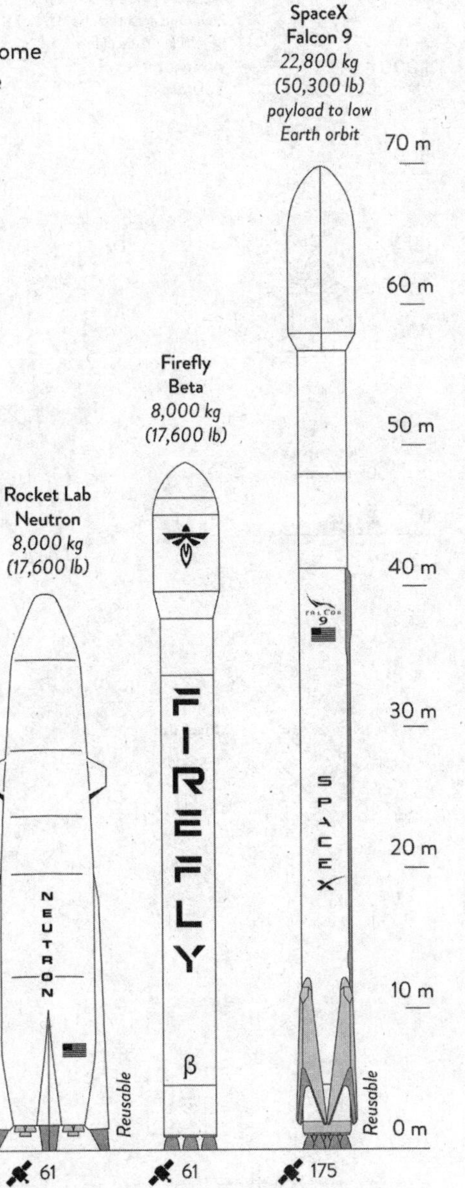

CREDIT: EVAN APPLEGATE DATA: COMPANY REPORTS

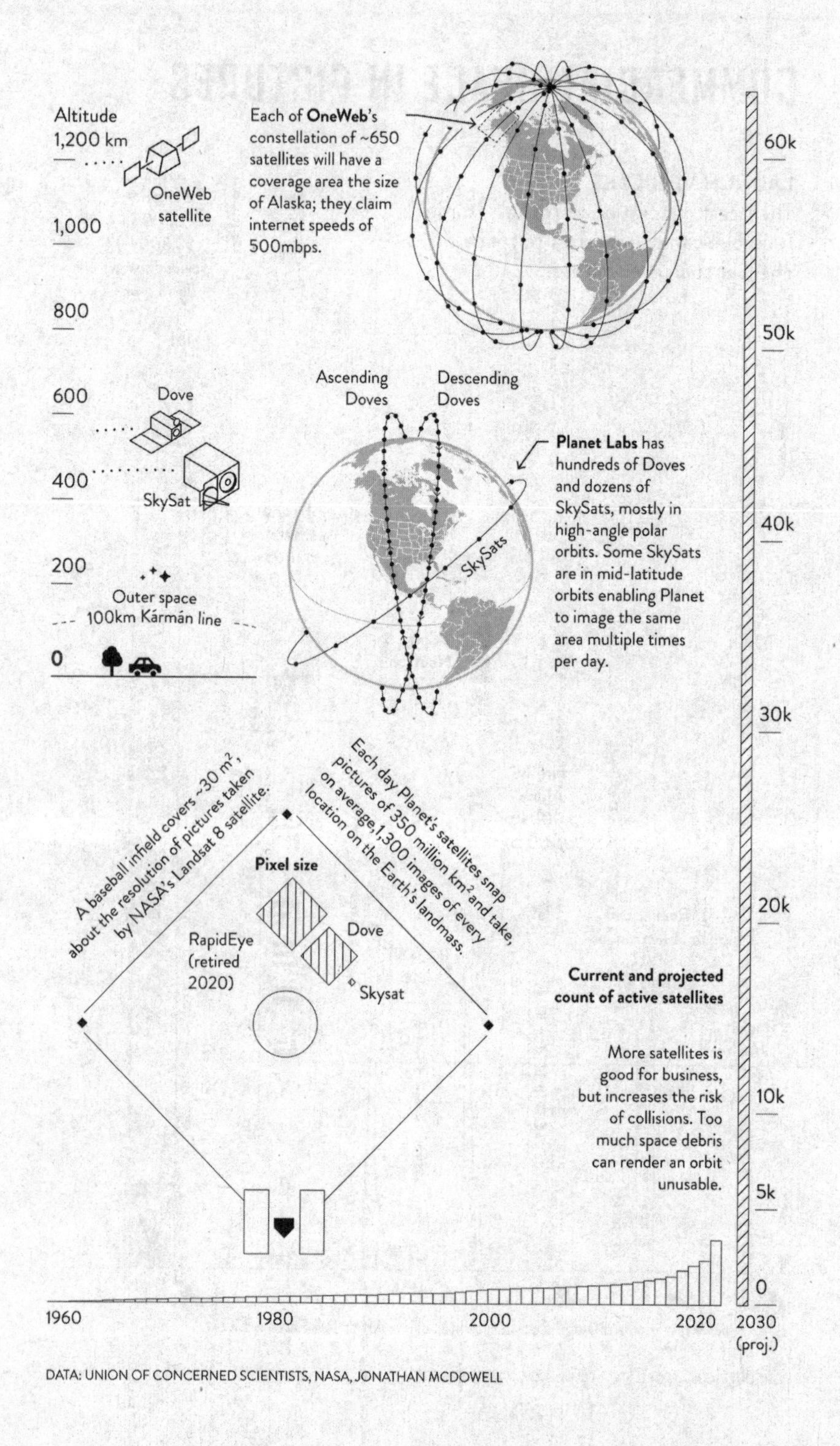

DATA: UNION OF CONCERNED SCIENTISTS, NASA, JONATHAN MCDOWELL

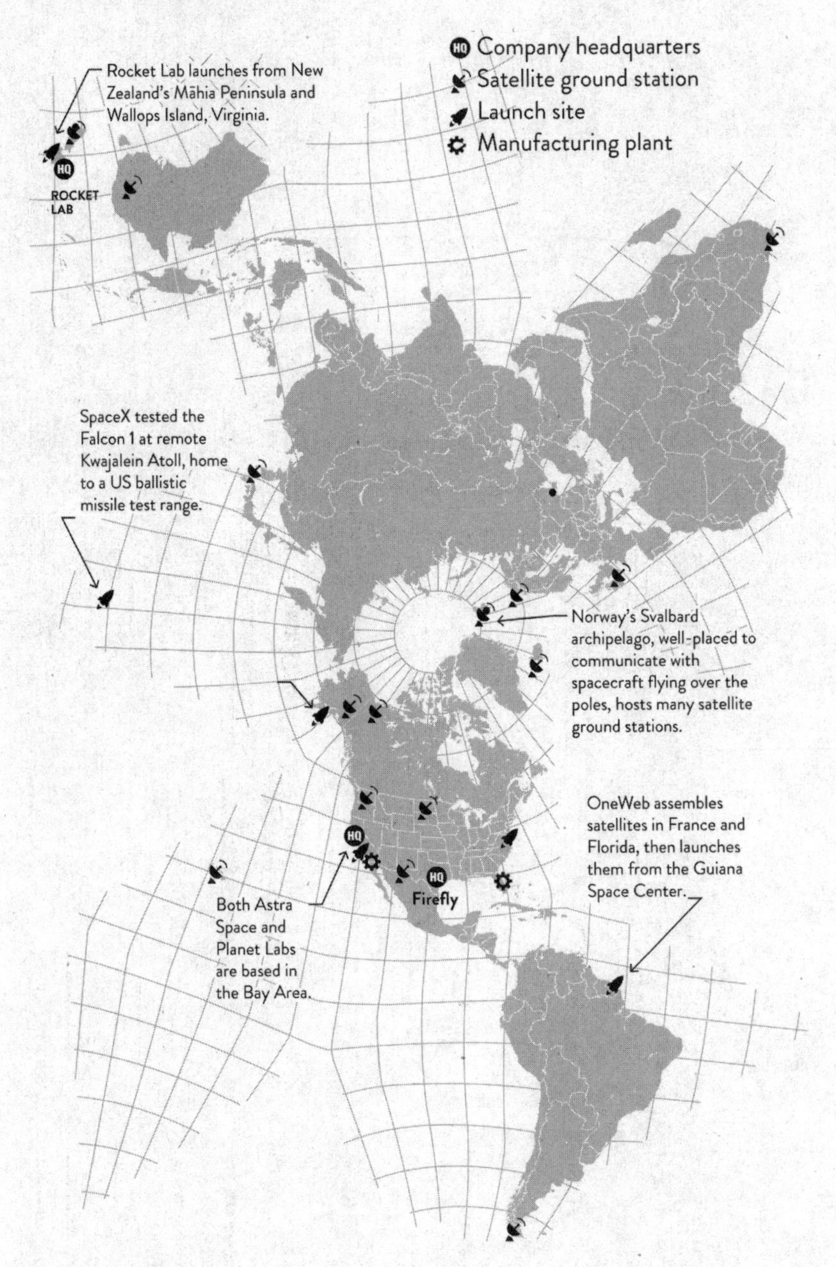

DATA: COMPANY REPORTS

THE MADDEST MAX

CHAPTER TWENTY-EIGHT

ON PASSION

Max Polyakov was in Texas, so he wanted to do Texan things. It was an afternoon in October 2018, and he hopped into a truck with a couple of other people to drive to the 5-Way Beer Barn in Briggs, about fifty miles north of downtown Austin. Because that was where you could pull up, roll down your window, be greeted by a nice woman who would ask, "What y'all need today?," and then have beer, booze, or a bale of hay plopped right into your vehicle.

Polyakov hails from Ukraine and displays next-level enthusiasm for many things. Upon approaching the large red barn, for example, he exclaimed, "Beer Barn! Look at that! Enter only! Look at that! Look at that! Look at that! We've come at the right time for the Beer Barn!" He reveled in the corn feed for deer. His eyes flitted from the chewing tobacco to the beef jerky. He was surprised and then excited that the Beer Barn accepted credit cards in addition to cash. He took it all in and savored the experience in a way that should make most of us jealous. "Culture is culture!" he said on the drive back, twenty-four-beer variety pack in hand.

In all likelihood, you have not heard of Max Polyakov until now, which is both surprising and not. After Elon Musk and Jeff Bezos,

Polyakov has risked more of his personal fortune on the grand gamble that is space than any other human: he has plowed $200 million of his own money into the rocket start-up Firefly Aerospace. For good reason, then, he had come to Texas to see how the money was being spent at Firefly's vast rocket engine test site and manufacturing facilities (located about a half mile from the Beer Barn) and to make his presence felt at the company's office headquarters in nearby Cedar Park.

I'd first met Polyakov the year before out in Silicon Valley, where we both lived at the time. People in the aerospace industry had been talking about a mysterious rich guy from Ukraine who had gotten into the rocket game, and I wanted to know who he was. According to information on the internet, Polyakov ran something called Noosphere Ventures Partners, although exactly what that organization did remained vague. He'd also been involved with some very successful online dating and gaming sites and business software companies. The public résumé, though, did not really explain how he'd made enough money to bankroll a rocket company or provide many clues to what type of man Polyakov was.

My initial visit to Polyakov's office in Menlo Park revealed a man in his early forties of slightly above average build with a dad paunch, closely cropped light brown hair, and a face that oscillated between cherubic and mischievous. Polyakov greeted me at the door and showed me around his building. The man clearly liked sci-fi art and had a sense of humor. He had a sculpture made out of dozens of small security cameras positioned over the entryway and another of a metal cyberpig by a conference room. Elsewhere, there were tons of space artifacts, like models of rockets and satellites and hunks of machinery. He also seemed fond of religious iconography and had paintings of saints on many of the walls.*

The varied decor matched the human speaking to me. Words and ideas blasted out of his mouth at incredible speed in a Slavic accent that took some time to understand.

* That visit took place years before the Russian invasion of Ukraine. The war would have a major impact on Polyakov and the other people and places that follow.

At the outset of the verbal onslaught, I learned that Polyakov had been lying low because he did not really trust the press, particularly the Western press. He'd agreed to meet with me on the recommendation of his chief lieutenant, a Belarussian named Artiom Anisimov, who had vouched for my reporting and offered the possibility that I might not be a terrible person.

After about a half hour of chitchat, Polyakov started to warm up and reveal himself. Firefly was in the midst of building its first rocket. It would be much larger than the machines made by Rocket Lab, Astra, and other start-ups but smaller than SpaceX's Falcon 9. If the small guys were sedans and SpaceX was an eighteen-wheeler, Firefly would be the minivan of space. It could fly a lot of cargo and do so at a reasonable price. But being fixated on the rocket was stupid, Polyakov told me, because he had much grander ambitions. He wanted to make thrusters, satellites, and software and more or less take over a huge swath of the aerospace industry. His go-to phrases to describe his business strategy were "full attack" and "with passion," and he delivered several soliloquies centered around those words and on how his energy and business acumen would combine to annihilate the competition and fuck them into oblivion.

Needless to say, I liked Max right from the start. Then things just kept getting better.

Polyakov told me that his parents had worked for the Soviet space program. Ukraine had been an engineering and manufacturing center for both ICBMs and spacefaring rockets. The country's economy, however, had languished since the fall of the Soviet Union, and the aerospace industry in Ukraine had collapsed, leaving its talent to go to waste. Polyakov's grand plan was to do something that would have been unthinkable during the Cold War era: combine the best of old Soviet aerospace engineering with the best of American New Space engineering. He would do so by setting up factories and research facilities in Ukraine that would employ the top engineers from the Soviet space program along with bright young up-and-comers. Those employees would develop technology and send it to Texas. Firefly would become a nexus of the knowledge of the two greatest space superpowers in history. Even Elon Musk would stand no chance against such a beast.

The plan sounded daft because it was daft. The United States puts strict controls on aerospace technology, and neither the government nor the military would likely warm to the idea of a random Ukrainian guy creating an intellectual property conveyor belt between the former Eastern Bloc and central Texas. Yet there was something romantic about the idea. The world had changed; why not find a way to make use of decades of knowledge and fabulous engineering instead of letting it sit idle or, worse, go to an enemy of the United States? Perhaps Polyakov was the one person who could make something like that happen.

Based on the same internet fumbling I'd conducted, people in the aerospace industry who'd heard of Polyakov were already peddling the idea that he was either dodgy or a spy or a dodgy spy. His record of owning dating sites coupled with his Ukrainian background made it all too easy to suggest that he must be up to something suspicious. Maybe he was going to siphon US technology back to Ukraine or—even worse—Russia. Maybe he'd made his money in unscrupulous ways that we didn't really know about. Plus, he talked funny.

In all honesty, I didn't care about any of that at the moment of our meeting. The man before me was bombastic and hyperbolic, but he was obviously also very intelligent and full of a joy for life that made you want to be around him. I'd never heard someone interlink talk about space, business, and his homeland with such zeal and romanticism. Polyakov said he had come to the United States to show that an immigrant could do the grandest of things and that he wanted to make great rockets to benefit this great nation and hopefully help his people back home along the way. And a big part of me believed him.

THE FULL PICTURE OF FIREFLY'S history and present is complicated, and we'll get into both later on, but what you need to know for now is that the company was founded in 2014 by a man named Tom Markusic and two other people. Markusic ran Firefly for a couple of years until it went bankrupt in 2017. That was when Polyakov appeared almost like magic and rescued the company. He even let Markusic return as CEO.

Markusic boasts serious space credentials. He's worked at NASA, SpaceX, Blue Origin, and Virgin Galactic and has done pioneering research into futuristic areas of rocket propulsion. Though not a Texan by birth, Markusic has learned to play the part well. He often dresses in boots, jeans, and a short-sleeve collared shirt. He sports a mustache and goatee. He likes guns and beer and trucks and combining all three when possible. Perhaps most important for anyone method acting Texan, Markusic goes about his day full of maximum confidence and swagger. He says things to his team such as "One thing that I hammer home to all of you every week is 'Work smart,' and the second thing is 'Get back to work.'"

The magic of pairing Polyakov and Markusic was that it created a stage on which the theater of the absurd could play. An eccentric rich guy from Ukraine owned a rocket company that no one really wanted him to own. He'd allowed Markusic to hang around as the boss of Firefly in the hope that Markusic really did know how to build rockets and because it made the rest of the world feel more comfortable about Firefly's existence. Neither man, it would turn out, cared all that much for the other, but they were united by the same New Space fantasy as everyone else: Maybe we can do this? Yeah, that would be so cool!

As proof of how wonderfully bizarre those two men were in concert, may I present to you the moments that led up to the Beer Barn run. Polyakov was surveying Firefly's grounds out in the Texas countryside, while Markusic sought to prove that he was a generational talent who wasn't about to let his rocket company go bankrupt for a second time.

POLYAKOV: Texas is good! Good! Not like California. Real America. True stuff. Good! When I came here, I could feel that passion, that energy, everything being synergistically aligned and moving together toward the stars. It's like finding a wife. You just see the beauty at first, and then all of these other good characteristics open up. You just wake up, you feel energy that you shall do something, right? Yeah. Yeah. Full attack—attack. Attack! You hear me? Yes. That's why we build rockets!

MARKUSIC: Here's our test site. The most important thing about a test site is that you have good weather, because you're working outside a lot. Texas is usually dry year-round. It gets hot in the summer, but it's bearable. The winters are quite mild. You need lots of space because you're making lots of noise. You need the space where if things go wrong and blow up, you're not affecting the land or anyone around you.

There's a lot of oil industry in Texas, and a lot of the parts of building a test stand involve building structures and welding and piping and plumbing and valves and machine shops and things like that. So there's actually a lot of supporting industries here.

We looked all around Austin, and this place had the right combination of low cost and low regulation. We could build anything we want out here without any permits or anything like that, which is important to make things low cost and fast moving. There aren't any ordinances for noise or really anything that would prevent us from doing whatever we need to out here.

POLYAKOV: Our land! Look at how beautiful it is! Firefly Farm.

MARKUSIC: Rocket Ranch. We pay like three hundred dollars a year in property taxes because this is a farm. Max is a farmer.

POLYAKOV: I look like a farmer, but I have a rocket hobby. Oh! Look over there. My lovely cows!

MARKUSIC: Yeah, there's probably a hundred of them out here. These cows have seen some things.

POLYAKOV: Lovely cows! Look at my lovely cows! Lovely test site! More cows! Look how many cows we have! We are like space oligarchs in cows.

MARKUSIC: Beef.

POLYAKOV: Beef!

MARKUSIC: Beef and rockets.

All right. We'll go look at some hardware in a minute, but first I want to show you the new test stands. We're building the second stage of the rocket up on the stand.

POLYAKOV: All from scratch! Look at that cow. Is that the dick of the cow?

MARKUSIC: No, it's probably an udder. Ha. That cow has four dicks! Let's go over here. I think we can see things better.

As Max said, we built all this stuff from scratch. This was just a cow field. Underneath the ground, there's caliche, which is the stuff they use for road base. We mined a lot of it ourselves out of the ground to make the roads here. We do everything in-house. All this welding, everything you see was done here in our little buildings. At SpaceX, I ran the test site early on, and I really learned a lot there. I think this is going to be what distinguishes us.

Over here, this is our composites shop. We do a mix of things. We have clean rooms for the engine assembly and fluid system assembly. That's an oven where we cure all the composite parts.

POLYAKOV: Heavy metal. Not Silicon Valley. You see? Heavy metal. How many tests will happen today? Will there be more engine burn tests?

MARKUSIC: I tell them do it twice a day. Run a test. Take the engine down. Fix it. Put it back. It should take them about five hours to get it ready for another one. We want to build another facility over there back in those trees. We can build a wall around it like the Alamo or something and make a safe house for you, Max, a fortress.

POLYAKOV: Some bloody Russian spies come and try to kill me.

MARKUSIC: Yeah, so when the Russians come, you can just go in there, and it's like "Fuck you."

POLYAKOV: Shall we go to the Beer Barn?

MARKUSIC: We've got plenty of beer at the office.

POLYAKOV: Come on, that's not fun. Beer Barn!

MARKUSIC: Okay, we'll get some beer. You think they're open at two o'clock in the afternoon?

POLYAKOV: Of course they're open at two o'clock. What else are they doing? Beer Barn!

CHAPTER TWENTY-NINE

GOD TOLD ME TO DO IT

Tom Markusic's story was simple. He was on a mission from God to spread human intelligence throughout the universe.

Markusic grew up in Mantua, Ohio, a blue-collar, one-stoplight town of about a thousand people in the northeastern part of the state. His father was an assembly-line worker for General Motors, and the family had a plot of land in the countryside where Markusic frolicked with his two brothers, trapping fox and muskrats.* Though he would eventually become embroiled in esoteric areas of physics, Markusic found that the people around him in his youth stressed labor and brawn over brains. "I grew up in a family of very humble means that was hardworking but did not have an academic bone in our history," he said. "In that part of the country, you didn't necessarily need that. There were good industrial jobs, and so that's what people did."

When Markusic turned thirteen, he got a job at Stachowski Farm, a family-run business famous for breeding and training Arabian horses, and started out tending to the horses and bailing hay.

---

* Instead of getting an allowance from his parents, Markusic made money by killing the animals, skinning them, and selling the furs.

Eventually, he worked his way up to repairing mechanical equipment. That gave the teenager his first taste of grease and tools and, more important, forced him to understand how the machines functioned from the ground up—or, as he put it, to learn the "essence of engineering." "If you've ever torn apart a hay-baling machine, it's like a garden of wonders inside," he said. "It's like a Rube Goldberg device with little pieces of metal and tiny little knots and all these things sliding together. The mental part of dealing with this stuff was learning to push through problems and persevere even when you feel like you don't really want to do it anymore."

An unexpected bonus of working at the farm came from the circles of people Markusic encountered. Some of the Arabian horses were worth up to a million dollars. Rich buyers and their rich kids would stop by during the summer to inspect the animals and ride them. Markusic would also head out on the road to horse shows held around the country. Through those excursions, he discovered a stratum of society that impressed the small-town midwestern boy and made him aware of the opportunities beyond the confines of Mantua.*

Markusic's parents rarely paid attention to his schoolwork, and they didn't really need to. The young man felt a natural draw to academics, going so far as to describe school as both "liberating" and "seductive" because it was a realm that he controlled and that came without any nagging expectations. He did well in the required subjects and also dug deeper into things that caught his interest.

By the time high school ended, he did not know what he'd do next. Yes, he'd done just fine academically, but no one in his family had ever raised the idea that he might attend college. Markusic, though, was dating a schoolmate named Christa English, whom he'd been friends with since they were both ten years old. English's family was in the real estate business and saw attending college as a must. Mr. English made it clear that if Markusic hoped to keep dating his daughter, he'd need to pick a field of study and a place to pur-

* Markusic remains in touch with the farm owners and hangs out with them when they visit Austin to see one of their customers, Michael Dell.

sue it. Markusic thought about the question for a bit, remembered that he'd always liked rockets and planes as a kid, and figured he might as well go to school to study them.

The weird thing is that Markusic has a second version of his aerospace origin story that has nothing to do with chasing a girl or trying to placate her parents. In this far more mystical tale, Markusic is back in fifth grade. He's just received a model rocket kit for Christmas. He rips open the package, puts the thing together, and heads to the two-hundred-acre cornfield near his house. The sky is blue, the air is still, and there's a vast expanse of virgin snow before the eager kid. He slides the rocket down onto the metal pole coming out of the launchpad and lights the fuse. Then things get heavy.

"The sound kind of shocked me," he said. "I remember looking up at this blue sky and just seeing the speed of it going up and then just sailing. It separates, the parachute comes out, and it's coming down. By that time, I can smell it—the smoke coming out. If there was a camera on me, I would have a grin ear to ear, I'm sure.

"And that was, well, it was kind of metaphysical. I felt like a meeting with destiny there. Like 'This is living. This is what I was built to do.'

"I'm a Christian guy, and I do believe in Providence, and I believe that we all have a purpose and there's a plan for all of us. I didn't know it then, but I think, you know, that I was kind of locking into the resonance of the universe of what my thing is. And so I think I'm built to build rockets and that was the first emotional response I had to the whole thing. At fifth grade, even though I didn't have any guidance or know what I wanted to do, it was kind of set right there. That was a deep passion for me, and that was what I was going to do with my life."

Take your pick as to the cause: Christa English, the divine model rocket, or some combination of the two. Whatever the inciting incident was, Markusic chased it with a hell of a lot of determination. He first enrolled at Ohio State University and received an undergraduate degree in aeronautical and astronautical engineering and then got a master's degree in aerospace engineering and physics from the University of Tennessee before heading to Princeton to pursue a doctorate in mechanical and aerospace engineering. "I ended up going to school until they ran out of degrees," he said. "It was pretty extraordinary to get a PhD from Princeton given where I came from."

Now, if you're on an intergalactic mission from God, you aim high, and that was exactly what Markusic did. He dived into the ultracomplex field of plasma physics, or the study of the "fourth state of matter."

There are gases, liquids, and solids, and then there is plasma, which can be thought of as an electrified gas. Crank a molecule up to a high enough temperature, and its atoms will begin to fray. Electrons will start separating from nuclei, ions will form, and the body of energetic chaos enters the plasma state. Lightning, the aurora borealis, the cores of stars, and nuclear weapons are all examples of plasma. For Markusic, plasma was a promising source of fuel for a new generation of rockets. Forget going to low Earth orbit or the moon; Markusic wanted to make rockets that could travel past Mars and even to the farthest reaches of the solar system. To do something like that, you need to replace the chemical explosions taking place in conventional, fire-breathing rockets and turn to using the almost unlimited potential of plasma, where in essence you're putting electrical energy into matter and then accelerating it.

As Markusic explained it, "Instead of just heating things up, we interact with the nuclear structure of things. So we use electrostatic and magnetic fields to interact with matter. If you use these electromagnetic interactions, you can push on matter and particles and accelerate them to very high velocities, much higher than you would get out of a thermal type of device."

The air force and NASA were very interested in this type of technology both for military operations and deep-space exploration. The two organizations paid for Markusic's schooling, and in return, he first did a stint at Edwards Air Force Base from 1996 to 2001, working on plasma thrusters for satellites. The military hoped to use the technology to readjust imaging satellites on the fly, moving a satellite focused on India, for example, into an orbit over Iraq if a conflict or other urgent situation arose. After that, he ended up at NASA's Marshall Space Flight Center in Huntsville, Alabama. He became one of the first members of an advanced propulsion research team, working among large vacuum chambers and varied propulsion experiments that tested the properties of such things as antimatter and nuclear energy. There were hopes that the technology would be used to explore Jupiter's icy moons and perhaps make it possible to reach Mars in record time.

During his five years at NASA, the agency began to shift its priorities away from basic research and toward human spaceflight. Even at a place like NASA that's full of nerds, Markusic fell into the supernerd category, and the funding for many of his types of projects dried up. To placate him and keep him interested, NASA tapped him to be the head of a quasi-*X-Files* unit that investigated the weird letters and emails that arrived from individuals claiming to have hit on major breakthroughs or mysterious phenomena. "Some guy would call and say, 'I've got this thing floating in my garage' or 'I've invented a perpetual motion machine,' and I would hop on a NASA private jet and go investigate it," Markusic said. "I'm a PhD. I understand physics well. I understand engineering well. So I could go evaluate these things and come back and say if there's something real here or not. There was never anything real."

On one occasion, he received a letter from a company based in a strip mall. He flew out to visit the operation and found a guy with a triangular object before him and several senior citizens lining the wall of the room, who were there to observe the proceedings. The object was a foot long on each side and had a couple of wires hooked up to its body and six-foot-long strings attached to its sides. After some chitchat, the man flipped a switch on a power supply, and the object suddenly flew up and hovered a few feet in the air with the strings now stretched out, holding it in place. "It's just basically floating," Markusic said. "And as I got closer to this thing, lightning bolts start shooting out of my toes into the ground. It stung a little. It was kinda disturbing." Following the demonstration, Markusic started chatting up the inventor and realized that the older people in the room had sunk their life savings into what they believed was an antigravity device.

Perplexed, Markusic returned to his hotel room for the night and began pounding away at the internet, looking for explanations. After doing some digging, he hit on something called the Biefeld-Brown effect, in which a high electrical voltage (60,000 volts or so) is applied to a pair of wires, which causes the air to break down into a plasma, which in turn creates an ionic wind strong enough to lift an object.

The next day, Markusic returned to the inventor's office, where the investors were also present and anxiously awaiting the verdict of the man from NASA. Markusic explained that the device was not a breakthrough discovery but rather a replication of a well-known experiment. It was not an antigravity machine, and it was of no use to NASA because it required air to do anything interesting, and there is no air in space. Markusic added that NASA would not be contributing the $100,000 the inventor sought to further his research. At which point he left the office, and the investors' stomachs turned in unison.

Though the *X-Files* stint served as a weird diversion, it did not really satisfy Markusic. He spent his downtime reading management books and going through payroll charts to try to learn how the business world differed from government. The self-education program was okay, but he thought most of the books were full of information that should be self-evident instead of brilliant insights. He grew bored and felt that he was dying on the inside. Just as he began to entertain the idea of leaving NASA, something remarkable happened.

In the middle of 2006, David Weeks, one of Markusic's *X-Files* colleagues, strolled into the office and said he'd been tasked with an intriguing mission. A multimillionaire or billionaire or whatever was out on a jungle island in the South Pacific trying to launch a rocket. NASA wanted to observe that rich guy and his crew and see if there was anything to learn from the operation or any cause for worry. NASA's management wanted Weeks, a veteran, to take along someone young, and Markusic ended up as the Scully to his Mulder.

Markusic had never heard of Elon Musk or SpaceX and didn't find the mission all that interesting. From what he'd been told, SpaceX had a small rocket called the Falcon 1 that relied on the typical small-rocket propellants of liquid oxygen and high-grade kerosene. It was a fire-breather, not a plasma dynamo. Boring. "I was about advanced technology, not some mundane thing where we'd already figured out all the problems," he said. Still, a trip to the South Pacific seemed better than another trip to a crackpot's lab in a strip mall. Markusic and Weeks flew from Alabama to Hawaii and then hopped aboard another plane and headed out to the middle of nowhere—to Kwajalein Atoll, part of the Marshall Islands.

The main hotel on Kwajalein looks like an army barracks, and that was where Markusic set down his luggage and stacks of management books. The rest of the surroundings were anything but Huntsville. Ronald Reagan–era military complexes and top secret weapons systems. Japanese pillboxes from World War II. Sharks. It was more of a scene than Markusic was prepared to deal with. "It was just a wild place," he said. "I'm sitting there on the beach praying 'Be one with nature. Be one with nature.'"

Each day, Markusic and Weeks boarded a catamaran and sailed to Omelek, the island in the atoll where SpaceX had set up its operations and built a launchpad. A couple dozen people from the SpaceX crew would be there banging away on the Falcon 1. Many of them were in their twenties, wearing T-shirts, sweating, swatting away bugs, and generally not looking anything like NASA rocket scientists and technicians. As the sun set and night came, they would light huge bonfires, bust out their stashes of alcohol, and party.

A month of that went by. It should have repelled Markusic and his NASA sensibilities, but it intrigued him. He went home each night to books full of corporatespeak and jargon and platitudes. Meanwhile, those young men and women were getting their hands dirty and doing something pretty incredible. Everything Markusic had been chasing felt fake and synthetic. The SpaceX work felt real, and a spiritual wave washed over him just as it had when he had been in the snowfield as a kid.

"I mean, these guys were one with the machine," he said. "They were putting the rocket up. It was cutting them. It was disappointing them. Things were going wrong. There were people drinking. People were teabagging* each other. I mean, very unprofessional behavior, but at the same time strong teamwork and behavior that seemed pretty advanced from the management stuff I was reading.

"At some point, it just clicked to me that, man, there's a whole world about pretending that you're doing things, and then there's a world about really doing things. So I started working with them more day by day, picking up wrenches and actually getting involved. And then at night, we'd go fishing and camp on the island with the bonfires and coconut crabs crawling all over us."

Markusic realized that, yes, a lot was known about liquid-fueled rockets and how they should work, but actually making them fly proved tricky. The practical engineering brought challenges. So, too, did the elements and the pressure of the situation. The bugs. The heat. The wind. The salt air doing its best to rust anything it could. The struggle seduced him.

What's more, he came to see that people would not start chasing after Mars or the exotic new technologies needed to explore deeper into space unless some real energy was injected into the industry. Getting things into space had to become a regular, familiar experience and an affordable one. Private companies needed to lead the charge. It would take SpaceX and dozens of other New Space companies to perfect the basics, which would, in turn, open a path to pursuing the more advanced technologies that had already consumed about a decade of Markusic's life. "I basically went native on that jungle island," he said.

After returning to the United States, it took him only a few

---

* For those who are unfamiliar with the term, teabagging in this context refers to the practice of placing one's testicles onto the face of an unsuspecting victim in a manner meant to cause general amusement for others in the vicinity. Though such an act would not be condoned under NASA protocols, SpaceX had no protocols.

weeks to ingratiate himself with Musk and a few key members of the SpaceX staff and receive a job offer. By that point, he had married his high school sweetheart, Christa, and they had three children with a fourth on the way. SpaceX didn't pay well at the time; it mostly dangled the promise of future riches in the form of stock options to employees to keep them interested in working borderline inhumane hours. The young employees, who made up the bulk of the workforce, could cope with that, but Markusic had a family to support and a mortgage to pay. Christa had little desire to uproot the family from their nice redbrick Alabama home, move into a smaller, more expensive place near SpaceX's Los Angeles headquarters, and suffer through soul-sucking traffic. Markusic presented those concerns to SpaceX, and it offered him another path: to head to the bush country of central Texas and build out the company's rocket engine test site. Big houses come cheap in places where few people want to live.

Markusic took the job and with it placed a massive wager. The bureaucracy of NASA made it almost impossible to fire someone like him. He could have spent the next three decades reading books, tinkering, and hoping that NASA would get back into the wonders of plasma thrusters, all while living a comfortable existence.* Instead, he opted to hop aboard the World's Riskiest Company Express. After all, SpaceX's first Falcon 1 had launched in March 2006 and promptly come right back down onto the launchpad, wrecking the start-up's island facilities. No one really knew how much money Musk had to see the venture through. It barely made sense to take out a mortgage on one of those big homes in Texas because the whole adventure could end in a few months.†

"When you work in government, you see one program after another being canceled due to political influences or whatever," Markusic said. "It can be very demoralizing because you work hard, you get knocked down, you work hard, you get knocked down. You

* In NASA terms, Markusic was a "gold badger," meaning a full-time government employee as opposed to a contractor.

† "You had to be young and crazy to do it," Christa said. "But the problem," Tom said, "was that I was older and crazy."

can work your entire life going through cycles of things that start with big fanfare and never get finished. And you can say, 'Well, gee, I'm going to have a very secure retirement and everything, so it's okay, I don't give a shit.' Or you can join up with some crazy-ass guys that put their own money on the line and have very audacious plans and visions. And you can get on board with that and actually try to finish something. This New Space is an opportunity to finish things."

In June 2006, Markusic turned up at the McGregor, Texas, test site, where only a handful of SpaceX people worked full-time. The site had a storied history as a place where people blew things up. During World War II, the military had produced TNT and bombs there, and in the decades following, the land had been used off and on by chemical and munitions manufacturers and aerospace companies testing their hardware. The tenant right before SpaceX had been Beal Aerospace, a private company founded by the billionaire Andrew Beal that had managed to churn through many millions of dollars in three years before folding in late 2000.

The immediate surroundings did not inspire much confidence. There were cornfields and cows, a little black house with a boat parked out front, and winds whipping around in 110-degree heat. One of the workers at the site, Joe Allen, had been there for decades—in the military days, the Beal days, right up to the SpaceX days. He was beloved for his spirit and his intimate knowledge of the land—"Don't dig there. We did that back in '78, and it ended bad."—but he also reflected some rougher realities. Allen's first wife had shot him once, and he'd been served with paternity papers three times over his years at McGregor. "But only one of them was mine," he liked to say.

SpaceX employees had been commuting back and forth between Los Angeles and McGregor to test new hardware, treating the place like a hard-core engineering getaway. Everyone knew that the site had to be built out and turned into a more serious, permanent operation, which was part of Markusic's job, along with refining SpaceX's propulsion technology. Just a few weeks into his taking the gig, Tom and Christa had their fourth child. Christa sensed that the task of raising all those kids would largely fall on her for the foreseeable future. Tom had regularly been working at McGregor until 2:00 or 3:00 a.m. The cushy government job had given way to start-up hell. "I didn't want to coast to my death, you know," he said. "I wanted some adventure. But life is complicated because sometimes adventure is just stupid, a stupid risk."

Markusic got lucky. He arrived at SpaceX as employee 111 and would spend the next five years trying his hand at a wide variety of projects. His immediate focus was, of course, the Falcon 1 and any tests that needed to take place for the rocket to finally reach orbit. In September 2008, after a long, hard slog, a SpaceX rocket did just that, the company having spent about $100 million along the way. Christa penned a blog post to mark the occasion, playfully nodding to the couple's Christian sensibilities in the process:

> Sept. 29, 2008—And it came to pass . . .
>
> And it came to pass upon the Earth that the Falcon soared. And there was great rejoicing, revelry and relief all around. To our west, the Great Man* himself was overcome. . . . And grown men embraced and even wept. There was much kissing and much joy. As the tidings of success reached far away places, offerings of praise and congratulations were dispatched from the corners of the world.
>
> And it was good.

* The "Great Man" descriptor used by Christa can be read two ways: as others saw Musk and as Musk saw himself. Christa never really enjoyed seeing her husband subjugate himself to another man. During one of our interviews, she reported that she had recently been reading my biography on Musk and had almost finished with it when the family tomcat had decided to pee on it. She had found that hilarious and appropriate. "Little male cats are really obsessed with who is the alpha in the house," she said. "It was like Elon is in the house. It's like this testosterone-fueled kind of like territorial thing. He was drawn to the Elon book, like this big alpha-testosterone, brimming book. And squatted right on top of it. It was crazy. I thought, 'Well, I've got to finish the book.' So I took out a blow-dryer and tried to make it readable."

Many SpaceX employees have a love-hate relationship with Musk. They admire Musk's drive and all the opportunities that the man's unmatched vision and relentless spirit afford those around him. They also eventually find Musk's demands and wrath deflating and grow weary of the long hours. Markusic, though, loved working for Musk and appreciated his directness and trust in people to do their jobs. "He's like 'I want this, and this is when I want it, and now go make it happen,'" Markusic said. "I thrived on his demanding nature."

During the latter part of his stint at SpaceX, Markusic commuted between Texas and Los Angeles, where the company bought him an apartment. He did studies on what the market would look like for much, much bigger rockets and some early work on Raptor, the engine SpaceX ultimately designed to fly its giant rockets. The commute took him away from his wife and four young children for much of the time. By 2011, SpaceX had grown to about a thousand employees. Markusic began to think that it might be time for a change. "It had become clear that SpaceX would be successful," he said. "And this thing called the New Space Movement was real. To keep that going, there needed to be more SpaceXs, and I could help make that happen."

Markusic dropped a couple of emails to friends, putting out feelers to let them know he would consider new opportunities. In minutes, emails came back offering him jobs. Tom and Christa soon found themselves in a meeting with Jeff Bezos at Blue Origin's offices in Kent, Washington.

The trappings of Blue Origin's office dazzled the couple. Bezos had collected space artifacts such as cosmonaut suits, a model of the *Star Trek Enterprise*, and mailboxes that had been dinged by asteroids. Right in the center of the office stood a giant bullet-looking thing—a steampunk-styled spaceship that recalled Jules Verne. Bezos struck Markusic as smarter and kinder than the other tech billionaire types he'd met before, and Markusic agreed to take a job as a senior systems engineer.

Things did not go well, though. Unlike the frenetic, hard-charging SpaceX, Blue Origin moved slowly. The company's logo has a pair of turtles standing on their hind legs, reaching up toward the heavens and harkening back to the story of the tortoise and the hare. Bezos used the logo to try to stress that Blue Origin would focus on steady progress and winning a very-long-term race to industrialize space. The culture that resulted, however, was not to Markusic's liking. He would roll into the office and receive an email about a companywide bike ride later in the week or perhaps someone's favorite oatmeal cookie recipe instead of an order to work harder, faster, and longer. "I'd get fired for sending shit like that somewhere else," he said. "There was an emphasis on work-life balance that I was not accustomed to. It's wrong, but I found it kind of offensive. I felt like I was part of the museum, part of the collection. 'Over on your right, we have Tom Markusic, formerly of SpaceX.'"

Markusic lasted all of two weeks at Blue Origin.

Lurking in the background had been Richard Branson and his space aspirations at Virgin Galactic. Like Musk and Bezos, the lion-maned Branson had set out in the early part of the 2000s to create a business in space. Ever the showman, he focused first on space tourism, seeking to build a spaceplane that could carry people for a few minutes to the edge of space, where they would experience weightlessness and have the privilege of observing Earth from a rare perspective for $250,000 a pop. Virgin Galactic started in 2004 and had made some progress in the years since, although by 2011, it remained a decade away from having a real spaceplane and farther from having a real spaceplane business. Nonetheless, it decided to take on even more and to build more stuff.

Branson had been calling Markusic, telling him that Virgin was getting a big chunk of the old Falcon 1 gang back together and he should come join the group. Their plan was to build a small rocket similar to Rocket Lab's and later Astra's for carrying satellites instead of humans. The big difference would be that Virgin would fly the rocket up into the atmosphere on a plane and then let it go and have the rocket engines fire and do the rest. The pitch sounded wild, but it was better than being a relic in a space museum, so Markusic, Christa, and the kids set off for the Mojave Desert, where Virgin had its headquarters.

The heart of Mojave is the local airport. It has a plane parking lot where commercial carriers drop off aircraft by the hundreds to sit in the dry conditions and degrade as slowly as possible while they await repair, calls back into action, or their eventual scrapping for parts. There's a control tower, and there's also a long runway, which makes Mojave one of the favored spots to put prototype and unusual aircraft through their paces. Best to have some room for error.

Hangars surround the runway and are inhabited by that strange breed of inveterate tinkerer. Most notably, this includes space start-ups—three-, four-, five-man operations that operate in the desert for years, feeding off the occasional government or research contract to pay the bills while ever so slowly building the rocket of their dreams in the back room.*

Only a couple of companies in Mojave have facilities that stand out as more polished than the rest. Virgin has long been one of them. It had the money to buy a giant hangar well away from the hoi polloi who cling to the buildings by the airport runway. The floors at the Virgin factory† are coated with epoxy that makes them shiny and filled with lots of fancy machines. The hallways have life-size cutouts of Branson that you can take your photo next to. And there are real cubicles where people can work behind serious computers.

A short drive away from the headquarters, Virgin also has a large engine test stand that's a labyrinth of horizontal and vertical metal painted a mix of red and white. Just as in Texas, Markusic had to build out the site, going from a concrete pad and one tower to a world-class facility full of instruments capable of taking myriad

* Masten Space Systems is the canonical example.

† The building is named "FAITH" for "Final Assembly, Integration and Test Hangar."

measurements, firing off engines at will, and sending out roaring forty-foot flames. His main mission, though, was to figure out what Virgin's rocket—dubbed LauncherOne—would look like and how it would perform. The company came up with a target of putting about 450 pounds of cargo into orbit for less than $10 million per launch.

At the time, both the spaceplane business and the satellite-launching business were run by George Whitesides,* another Princeton grad, although with a degree in public and international affairs rather than anything to do with engineering. Whitesides's most legit space experience came from a one-year term in 2010 as chief of staff to then NASA administrator Charles Bolden, Jr., and, on the fringes, from his deep friendships with Will Marshall, Chris Kemp, and the Schinglers. Whitesides is a tall, skinny stick of enthusiasm and wide-eyed optimism, and for a while, the relations among him, Markusic, and the rest of the Virgin crew were just fine.

By 2013, however, Markusic's views about the future of LauncherOne began to diverge from those of Whitesides and others. They fought about the type of propulsion the rocket should use, how much payload the rocket should be expected to carry, and the distraction that Virgin's space tourism efforts posed. "What they wanted to do was not what I wanted to do," Markusic said. "It's not to say they were wrong, but our visions weren't aligned."

Markusic had cycled through the complete suite of space billionaires, which left no other option than the most painful one—DIY—if he wanted to chart a different course. As the good Lord would have it, Markusic's SpaceX stock had become quite valuable during his years hopping across the rocket industry. He liquidated most of his shares on the private market, minting a small fortune. He also made friends with a couple of businessmen turned investors, P. J. King and Michael Blum. The three men sat in Blum's hot tub one evening, drank several bottles of wine, and then made the

* Virgin later split the two businesses into two companies, with Whitesides remaining in charge of Virgin Galactic and new leaders for what became known as Virgin Orbit.

type of poor life decision that follows that much booze. "We put our heads together and said, 'Let's do this,'" Markusic said.

THE FIREFLY SPACE SYSTEMS JOURNEY began in January 2014.* Once again the Markusics played the role of rocket industry vagabonds, packing their bags and returning to Texas. This time, they traded the boonies of McGregor for something closer to civilization, with the company setting up its headquarters in Cedar Park, a strip mall–filled suburban town about twenty miles north of downtown Austin.†

Markusic knew that Rocket Lab had been around for years, but he considered the company more of a research-and-development organization than a full-fledged rocket maker. True enough, Rocket Lab had been doing work for the US military to develop new kinds of propellants and very small rockets that could be launched at a moment's notice. Its future workhorse, Electron, would not be revealed until August 2014. Markusic also clearly knew of Virgin's plans to enter the small-rocket market and considered its approach to be flawed. Other efforts here and there were described by Markusic as "Mohobby grade," referring to those dreaming tinkerers in their hangars back in the Southern California desert. "It's not too hard to get fire to shoot out of the end of something, but there's a world of difference between that and an orbital launch vehicle," he said.

Overall, Markusic saw the small-launch market as an open playing field and a race to see who could get something real into orbit first. And once some company did figure it out, the skies would soon fill with rockets carrying hundreds, if not thousands, of satellites into space every year. Markusic picked the Firefly name to conjure up that very notion after watching lightning bugs twinkle around his backyard. Why couldn't space look the same way? Myriad rocket engines burning their way through the void.

---

* For a few months, Firefly had its offices in Hawthorne, California, right next to SpaceX's headquarters simply because P. J. King, one of the cofounders, had an office there.

† Christa and the kids were happy with the move. Their new house was on three acres next to a nature preserve. Herds of deer slept in the yard. Wild boars ran through the land, too.

Along with the Cedar Park headquarters, Firefly acquired a two-hundred-acre plot of land in Briggs, a smaller, more remote town in the central Texas scrublands about a half hour north of the main office. There Markusic would build yet another test stand and a manufacturing facility for the engines and rocket bodies. He'd picked Texas precisely because of the state's "do whatever you want" culture and cheap land. Firefly could blow things up as needed and have the room to locate all its major operations right next to each other. No more building rocket engines in California and carting them out to Texas for tests, as he'd done at SpaceX. Engineers and technicians would be able to make tweaks to their products, put them into action quickly, and then take them right back into the manufacturing facility to make more tweaks and do more tests and on and on. Best of all, Cedar Park and Briggs were both close enough to Austin to attract talented engineers who wanted a low cost of living compared to California's while maintaining easy access to a colorful, thriving city.

In a moment of foreshadowing, Markusic also planned for the worst, while hoping for the best. "We knew that if this didn't work out, the employees would be able to find other jobs, unlike bringing people to Mojave or McGregor," he said. "We knew that if Firefly failed, they could sell their homes for a profit because this is one of the fastest-growing cities in America."

For a while, things at Firefly went really well. Markusic pulled some people away from his former employers and found numerous young engineers coming out of the University of Texas and other nearby schools. Trusted old hands such as Les Martin, who had built test sites for SpaceX and Virgin,* turned up at Briggs to lay cement and bend metal. Markusic tried to find his footing as the big boss, learning how to run a company and keep this complex operation on track.

In media interviews at the time, he disclosed that Firefly's first rocket would be called Alpha and that it would be a new take on

* And then Astra.

this type of machine. Instead of using aluminum or an alloy for the rocket's body, Firefly would build it out of carbon fiber.* The material cost more and required serious expertise to know how to make it, cook it in an oven, and shape it, but it also delivered major benefits in that the rocket would be lighter weight and stronger. The company also had some tricks up its sleeve in terms of propellants and engine design.

If all went right, Alpha would be able to carry a thousand pounds of satellites or other cargo into low Earth orbit by the end of 2017 for $8 million per launch. After that, it would build a larger rocket called Beta that could carry 2,500 pounds to orbit. And then, after that, it would build—get this—a reusable spaceplane called Gamma that would have rockets on its sides to take it off and help it reach orbit. The plane would carry out its mission and then glide back to Earth. You can tell that Markusic was really feeling it back in these early days of Firefly, because, like all rocket start-up CEOs before him, he presented the press with a litany of overly optimistic promises. The 2017 launch target was a doozy. A suggestion that Firefly could be profitable by 2018 was downright Muskian.

The Falcon 1 had taken about six years to build and fly into orbit. No doubt people like Markusic had learned lessons from the experience and could apply them and maybe speed up the process. To go from nothing to space in three years, though, would be one of the great engineering feats in history. Technology and start-up people always make unrealistic promises. It's part of the game, something that keeps the troops moving and investors feeling as though they've gambled on something tangible. But rocket start-up CEOs appear to suffer from a unique form of self-delusion more spectacular than their peers'. Perhaps it's just that the rocket game makes so little sense compared to other parts of the tech and business world that the ante must be upped. Rather than go more conservative, you go more ridiculous. Then the investors, who almost assuredly will lose all their money, can tap into the immense power of that gargan-

* Rocket Lab was clearly already doing so but was also considered an anomaly at the time and a not well-known one at that.

tuan lie and use it to overwhelm their rational neurons crying out for attention.*

Anyway, off went Markusic, spinning the Firefly story and enjoying the limelight as the CEO after being some billionaire's resident propulsion nerd for so many years. And the Firefly troops actually made their big-talking leader look good. In September 2015, the company issued a press release celebrating the completion of its first forty-foot-tall test stand and the construction of a nearby ten-thousand-square-foot manufacturing and control center. By the end of the month, Firefly got to use its shiny new stand to conduct the first test of its engine, and, by God, the engine worked. The company had grown to sixty employees, and Markusic decided to promise that it would hire another two hundred people by 2019, by which time, he said, Firefly would be producing fifty† of its Alpha rockets per year and likely turning a profit.

Markusic mastered the rhetoric needed to transition from futuristic plasma engine specialist to swaggering industrialist. He described Firefly's mission as building the "Model T" of rockets—something that could be made cheaply and consistently. The company would transform itself into a delivery company that would supply a burgeoning economy in low Earth orbit. "I like to think of SpaceX and Blue Origin as being like the Netscapes—the first wave—and we're going to be like the Google," he said. "I am very happy for Elon and Jeff to duke it out to be the human exploration pioneers to Mars. Like 'Y'all go do that. I'll just stay here and make a few billion dollars.'"

By the start of 2016, Firefly was the talk of the new aerospace industry. More than twenty-five companies around the world had announced plans to build small rockets and fly them into orbit. Elon Musk had convinced engineers in every corner of the globe that they

---

* One downside of being in Texas was that Markusic and Firefly could not wow investors with their eye candy as easily as if they'd been in California. "Rich people would really like to drive their friends over and show them the rocket they're developing," Markusic said. "Their investment is like part of their identity, and they want to show it to people." In my experience, this is comically true.

† Why stop there?

and a few friends could conquer space if they were just willing to try hard enough. Very few of the efforts had real money behind them, and real space people considered them jokes. Rocket Lab looked like an actual contender since it had venture capital money and non-imaginary rockets sitting in its Auckland factory. But it was run by a guy who had never gone to college and had zero aerospace experience at a company like SpaceX or Blue Origin or even an old aerospace giant. Firefly seemed like perhaps the best bet, given the pedigrees of its founder and engineers. Just as important, its operations were all in the United States, which meant easy access to investors and government contracts.

To that end, the money flowed into Firefly's coffers. The start-up had raised a few million dollars from a small group of investors that included the cofounders, their friends, and a handful of other wealthy individuals. In the middle of 2016, it raised another $20 million, increasing the company's value on paper from a start of around $2 million to $110 million. Meanwhile, it picked up a NASA contract worth $5.5 million to fly satellites for the space agency and had $20 million in signed contracts from other government bodies and companies. Markusic figured it would take about $85 million to get the first rocket into orbit, and Firefly appeared well on its way to achieving that goal.

In the background, however, all was not well. In 2015, Virgin Galactic began taking legal action against Markusic and his cofounders. Virgin accused Markusic of swiping Virgin's intellectual property when he left the company and using the information to help Firefly. Although Firefly had just raised a pile of money, it needed more—but one of its key European investors turned skittish as the specter of Brexit loomed and declined to keep pouring cash into the rocket start-up. Making matters worse, a Falcon 9 rocket blew up on the pad in September 2016. That dose of reality about the risks of the rocket business proved too much for another of Firefly's investors. All of that meant that Markusic had to find new supporters who would be willing to fund a rocket company with a lawsuit hanging over it. "The litigation thing was not a problem for the previous investors, but now we were scrambling," he said. "We're coming into meetings desperate, and there's this lawsuit going that made it more of a challenge for new investors, and it felt like the perfect storm."

Shit got real bad for Firefly real quick. The start-up burned through $1 million a week as Markusic went on a madcap scramble to try to talk anyone into funding his space dream. During the flights to see investors and hours spent in rental cars, he realized that he had perhaps not been the most forward-thinking CEO.* He should have kept better track of the company's spending and maybe laid some people off or shut down nonessential work. Instead, Firefly had kept right on cruising full speed ahead until it had run smack into an empty bank account. "Part of the reason we were burning money so fast is because I wanted people to come in here—investors to come in here—and feel the energy," he said. "Like 'These guys are flying!' I wanted them to feel the urgency and the need to get on board now because this thing is moving. To do that, you have to spend a lot of money, so it's a double-edged sword. We had a lot of momentum, and then it all came to a screeching halt."

In late 2016, Firefly furloughed most of its staff. Markusic told the employees that he still hoped to find new investors and bring them back soon. Some of them even kept showing up at the office, believing that money would arrive any moment. No such luck. In April 2017, Firefly filed for bankruptcy, having burned through $30 million.

---

* Markusic once gave me a rousing speech on planning ahead and how it relates to the state of America: "I think culturally in America, we plan things out way too much. When my parents were young, the idea was 'Do what your impulses tell you. You get a job, you have lots of sex, you have babies, things work out.' My kids today, millennials today, it's like they've got bucket lists and they've got a plan for 'I'm going to go to college and then my income is going to be this much and at that point I'm going to get a spouse and I'm going to have a baby.' By the time they're in their thirties, they're like 'Oh, boy, the biological clock is ticking.' I don't think you can take things for granted, and I don't think you can plan too far out ahead. I think you're fooling yourself. And you may actually miss opportunities today by trying to orchestrate your life. And so I think that applies to organizations, too. And it's one thing I think I really learned at SpaceX. Elon always had us very focused on the next thing. And it's like 'Look, the mentality we've got to get into is if we get the next thing right, then we'll have an opportunity to do the next thing after that.' But if we are looking like ten steps ahead, you lose focus and can't get it done. Yeah, so I encourage everybody to have babies when they're eighteen."

"I don't normally sleep that much, but I was down to like three hours a day because you wake up and the problems of the world are just beating down on you," Markusic said. "It was incredibly emotionally draining. I was on my knees praying for strength to get through it. If we were doing something wrong and we failed or we just weren't smart enough to get it done, that's one thing. But we were kicking ass."

To outside observers, the bankruptcy made little sense. Investors all over the world were funneling money into New Space at record levels. Questions, though, had started to be asked about Markusic's leadership. People in the tight-knit rocket-making community spread rumors that Firefly was disorganized and way behind schedule. Where Markusic saw a company that had done most things right, others saw yet another rocket start-up bumbling around and gobbling up cash.

During his darkest hours, Markusic prowled around Firefly's empty offices by himself. He reminisced. He shed tears. He started making spreadsheets of Firefly's assets and trying to figure out who might buy them during the bankruptcy proceedings. Most of all, he kept on praying that someone would appear out of nowhere to save the company. After all, God had created him to build rockets, and God would send some kind of help now.

CHAPTER THIRTY

FULL ATTACK

Artiom Anisimov had been watching Firefly's implosion. And he had a plan.

Anisimov was born in 1986 in Osipovichi, a small town in the center of Belarus afflicted by the waning fortunes of the Soviet Union. As a young child, he moved to Mongolia with his family. His father was serving in the military and participating in the Soviets' war in Afghanistan in the hope of securing a better life back home. After four years of service in Mongolia, the Anisimovs received their reward and were granted a one-bedroom apartment back in Belarus.

Osipovichi did not offer much in the way of opportunity. The criminal element had a big presence in the city, the economy had collapsed, and the schools were run down. Anisimov, though, was smart and industrious, and in his teenage years he signed up with a student exchange program that sent him to live with a family in Tennessee. The father of the family was a surgeon, the mother was a substitute teacher, and the kids were sporty and popular. The visit provided Anisimov with a taste of the comforts the United States could offer. "It's hard to explain," he said. "You learn that there are places where people live differently and the different things might be better."

Anisimov went to university first in Belarus, then in Lithuania, and then in Nebraska, obtaining a couple of law degrees along the way. At the University of Nebraska, he studied under Frans von der Dunk, one of the world's foremost experts on space law. The professor convinced Anisimov that space law was about to become a very big deal as the industry changed, and Anisimov decided to pursue it as a career.*

After graduating, Anisimov traveled to Washington, DC, and tried to find a job in the aerospace industry. Naive and not sure how one would go about such a task, he'd often show up at a company headquarters unannounced and ask to speak with a recruiter. On at least one occasion, he had to hitchhike to a company campus in Virginia because he had neither a car nor the money for a taxi. Due to visa issues and a series of swings and misses with corporate jobs, he found himself working as a parking valet at a grocery store for almost two years. He used the time to study for and pass the bar exam and make contacts in the space industry.

Anisimov evolved into a space junkie. He sneaked into space conferences and talked to as many people as possible. Conversation by conversation, the strategy worked, and he managed to find a couple of jobs doing legal work for one space start-up and then another. In 2013, he moved to Silicon Valley and, after a couple of twists and turns, ended up working as Max Polyakov's right-hand man on all things space.

Anisimov possessed a near-encyclopedic knowledge of space history, the politics governing the industry, and the major space players. He's a relentless networker and built up a long list of contacts that were valuable to Polyakov. He also had a good sense of the undulations in the industry, charting who was up and who was down and how someone else's weakness could be exploited for an advantage.

Back in 2016, Polyakov had dramatic space ambitions but not

---

* When he arrived in Nebraska, Anisimov had $160. His exchange family in Tennessee was generous enough to sign for a $75,000 loan that let Anisimov stay in a student dorm, buy food, and pay tuition. He paid the family back several years later.

much experience to back them up. He'd made his money through internet sites and business software, and space was something of a sideline. He funded a company called EOS Data Analytics that took satellite images and analyzed them more or less in the spirit of Planet Labs, and he funded a few engineering projects in Ukraine. But that was the extent of his space empire.

As stories of Firefly's financial issues became public, Anisimov saw an opportunity for Polyakov to go big. He began sending feelers to Markusic and, after making contact, asked if he'd like to meet Polyakov and talk business. Anisimov sensed that the vulnerable Firefly could catapult Polyakov right into the rocket business and do so without needing to start from scratch. Markusic gladly met Polyakov in the fall of 2016 and then kept chatting with him through the end of the year. In January 2017, he found himself on a first-class flight to Ukraine to visit Polyakov's homeland and hash out more details of what might be possible.

There are two versions of what happened next and of how Polyakov came to own Firefly.

In one accounting of the events, Polyakov played the role of a white knight. He stepped in with a ton of money during Firefly's darkest hour and saved the company from certain demise. Markusic had tried his hardest to find other suitors, but they had never materialized. Partnering with Polyakov helped make sure that Firefly's technology could live on and that all the people who had been involved with the company to date would come away with something rather than nothing. Simple.

The other version of events has a more cynical and sinister plotline. In this tale, Markusic met Polyakov and sensed an opportunity to push his cofounders and existing investors out of the company and to restart it with a clean financial slate. Rather than trying his best to keep Firefly alive in the latter part of 2016, Markusic basically let it die and go into bankruptcy, which devalued the positions of the existing stakeholders. It also made it possible for Polyakov to swoop in and buy Firefly's assets on the cheap in an auction that was rigged so that only he could win it.

Firefly Space Systems disappeared, and so, too, did the ownership stakes of previous investors. Firefly Aerospace was born with Polyakov and Markusic as the majority owners.

The people subscribing to this version of events were Firefly's other cofounders, who eventually sued Markusic and Polyakov for allegedly screwing them out of the company they'd help build. Polyakov and Markusic have refuted any such nefarious claims and said they were just businesspeople doing business under desperate conditions.*

Whatever the case, Polyakov ended up with a rocket company and immediately put about $75 million into the venture. That allowed Firefly to rehire many of its employees, restart production of its rocket, and expand its facilities. It's rare that someone such as Markusic would be allowed to stay on as CEO in a situation like this. Usually when a company goes belly up, the new owner brings in new management, partly because they'll be loyal to the new owner but also because they're supposed to be better at running things. In this case, however, Markusic remained the boss after talking a fresh investor into funding his heavenly ordained mission to space.

There's nothing like buying a rocket company to get your blood flowing, and the Polyakov I met following the deal was filled with optimism. He had a company that analyzed satellite data. But now it was time to get serious and start making hardware. Polyakov wanted to build satellites and build the rockets to fly them. Other companies were focused on their individual slices of the market. Planet made satellites. Rocket Lab made rockets. Firefly, by contrast, would be a one-stop shop, which would give it economic advantages. It could fly its own satellites into orbit at cost, instead of paying a premium to another rocket company. And it could prioritize its satellite launches and make other customers wait for its excess rockets.

According to Polyakov, the first round of commercial space companies had all made major mistakes. Rocket Lab, Virgin, and

---

* At the time of this writing, the litigation was ongoing.

Astra were building their rockets too small. The satellite companies were building shoddy machines that broke too quickly in orbit and were beholden to the schedules of the rocket companies. Some of the companies were ahead of Firefly with their products, but their time advantage could not make up for their strategic and technological blunders. "You just sit back and smile because it's all fucked up," Polyakov said.

Polyakov calculated that it would take him "double-digit millions" to finish building Firefly's first rocket and predicted that it would go up by the middle of 2019. "We will spend much less to get there than Rocket Lab," he said. Firefly planned to keep its costs down, in part, by turning to the Ukrainian aerospace expertise. Polyakov had access to designs for very complex parts that had been perfected over decades and could be transferred to Texas. "We will bring this Ukrainian heritage to the USA just like SpaceX used some of NASA's heritage,"* Polyakov said. "We have good materials. Precise ballistics and guidance systems. Heritage should be reused." The Ukrainian engineers came cheaply, too, which would help lower Firefly's labor costs. "It's about discipline and process," Polyakov said.

Firefly's rocket would also carry about ten times as much cargo as the small rocket makers' machines. "Virgin is fucked," Polyakov said. "Peter Beck with his 150 kilograms is no good. We look at this industry so far as a cynical business. It's all hype. We don't want to fly to Mars. Fuck it. We're here to make a lot of money."

The main source of Polyakov's confidence in the aerospace industry appeared to derive from his upbringing. He had come from humble beginnings and battled through the chaos accompanying the fall of the Soviet Union to make a fortune. He would use the wisdom gained from his other business dealings and trample people such as Peter Beck and Chris Kemp. "Most of the people in the space business are like children," he said. "They don't understand what a dollar means. They never made their first hundred dollars and cried. It's a show. It's a circus. I love the space market.

---

* This is very true. Throughout its history, SpaceX had partnered with NASA on technology that the space agency had developed over the decades.

"What's happening right now is a bubble funded by big government money. There will be many businesses, which will die, and because we control the satellites and the data and the rockets, we will buy them and consolidate the market. Then things will continue because humanity has a passion for space. It is the final frontier."

MAXYM POLYAKOV GREW UP IN Zaporizhzhia, a city of about 750,000 people in southeastern Ukraine. Like much of the country, Zaporizhzhia's economy and daily life had revolved around farming for centuries. The formation of the Soviet Union, however, reshaped Zaporizhzhia into an industrial powerhouse. First came the railroads. Then a dam. Then one factory after another. The Soviets loved to create model cities that would celebrate various capabilities, and they forged Zaporizhzhia to represent industrial might. Young, strong men were hauled in from all over the Soviet empire to build the city and work in its steel, aluminum, and heavy machinery factories. They were lured by solid wages but found day-to-day life grim. Most of them lived in barracks that lacked toilets and running water. The remnants of those bustling days remain, although in tarnished form. The factories are rusted and crumbling and function now as canvases for graffiti artists. In a park that runs alongside the prominent Avenue of Metalsmiths, a statue of a muscle-bound worker with his shirt open and tools in hand stands watch over a weed-strewn pathway.*

Polyakov's parents were of a different class of worker who arrived later in Zaporizhzhia's history. They were scientists in the Soviet aerospace program, which occupied a major position of prominence in this part of Ukraine. Polyakov's father, Valeriy, wrote software that interconnected various systems in rockets and spacecraft, serving, in effect, as an operating system for the machines. The code ran on some of the most ambitious aerospace systems ever conceived, including the International Space Station and Mir (the

* For anyone interested in the backstory of Dnipro, I recommend Roman Adrian Cybriwsky's *Along Ukraine's River: A Social and Environmental History of the Dnipro* [*Dnieper*].

Russian space station) along with the massive Energia rocket, which could carry a hundred tons of cargo into orbit, and the very short-lived Buran space shuttle. Polyakov's mother, Ludmila, worked in the same office, helping create hardware systems that would let parts of the Soviet rockets return to Earth smoothly and be reused.

The family lived in a small house that belonged to Polyakov's grandmother. During the heights of the Cold War and the Space Race, Polyakov's parents enjoyed themselves and felt energized by being part of an ambitious scientific community and among relatively free-thinking like minds separated from some of the bureaucracy and controls of Moscow and Kiev. There were even occasional perks when a scientist achieved something truly remarkable. After the Energia rocket flew for the first time in 1987, for example, the Polyakovs were awarded a 650-square-foot flat in a better part of the city.

The decline and fall of the Soviet Union took Zaporizhzhia and its space centers with it. Budgets were already dwindling, and then Mother Russia pulled the plug altogether. If Ukraine wanted to have a rocket program and make use of all its talent and assets, it would need to find a way to keep it going on its own. For the Polyakovs, the immediate effect of the transition was devastating. "After the Soviet Union crashed, my father received five dollars a month to feed a family of four," Polyakov said. "He said, 'If I find you doing space, I will beat you.'"

Valeriy kept clinging to the hope that someone would find a way to revitalize the Energia rocket or Buran shuttle, but years passed without any meaningful change. Ludmila kept the family afloat by buying roses and tulips in bulk from Holland and selling them during major holidays in Ukraine. "She came from a family that always found opportunities," Polyakov said. The family also had a dacha in the countryside, which proved to be a lifesaver. They grew potatoes, cucumbers, and tomatoes and stored them in a cellar to get through the winters. "Each family needed to get at least nine hundred pounds of potatoes, otherwise you didn't survive," Polyakov said. By 1994, Ludmila was bringing in $2,000 a year from her flower business. "My father was pushed to get out of space, get out of this shit, and make some more money, too," Polyakov said. "It was very painful. He'd put his whole life into it." Valeriy wound up earning $50 a month by traveling around the Middle East, Uzbekistan, and Tajikistan doing engineering work on industrial controllers used in old Soviet factories.

While his parents struggled, Polyakov excelled at school. He won national competitions in math and physics and tore through his requirements. At eighteen, he enrolled in medical school and spent six years studying to become an obstetrician gynecologist. After delivering a few babies and learning how much doctors earned under the national health system, however, he dropped out in 2000 just before completing his education.

Those were the go-go days of the dot-com boom, and Polyakov noticed that no one from Ukraine had seized the moment. Large American technology companies like Intel and IBM scoured the world for math-inclined people who could write code cheaply, often hiring teams of thousands of software developers in places such as Russia and India. While still a student, Polyakov conducted his first foray into business, creating a software outsourcing company in which Ukrainian engineers were farmed out as low-cost labor to the highest bidders.

After giving up on becoming a doctor, Polyakov threw his energy into bigger technology ventures. He learned how to build software companies that made their own products and also started developing internet services. He created start-ups like HitDynamics and Maxima Group that helped other companies track their online marketing and ad campaigns. He cofounded a number of online dating sites as well, including Cupid, and ran some shadier-sounding ones such as Flirt and BeNaughty. Founded in 2005, Cupid was the breakout success of the bunch. Over the next few years, it grew to have 54 million customers and went public in 2010. During all that, Polyakov obtained a PhD in international economics from Dnipropetrovsk National University.

Dnipropetrovsk, or simply Dnipro, a city about sixty miles north of Zaporizhzhia, had become Polyakov's stomping grounds. He'd found the talent for many of his businesses among the bright students there and had set up offices in the heart of the city. In August 2018, I made my way to Dnipro to see the city and Polyakov's operations firsthand.

ARRIVING IN DNIPRO FELT LIKE going back in time. My plane landed at an airport straight out of the Soviet Union's favored rectangles catalogue. There was a rectangular main terminal made out of rectangular cement blocks with rectangular windows and a rectangular roof on top. The dominant colors were white, gray, and acceptance. After hopping into a car and heading toward the city center, however, I discovered that Dnipro proper had many charms that its airport lacked. Yes, many of the buildings were brutalist Soviet-era hunks of stone, and decaying ones at that because of decades of neglect. But there were parks, markets, grand plazas, and the Dnieper River winding its way through the home to a million people.

Polyakov had offered to pay my way to Dnipro, and I had declined the offer. That, however, did not stop him from trying to exert some measures of control over my trip. He'd arranged for a group of people to accompany me based on the pretense that Dnipro sat close to the war with Russia in Crimea and I required protection. My travel companions were a pair of very pretty women named Tanya and Olga and a square-jawed bodyguard whom I'll call Dimitry. Throughout my trip, Tanya and Olga would don form-fitting dresses and five-inch heels, while Dimitry would cruise around in drab quasi-military garb with a satchel for his gun and other essential protection items. The writer, the supermodels, and the muscle. We were like an Eastern European version of the A-Team.

Dnipro has a long, glorious history that includes lots of industry and heavy manufacturing. When the Russians arrived after World War II, the industrial base really captured their imaginations, and Dnipro was selected as a city in which to build large military machines, planes, and cars. German prisoners of war were hauled in to create new factories, and everything went so well over the next few years that Soviet premier Josef Stalin decided that Dnipro should serve as a base for secret aerospace projects, too. And so around 1950, workers refashioned a large car plant into a factory for ICBMs, and Dnipro became a closed city.

Polyakov wanted me to learn about that history, so he first sent me and my companions to the local aerospace museum. One might expect such a facility to have modern flourishes and showstopping multimedia displays celebrating the glories of the Soviet and Ukrainian weapons and space programs. That was not the case. No, the two-story aerospace museum continued the study in uninspired uses of rectangles and gray on the outside and explored the aesthetic of dark, cavernous, and dusty on the inside. There were quite a few artifacts—satellites, nozzles, fuel chambers—resting in largely empty rooms with walls lined by portraits of stern-looking former aerospace bureaucrats and engineers dressed in military uniforms.

What the museum did have going for it was my tour guide, a sedate but knowledgeable white-haired gentleman heading toward his eighties. He explained that when World War II had ended, the Americans had grabbed all the German missile experts and taken them stateside, while the Russians had obtained all the schematics and plans for the projectiles the scientists had liked to build. The plans provided the Soviets with a running start that people in Dnipro actualized. The city's ICBM factory opened in 1951, and by 1959, it could produce a hundred missiles a year. As time went on, it would become the busiest ICBM factory in the world, producing a smorgasbord of flying tubes of death, including the SS-18, or "Satan," as US officials called it. Dnipro would eventually manufacture about 60 percent of the Soviets' land-based missiles. The prodigious output prompted Soviet premier Nikita Khrushchev to brag, "We are making rockets like sausages." The next four decades would be spent making ever-deadlier ICBMs that could go farther and farther. "Ultimately, both the Soviets and the US got to the point where they could obliterate each other many times over," my tour guide said.

"We could reach any part of the US in eighteen minutes and turn a four-million-person city into a desert." In other words, success.

From there, my man got into the space side of Dnipro's history. In 1962, the first Dnipro-made satellite went into orbit. Three years later, the country produced twenty-four satellites. One of the most famous was an imaging system that could snap surveillance pictures of the earth and get clear pictures of objects sixteen by sixteen feet in size. Following on those achievements, the engineers in Dnipro turned to rockets and produced some of the Soviet space program's workhorses. The most famous rocket to emerge from Dnipro was the Zenit, a two-hundred-foot-tall beauty that appeared in the 1980s. Elon Musk has hailed it as one of the finest machines ever created, and the locals will remind you of this with glee. Polyakov's parents had worked on much of the technology that was celebrated by my guide.

If you want to see some of the ICBMs, rockets, and engines, you need only go to the museum's parking lot because they are laid out right there on the asphalt in a totally haphazard fashion like the most neglected of aerospace relics. Still, to stand among them was cool.

After the museum, my companions and I hopped into a van and continued our historic space journey. We drove to the outskirts of Dnipro where the city gave way to forests. We pulled off the highway and traveled down a bumpy side road that led to a checkpoint with an electrified, razor wire–topped gate. A couple of guys in uniform materialized and looked at my passport. Their half-assed approach to security made me think that either they didn't love their job or that some strings had been pulled to make all of it possible, because in a matter of moments I went through the gate and right into the former top secret Soviet rocket engine testing site hidden amid four hundred acres of trees.

The star of the show was a test stand that had been used to break in some of the best rocket engines ever built. It was several stories high and about a hundred feet wide, a snarling mass of scaffolding that really did look like the brainchild of a mad scientist performing rocket experiments in the woods. Huge metallic exhaust tubes extended out from the main structure and into a cement-lined reservoir created by clearing out a couple acres of trees. During tests, technicians bolt an engine high up in the test stand, click their buttons, and then blast flames down the elephant trunk–like tubes, pushing a wall of fire and thunderous sound into the forest.* The resident squirrels, foxes, and rabbits scatter.

In its heyday, the compound must have been spectacular. More than a thousand people would have inhabited the aerospace fortress, which includes propellant production facilities, bulky water storage tanks, and a railway that leads straight to the Baikonur Cosmodrome, a preferred Russian rocket launch site in southern Kazakhstan. But although it was still impressive in its scale and all-out oddness, the site now looked run down and from another time. All of the metal structures at the test stand were rusting. And its innards—an intricate, shockingly complex maze of wires and tubes—looked not to have changed in fifty years. There was a bunker about thirty feet from the test stand where engineers sat behind their computers and conducted the tests, and it seemed like a forgotten World War II prison with its six tiny windows poking out of a blackened, scarred cement exterior—a data center with a gulag aesthetic.

My guide there was a stocky scientist who had worked among the test equipment for more than thirty years. He was generous with his time and knowledge. The number of employees, he said, had dwindled from a peak of 1,200 to 250. There used to be three tests per day to support the Soviets' nonstop production of missiles and spacefaring rockets. Such tests, though, had become infrequent and performed only on an as-needed basis if some country or company wanted to test out a new engine and learn from the Ukrainians. "It was a little more fun in the old days," my guide said. "It was full of young people. Many of the former workers have since gone into business. We hope our experience will be

---

* Each year, a hundred trees are planted in the surrounding area as a type of renewal project for the ones cut down to make this facility.

in demand again, as maybe start-ups come here. We are open and will work with anyone."

When the Soviet Union fell, the demand for Ukrainian rocket technology and engineering collapsed. Russia turned inward, opting to use its own rockets such as the Soyuz rather than a Zenit made in a country that it no longer really cared for or controlled. In a bid to fill the void, US officials rushed into Ukraine to try and stop decades of aerospace knowledge from falling into enemy hands. They dished out green cards to the top aerospace whizzes and set them up as professors at MIT or Caltech or tucked them away in government research labs. Ukraine, however, used to employ fifty thousand people in its aerospace industry, and that figure is now down to seven thousand. Though the United States found jobs for some of those folks, the vast majority were forced to find other work or lend their talents to any number of countries, such as India, China, and two others that people speak about only with winks and nods: Iran and North Korea. Even the engineers who still hold aerospace jobs are often suspected of selling the trade secrets learned in the Dnipro forest on the side because their wages are a fraction of what they used to be.*

The company in charge of many of Ukraine's aerospace assets

* The most dramatic bid to reenergize Ukraine's aerospace industry began in 1995 with the formation of Sea Launch, a consortium created by companies from the United States, Russia, Ukraine, and Norway with the goal of launching rockets from platforms in the ocean. This will sound nuts, but Sea Launch happened. The consortium pulled together $500 million and obtained a 660-foot ship called *Sea Launch Commander* to hold equipment, a mission control center, and 240 people. It also bought *Odyssey*, a 436-foot-long, 220-foot-wide former mobile oil-drilling rig, to use as a launch platform. The two craft worked together to ferry a Zenit rocket with Russian-made engines to launch sites in the ocean near the equator.

The Sea Launch idea was brilliant for a couple reasons. First off, it gave the Russians and Ukrainians money for their aerospace industry and kept their engineers busy and gainfully employed. Boeing stepped up as the US investor in the project, owning 40 percent of the consortium with the government's blessing. Second, the mobile launch operation meant that Sea Launch could travel to ideal spots from which to hurl explosives into space.

The first launch from the platform took place in October 1999 with the Zenit rocket carrying a communications satellite for DirectTV. Over the next fifteen years, Sea Launch would fly close to three dozen more rockets for a variety of commercial customers, including EchoStar and XM Satellite. But the operation fell into disarray in 2014 when Russia moved into the Crimean Peninsula and effectively went to war with Ukraine, ending any chance that the two countries would cooperate on rocket launches.

is Yuzhmash Machine-Building Plant. It runs the rocket test site in the forest. It also runs a two-thousand-acre manufacturing facility at the edge of Dnipro, the very facility where the nuclear missiles and rockets were made for decades.

If you know the right people, as Polyakov does, even this facility can be made accessible for a rare visit by a reporter. In much the same drill as before, the supermodels, the muscle, and I hopped into a van, drove to the ICBM factory, were waved through a razor wire–topped fence, and presented our passports to some guards. Those guards dug in more than their friends at the test site had and dragged the process out. Hardly any journalists have ever made it inside those walls, and the security officers can decide to end a visit on a whim. I actually had to get out of the van and stand around for ten minutes or so while my Ukrainian companions and the guards talked and pointed at various papers. We were parked in the middle of an asphalt expanse with some factory buildings on one side and a row of trees on the other. As mentioned, the place was fucking huge, but it appeared almost deserted. I looked around and saw a delivery truck, a fifty-foot-long missile on its side that was part of a display, and two guys in uniform driving around on what appeared to be a small tractor built in the 1960s. That was it.

Having grown up during the tail end of the Cold War, I had consumed enough anti-Soviet propaganda to be both awed and disappointed by the visuals. A totally ridiculous set of events had placed me there at the factory, which had once had my possible annihilation as one of its main goals. I kind of wanted the surroundings to look more sinister and intimidating. But no. It was just another very large complex in a hollowed-out, tarnished state; less bringer of doom and more the depressing retirement home where doom was spending its final days.

I never got to see the old ICBM factory lines or whether there was a supercool death-and-destruction roller-coaster ride through the complex. That's because I was escorted from the main entrance to the space rocket–manufacturing areas and introduced to another, older tour guide. He explained that Yuzhmash makes—or at least *could* make—Zenit rockets; the first stages of the Antares rocket flown by America's Orbital Sciences Corporation (since acquired by Northrop Grumman); the Cyclone-4 rocket, once meant to fly satellites into orbit from a Brazilian launchpad until that deal fell apart; and a variety of rocket engines.

The optimistic plan for Yuzhmash called for the production of twenty rockets per year. "We do not have so many orders now," my guide said. Over the past few years, Yuzhmash had gutted the rocket-making part of its workforce, and the engineers and technicians left over often saw their workweek shrink to two or three days to save money or simply went months without pay. To try to make up for the lost aerospace revenue, Yuzhmash had people producing everything from tractors and electric shavers to airplane landing gears and tools. That helped explain why, during my tour, there were rocket bodies at one end of the factory floor and buses being built at the other.

Despite the hard times, my guide exuded pride in the factory. He showed me some argon arc welding equipment and X-ray machines used to examine the precision of the welds. He loved that Elon Musk adored the Zenit rocket. He made a joke about the North Koreans getting their hands on Ukrainian engine technology. It related to an old *New York Times* story and subsequent reports, which noted that the North Koreans seemed able to improve their missile technology at an alarming rate and that their rocket engines resembled the RD-250s once produced in this factory. "Don't write about North Korea," he said with a laugh. "There is wrong information on the internet. It made inspectors come here."

We went from one large factory floor to another, and there were rocket bodies, interstage rings, and fairings all over the place just waiting for something to do. Most of the people making their rounds in the building were in their fifties, sixties, and seventies. Along the side of each factory chamber, older women in white lab coats sat behind wooden desks flanked by metal filing cabinets and watched me meander from spot to spot. A sadness hung over everything. Without question, Ukrainian engineers are exceptional.* Many other countries and companies have struggled to build the products that once tumbled out of the factory with frightening efficiency. But all of the knowledge and potential had been paralyzed by politics, corruption, and the steady march of progress outside the factory walls.

WHEN I MADE MY TRIP to Dnipro, Russia had already annexed Crimea in the south. Vladimir Putin's troops had been encroaching across the eastern part of Ukraine for years with the tension tumbling over to Dnipro, which was only about a hundred miles from the edge of the conflict. People in the city were worried about what might happen next, as Putin had long made it clear that he would like to obtain Dnipro and the surrounding area and make it part of Russia. Many of the Dnipro locals and Ukrainians in general felt as if they'd done what the West wanted by disposing of their nuclear arsenal, pursuing democracy, and trying to form deeper ties with Europe, only to be abandoned by the United States, in particular, during their moment of need. Meanwhile, the local politicians were doing the country few favors through their corrupt actions that siphoned off much-needed money meant to revitalize the Ukrainian economy.

The combination of Russia's military presence and Ukraine's dysfunction gave Dnipro some Wild West vibes that bordered on comical during those prewar times. I went to a restaurant with my companions, for example, and there was a metal detector at the entrance. That meant that my bodyguard had to wait in the lobby. He sat there with the other bodyguards, all of whom had their gun satchels resting on their laps. They shot the shit in their bodyguard zone while the rest of the clients and I ate. If, say, you wanted to smoke a joint

---

* Along with missiles and rockets, Ukrainians also make stunning planes. At a factory not far from there, in Kiev, engineers and mechanics built the Antonov, a giant cargo plane with a three-hundred-foot wingspan. When the Russians invaded in February 2022, they went right after the Antonov plant.

and not think about Putin coming for your land for a bit, you zipped over to the dark-web site called Hydra, picked out your weed, paid with Bitcoin, and received some GPS coordinates. You then traveled to the specified location and dug in the ground and—boom—there was the contraband with a "buried on" date and all.* Or maybe you felt like shooting a pig with a bazooka. That was also possible, especially if you knew Polyakov.†

Be it nurture or nature, Polyakov thrived in the environment and seemed to know exactly how to navigate it. His company had taken over two of the tallest buildings in Dnipro, and he occupied an office on the top floor. From his perch, he could look out and see the river, the forests, and his old university buildings. He often worked and drank late into the night and slept at the office, which had been equipped with a bed. The office also had security cameras in every corner and thick, soundproof doors to keep anyone from snooping on conversations.

While at his Dnipro office, I learned more about Polyakov's companies. He'd done well with the dating sites and business software to the point that he'd built seven ventures that brought in at least $100 million in revenue each per year. He also had major plays in things such as online gaming, robotics, and artificial intelligence software. Much of the dating and gaming stuff seemed standard. Other operations in those areas were more ethically dubious. "Gaming" sometimes meant gambling. And "dating sites" sometimes meant places where fake accounts of beautiful women were created to entice men to give up their credit card numbers for subscriptions that were borderline impossible to cancel. The ownership structures of the companies were complex, and their financials nested behind offshore accounts. At the time of my visit, Polyakov's companies employed almost five thousand people and were cash machines.

One evening, we went to a nice restaurant in Dnipro. About a dozen of Polyakov's top lieutenants, including Anisimov, were there.

---

* I may or may not have done this.

† See previous footnote.

A rotund, boisterous man who had been in the Ukrainian army seemed to have been responsible for much of my access to the old Soviet sites. A couple of women who were present headed up some of the online operations. It was hard to keep track of what everyone did, mostly because the waiters at the restaurant filled our glasses with endless streams of Oban scotch. The guests commented that Polyakov had once acquired the entire Ukrainian supply of Oban and had almost acquired the distillery for $19 million before thinking better of it. Those stories may have been apocryphal, but they seemed plausible in the moment and felt all too true the next morning.

Over the course of three hours of eating and drinking, it became clear to me that these people were very loyal to Polyakov. Most of them had worked for him for decades and helped turn fledgling companies into giants. He obviously rewarded those who succeeded well, and during the meal, he called out the top performers, while also noting who could be making more money for him. Many toasts were made, as Polyakov reminded his key executives that the profits gained from software, gambling, and horny men were now being funneled into something great and glorious. Their efforts were the key to making Firefly's rockets work.

"Sometimes I am a huge believer that you're already built with ideas that are preprogrammed," Polyakov said. "And then there are times that come along, and you open up and start to feel them. You should feel that passion. You should feel that passion for the idea. That's why I came to America and didn't touch anybody else's money. It's about how do you have your own money and explore your passion properly."

AS A CHILD, POLYAKOV HAD experienced the might of Ukrainian engineering on a daily basis, and then he had felt the sting as the country had tumbled into chaos. He'd made it a personal mission to try to save what was left of Ukraine's aerospace knowledge and inspire a new generation of engineers to feel that they could think big again. To help fulfill that quest, he had created a branch of Firefly in Dnipro with manufacturing and research-and-development facilities. He'd also been pumping large amounts of money into the local education system.

One morning, my travel companions took me to visit Firefly's Ukrainian factory. There were veteran engineers who used to work for Yuzhmash examining machines alongside younger engineers who had come straight out of school. Polyakov had spent millions on state-of-the-art manufacturing equipment, ranging from high-end 3D printers to laser cutters and turning and milling machines. The building was a former window factory, and its exposed metal rafters and brick walls harkened back to Dnipro's grittiness. But Polyakov had transformed the place. The factory was a bright, clean space with a palpable, youthful energy. More than a hundred people worked at the facility, and the theory was that they could produce rocket parts as cheaply as anywhere else in the world thanks to low labor and materials costs. More to the point, they could tap into decades of aerospace expertise to create technology that other rocket makers could not match.

There was reason to believe in Polyakov's strategy. Russians, for example, have long excelled at rocket engine making and have produced hardware advances that US engineers struggle to replicate. For example, United Launch Alliance (ULA), an American rocket company that sends top secret US military satellites into space, has comically depended on the Russian-made RD-180 engine for years.* Similarly, Ukrainian engineers have deep talent at making a key rocket system called a turbopump. This mechanism helps control the rate at which propellants mix as they're fed into a rocket's combustion chamber. Though straightforward to build in theory, good turbopumps have proved notoriously difficult to make and have slowed down many a rocket program, including SpaceX's Falcon 1. Workers at Firefly's Ukraine factory had designed the turbopumps that go into the company's rockets and taught engineers in the United States how to make them. Old Space meets New Space, or, as Polyakov put it, "We have the best of both worlds."

* Russia's takeover of Crimea helped trigger a slow end to that embarrassing situation. ULA is in the process of switching over to new engines built by Blue Origin.

At a nearby R&D lab, Polyakov had a team of people working on new thrusters for use on satellites. They were ion thrusters, in which a gas is hit with electricity to produce a beam of ions that can propel a satellite and help it change orbit either to avoid a collision with something or simply to be in a better place to fulfill its mission. The crew in Ukraine claimed to have built their latest thruster prototypes from scratch in about a year. The final product would cost about $200,000 versus millions of dollars in another country, they said.

During another part of my field trip through Dnipro, we went to visit the local university. Once again, the outsides of the buildings failed to impress. The insides, however, sparkled. Polyakov had donated millions of dollars to the school, upgrading its facilities and starting a variety of engineering, cybersecurity, artificial intelligence, aerospace, and robotics programs. He had paid to fix up the local planetarium and to run engineering contests. And he'd tried to use his money to keep good professors in town. "The professors still get shit salaries," he said. "So we are paying them and doing engineering schools and trying to improve the ecosystem."

Hoping to ape Silicon Valley, Polyakov attempted to encourage the university to set up a system in which professors and students could use the intellectual property they'd developed as the basis of new companies. He'd also tried to bring in fresh sources of venture capital and generally to coax people toward thinking in more entrepreneurial ways. With some luck, a few successes would lead to both financial rewards for the school and bigger dreams for the students, and the whole endeavor could move along without Polyakov's aid. "We are trying to build a sustainable model," he said. "If the situation is stable, then the passion starts to come and people have ideas. Passion and action must come together."

Beyond his swearing and bombastic tendencies, Polyakov has a deeply intellectual side. He's well read. He's thoughtful and introspective. Like many people from his part of the world, he can speak about science in poetic, almost religious terms. All of his philanthropic and educational endeavors were overseen by an organization named Noosphere. That was in homage to the Russo-Ukrainian scientist Vladimir Vernadsky, who created the concept of the noosphere in the 1930s and '40s.

After witnessing the first few decades of the twentieth century, Vernadsky realized that through technology, humanity had started to impact the earth on a scale similar to that of geological and other natural forces. He used the word *noosphere* to describe the phenomenon and wrote about it as "the energy of human culture."

One paper written in 1943, two years before Vernadsky's death, had the scientist enthusing in rapturous tones as he imagined humanity on the verge of fulfilling its potential. He looked past the horrors of World War II to marvel at the way in which people had built machines on such a large scale, created sophisticated communication systems, and started to unlock the secrets of atoms and nuclear energy. He felt as though humanity might soon reach something akin to enlightenment and unlimited possibility. "The noosphere is a new geological phenomenon on our planet," he wrote. "In it, for the first time, man becomes a large-scale geological force. Wider and wider creative possibilities open before him. It may be that the generation of our grandchildren will approach their blossoming." Later he added, "Fairy-tale dreams appear possible in the future; man is striving to emerge beyond the boundaries of his planet into cosmic space. And he probably will do so."

It's a shame that Vernadsky is not as famous worldwide as he remains to this day in Russia and Ukraine, where he founded the Ukrainian Academy of Sciences in Kyiv and fueled the country's interest in science. Because although impressed with progress and technology, Vernadsky also called for humanity to advance itself in concert with the environment—to use our wonderful new tools to create the purest of water and the cleanest of air. He envisioned all people flourishing equally and uniting across the continents in the shared quest to perfect the earth and then spread the human species throughout the universe.

Vernadsky's ideas provided some of the intellectual underpinnings behind the Soviet space program. It made sense, then, that

Polyakov would glom onto such writings and see himself as a vehicle for helping humans push "the energy of the human culture" beyond our planetary bounds.

But it was really the underlying ideas of expanding human potential that excited Polyakov the most. He believes that people are at their best when a critical mass of bright, passionate, well-meaning humans are gathered in one place and chasing a common goal. The Renaissance might be an example of this. So, too, might the early years of the Soviet Union, when people were unified and believed in the cause. Silicon Valley, Polyakov said, also used to exhibit such qualities until it got too distracted, money hungry, and shortsighted. "We've all become financially greedy," he said. "You start extracting and extracting, and it kills it."

Polyakov wanted to put his money into Ukraine to try to reenergize it and create "passion people who want to change things and do something," he said. The first step toward doing so was to find the people with deep aerospace smarts and have them pass their wisdom along to the next generations—not to let the skills gained from decades of toil fade. "We have already lost the grandads and some of the dads," he said. "We don't want to lose a third generation. It's about transferring the knowledge. It's about creating a different feeling. It's about nourishing the soil. If you lose something like this place, very often you cannot recreate it."

More broadly, he wanted to become an example to all young people for what passion could accomplish. Firefly's rocket would be a monument to his struggle and a sign that a kid from Zaporizhzhia or any other town could do great things. "You just wake up and you feel the energy that you shall do something, right?" he said. "But you shall do something for good. For better."

CHAPTER THIRTY-ONE

THESE ROCKETS: THEY'RE EXPENSIVE

A couple months after our 2018 trip to Ukraine, Polyakov threw an Oktoberfest party at his house in Menlo Park, California. Many of his business lieutenants were there. So were some space people. And so were some neighbors from what's one of the wealthiest suburbs in Silicon Valley—and thus one of the wealthiest suburbs in the world.

Polyakov had bought the mansion a few years earlier on a whim. He'd wanted to move his family to California and had hired a real estate agent to scout some Silicon Valley locations. After a handful of promising properties had been identified, he had flown out with his wife to see them. The agent had pushed hard on the house at Robert S. Drive and warned the Polyakovs that real estate moved fast in the Valley. Max had told him to buy it at whatever price seemed appropriate and then rushed to the airport to catch a flight back to Ukraine with his wife. A couple days later, the deal had gone through, and a story had appeared in *Palo Alto Weekly*: "Home Sells for $7.6 Million, Record for Menlo Park." The real estate agent had failed to tell Polyakov that he would be setting the market rate for an entire city.

The house looked like a minicastle. It had a huge backyard with a pool and guest quarters. Still, Polyakov wanted more privacy and convinced an older woman who owned the house next door to sell him her house, too. The woman had lived there for decades and initially rejected numerous offers. Undeterred, he had someone locate her children and told them how much he was willing to pay for the property. They convinced Mom to sell, so Polyakov and family now had a compound.*

That resulted in an even more spacious backyard, which Max upgraded by installing a *banya* (sauna) and an outdoor bar. On the night of the party, there was plenty of room for hundreds of guests to roam and consume the incredible amount of food and drink. Max's wife, Katya, was the perfect host, and the Polyakovs' four children were playful and well behaved. Dressed in lederhosen, Max sang and celebrated and insisted that people try the sturgeon that had apparently been smuggled into the country. Much fun was had by all.

As 2018 spilled into 2019, Polyakov still loved being in the rocket business. We would hang out off and on at his Menlo Park office, drink Oban, and gossip about the goings-on in the industry. Polyakov never had any kind words to say for old-line companies such as Boeing and Lockheed Martin, which he thought were corrupt and bilked US taxpayers out of money. He had even less kind things to say about Russians. Chatting with Max was intense because of the sheer volume and pace of his words and the energy he brought to every subject. That said, I found great pleasure in those talks. It's rare to meet someone who speaks about passion nonstop and can back up the talk with his personality. Max was unusual and intriguing.

* Not long after the Polyakovs moved in, Max's wife, Katya, asked a local Eastern Orthodox priest to come over for a visit so the family could form some religious ties in the area. Max had enjoyed a couple of drinks beforehand and decided to impress the priest by pressing a button that activated the pool's cover. Once the cover had slid across the pool, Polyakov walked across it and said, "See! I can walk on water! I'm like Jesus!" I'm sure the donation that followed eased any tension caused by that performance.

Polyakov remained convinced that all of the small rockets were pointless. They could carry only 200 to 500 pounds of cargo to space. Following Polyakov's investment, Firefly opted to make its Alpha rocket larger, so that it could carry 2,200 pounds of cargo. A second, even larger rocket called Beta was already in the works, and it would carry 17,500 pounds. Those sorts of payloads would let Firefly customers send dozens of satellites into orbit on each flight instead of just a handful. Since the aerospace business always moves more slowly than expected, Polyakov figured that the major demand for rocket launches was still a couple of years off. By that time, Alpha would be flying and then Beta would arrive and put the small-rocket makers out of business. "We know what the market will be in three to five years," he said. "We will be at the right time, at the right place, with a product that eats the market."

Once upon a time, Polyakov had thought he might have to funnel $50 million or $75 million into Firefly by the time the first rocket flew. If the rocket did well, his investment would be worth billions in an instant. He'd been burned by venture capitalists and investor partners in the past and did not like the idea of bringing on any financial partners.* His reluctance to take on outside investments plus delays with the tests on Alpha had already compelled him to put $100 million into Firefly to keep it moving. He insisted that the additional cash was all part of the bargain he'd made. "The first rule of aerodynamics is that without money nothing flies," he said. "There is nothing better than to put all your own money into a company. What you feel is your passion."

The money had bought an impressive amount of equipment. Back in Texas, Firefly had state-of-the-art facilities that could be the envy of any rocket company. It had a pair of huge test stands so that it could test rocket engines in both vertical and horizontal positions. It had a factory to make the large carbon-fiber rocket body. It had a mission control center for overseeing the tests. Hundreds of people moved about the complex, waving to the cows as they went about

* In fact, he thought that venture capitalists were the scum of the earth.

their work. On most days, the engines performed well, firing for minutes at a time. On others, things went wrong, and exploded hunks of metal were cast into a pile on the property—the graveyard of bad engineering. Some employees claimed that Firefly might launch in 2019 from a pad at Vandenberg Air Force Base in Southern California, although I doubted that anyone really believed in the date.

Polyakov's Ukrainian operations were doing well, too. The finishing touches were being put onto his multimillion-dollar planetarium upgrade, a gift to the children of Dnipro. The engineering schools were making strides toward creating more start-ups, and Polyakov had expanded his empire of technology businesses, particularly through a big move into financial technology. "It's my time," he said. "I can feel it. If I sell any one of these businesses, it could fund Firefly for five years."

Ukrainian rocket scientists were producing top-notch technology for Firefly, too. Much to my surprise, Polyakov and Anisimov had pushed a special agreement through with the US government that made it legal for Ukrainian aerospace intellectual property to go to Texas. It was a one-way deal. Firefly's engineers in Ukraine were free to teach their US counterparts just about anything, but technology cooked up in Texas would stay in the United States. Still, it was a step toward Polyakov's goal of uniting the two countries and revitalizing Ukraine's aerospace industry. Even Rocket Lab, based within the borders of one of the United States' closest allies, had struggled to do much better, as the US government tried as hard as possible to keep US aerospace technology out of the hands of foreigners.

Securing that special deal for Firefly had taken months and been incredibly difficult, and that had flummoxed Polyakov to a degree. Ukraine was supposed to be an ally of the United States, which should, he felt, have welcomed the opportunity to form a deep and broad technology partnership. So many immigrants had come to the United States to start technology companies in areas like software, internet services, and computer hardware. They were celebrated and often raised funds from foreigners. Beyond that, the Russian Yuri Milner had become one of the biggest investors in all of Silicon Valley, taking huge stakes in companies such as Facebook, Twitter, and Airbnb.* He had close ties to friends of Vladimir Putin, and the exact nature of his relationship with Mother Russia was opaque. Rarely, however, did anyone make much of a fuss about any of that, and he had free rein to fund what he pleased.

Polyakov rationalized that the aerospace business just came with extra baggage. "Space is one of those areas where foreigners are not welcome," he said. "It's probably the hardest industry to participate in. Hopefully, I can tell the story that there is one Ukrainian that is good."

At Firefly's headquarters in Texas, Markusic had always pitched Polyakov's arrival as an investor as "a friggin' miracle." In 2018, about a year after the investment deal had initially gone through, the two men gave a rah-rah speech describing what the future would bring.

MARKUSIC: Okay, all right, I'll get started here. So I just wanted to wrap up the week. We had the good fortune to have Max here, which is very rare, and so we should take a minute—

POLYAKOV: I'm always here.

MARKUSIC: We are getting close to doing this. We are really doing this, so I hope you have the same type of feeling. Having said that, there's just a lot to do in the next year to get us to launch. It's going to be very challenging.

POLYAKOV: We're back on track because we're not dead. It's a pretty good message for those people who had been fired. You've come back to the family, and now you will stay with the family. The family style is all passion, energy, very full of passion and energy.

MARKUSIC: We have high expectations of you. Max has high expectations of you, and as I tell you every week, if we stay on schedule and do what we said we're going to do, I think we're going to have support at all levels.

---

* He had also invested in Planet Labs.

POLYAKOV: For me, this is a hobby. I love it. If you do not succeed, I will not die tomorrow, right? But there are reasons we need to succeed. People outside hate us. Guys like Richard Branson. Well, he's an old-school guy, I don't really care about him. But lobbyists in Russia, China, and Ukraine, it's like all these people who want Firefly to fail. You need to feel it and flip it because every bit of negative energy goes against us. The Firefly family can flip it to the positive energy impulse.

We want to feel what Rocket Lab is feeling. And our rocket is four times bigger. We're all in the fight for money, success, and glory to some extent, right? To make America great again, right? Start to feel that.

Do not be afraid to take risks. You shall take risks and blow stuff up. We test, blow up, test, blow up, test, blow up. You're in America. In Texas, right? It's not California, right? So we shall blow them up, guys. We have a very tight schedule, but this is the only way to do it. You have the right to make mistakes.

I will tell you that the internet is beautiful. You invest much less money, you build seventy percent of a product and launch it, and you're still making crazy money. But science is beautiful, too. You chose this because it's hard. You chose it because it is your way. This will be the hardest job of your life. But you will get so much passion inside of you. Not many people can do what you can do. Try to respect it. Do not lose the passion energy. If things are not going well at the company, and you're becoming too rational, you can email me.

MARKUSIC: Please remember, all your emails are monitored.

These early feel-good vibes and Polyakov's overall sunny outlook on being a rocket magnate began to give way to something else entirely as 2019 passed without a launch and Firefly hit the early part of 2020 with no launch in sight. Firefly was going through the same engineering pains faced by every rocket company: its engines worked fine, but building the rocket body and getting all the electronics to function in concert was taking time.

The Ukrainian teams were sending designs for some of the rocket's most complex parts, such as the turbopump, but their US counterparts were struggling to build them and figure out how to use them properly. Polyakov felt as though he'd handed Markusic a wonderful gift and Markusic was frittering Firefly's advantage away. No Oban was required to elicit a monologue from Polyakov decrying Markusic as incompetent and slow.

Polyakov's frustrations were understandable and inevitable. He'd made the mistake of optimistically thinking that Firefly would hit its deadlines. Each month required him to put another $5 million or $10 million into Firefly. Whereas he'd once hoped to spend on the order of $50 million to build the first rocket, he was now approaching $150 million.

No one besides Polyakov knew his exact net worth, but he certainly did not have Musk's or Bezos's billions. Each check that he wrote brought with it some real measure of financial pain. Our discussions about Vernadsky and creating a tribe of passion people throughout the globe had given way to Polyakov's gripes that he could have spent all this money on a private jet, an island, or both. His wife, he said, certainly had questions about the rocket gambit.

Polyakov was particularly miffed that the US government did not show Firefly more love. The company had been passed over for NASA and military contracts. It wasn't so much that Rocket Lab, with its small rockets, was getting some of these deals. It was that the mom-and-pop operations out in Mojave were being awarded tens of millions of dollars to build things such as lunar landers. All the while, Polyakov was spending his own money on a huge rocket, but no one seemed to be taking his effort seriously. He noted that if he had been spending that much money in an Eastern European country, government officials would have been falling all over themselves to make sure he felt like a king.

There were times when Polyakov wanted to fire Markusic, but the two men were bound together by their business deal. Markusic gave Firefly credibility and an American face. Polyakov had also spent so much money already that he did not want to create more delays by overhauling the management. In his heart, he felt certain that Markusic cared only about Markusic and that he'd sell Polyakov out in a heartbeat if the opportunity presented itself. Sometimes Polyakov expressed such sentiments to Markusic's face. The CEO took the abuse because he needed Polyakov's money. Max owned about 80 percent of Firefly, which meant he owned Markusic's path to Providence.

THE PANDEMIC MAY HAVE SLOWED down lots of people but not Polyakov. Though he'd flown commercially in the past, he decided to deal with some of the annoyances of covid-19 by using private jets. It turned out that he really liked flying private and that the quick travel times made it more convenient for him to check up on Markusic. Where Polyakov used to visit Firefly's headquarters once a quarter or so, he was now taking trips to Texas every couple of weeks throughout the early part of 2020.

In August, I went on one of those jaunts with Polyakov and Anisimov. Over the past several months, Firefly had sent dozens of people to live in Southern California and work at Vandenberg Air Force Base. Their job was to overhaul an existing launchpad and prepare it for the first Alpha launch. Our itinerary had it so that we'd visit Vandenberg first and then zip over to Texas, where the rocket was being completed ahead of its journey via truck to California.

We arrived at the airport in Oakland around 6:00 a.m., took off, and drank some beer. It was a short hop down to Vandenberg,* and Polyakov was primed to be pissed off for the entire flight. As Polyakov saw it, Markusic had been futzing around at the launch site for half a year. How hard could it be? Firefly didn't have to build its pad from scratch; it just needed to do a retrofit. "Fuck," Polyakov said. "Fuck Tom."

Had it not been owned by the military, Vandenberg Air Force Base would be a resort. It occupies 98,000 acres of prime real estate nestled right up against California's breathtaking coastline. There are forests

* The onboard reading material included *Farm and Ranch,* which advertised large properties for sale, and *Executive Controller,* which advertised private planes for sale.

of eucalyptus trees growing amid fields of scrub brush with the crash of Pacific Ocean waves as a backdrop. As we drove up to that wonderland, a sign appeared amid the rolling fog that read "Welcome to Space Country" in a nod to the decades of missile and rocket launches that have taken place at California's main aerospace hub.

Since it's a military base, we had to check in at a security center and obtain badges that gave us access to the grounds. As I am a US citizen, my check-in went quickly, and the security officer went ahead and gave me a yearlong pass. Polyakov, by contrast, waited about an hour and then received a one-day pass even though he had paid more than $100 million to make the visit possible and would eventually pay Vandenberg $1 million per launch* as a customer of the site. He was treated like a contract painter or machinist, and a sketchy foreign one at that. Someone in our crew joked, "In another country, there would have been beers and prostitutes here by now."

With the security procedures done, we got into an SUV and Anisimov drove us to Space Launch Complex 2, which had a pair of launchpads. On the way, Polyakov rolled down the window and took in some of the cool air. "Oh, good, refreshing," he said. "There must be gold here." As we pulled up to the complex, he spotted a Firefly sign. "Serious shit, my friend!" he said. "It's all mine! Technically."

A guide of sorts showed up to greet Polyakov. He'd worked at Vandenberg for decades and was something of a historian. He recounted the great missiles and satellites, including all of the CORONA spy satellites, that had been launched from the pad Firefly would soon occupy. Back in the day, the United States could spin up an entire missile program and the requisite launch infrastructure in eighteen months. "We couldn't even write the environmental impact study in that amount of time now," the guide said.†

Polyakov soon grew tired of revisiting the past and sent the

* He even had to pay $500,000 for each scrubbed launch.

† If we were lucky, we had been told, we might see some deer, mountain lions, or black bears during our visit. If we were unlucky, we'd suffer from the "curse of Slick-6," which apparently was the result of the air force having built part of its launch operations on top of a Chumash Indian burial ground.

guide away, and we went down to the actual launch site and operations centers. Firefly had offices set up in a couple of low-slung beige buildings that appeared to be of Cold War vintage. The launchpads were a short drive away, marked by towers jutting up from the hills and scrub brush. Polyakov prowled around by the offices, looking for victims of his fury. A former SpaceX engineer passed us, and Polyakov asked for an update on some of the mechanical work down by the pad. The engineer relayed a hopeful timeline with dates that had been provided by his welders. "Welders are always bullshitters," Polyakov said. "Show me the pads."

One pad had a huge teal tower that had been used in the past to hold up NASA rockets and had since been retired. NASA had agreed to pay for the removal of the tower so that new customers could use the site, and some people from the base were pursuing that project. The second pad had been cleared already and turned over to Firefly, which was in the process of creating its own infrastructure on top of the concrete slab. In a building set between the pads was a data center used to monitor launches and receive information from the rockets. It was also of Cold War vintage. "If it smells like shit in here, it's because the sewer just backed up," said the gentleman giving Polyakov the tour. The mission control center for the launches was about eleven miles away and required a background check for entry, so we would not be visiting it.

By that point, Markusic had joined Polyakov for the survey of the grounds and operations. "Why does Rocket Lab already have a new pad built, Tom?" Max asked.

"It's a small rocket," Markusic replied. "It's a tiny-dick rocket. Plus, the entire country of New Zealand is behind them."

One of the Firefly employees wanted to show Polyakov that people were spending his money wisely. He pointed to a storage shed that had been acquired for a discount because someone had crashed a motorcycle into it. "He might've died," the employee said. "I'm not sure. We'll just restucco that corner and seal it up."

Another nearby storage shed was totally empty except for a single red metal cabinet about seven feet tall. A warning on its side read, "DANGER EXPLOSIVES, Keep Fire Away." It held the bomb that would blow up Firefly's rocket if something went wrong in mid-flight. About half the size of a shoebox, the bomb had two charges. One would go into the kerosene tank and the other into the liquid oxygen tank.

Polyakov went up to the cabinet and posed for a couple of photos as Max Danger. Otherwise, he did not take much pleasure in the tour. He ribbed Markusic nonstop for most of it to the point that it became uncomfortable, at least for me. Anytime we got into the car to move from one location to the next, Polyakov took a pull from an Oban bottle, which freed his tongue some more. Markusic tried to lighten the mood by telling Max that there would be a pig roast in his honor in Texas the next day. "I might not make it," Polyakov said. "My wife might not like it. She does not like people at Firefly spoiling me and treating me like a king."

Before we left for the airport, Markusic had all the employees gather in a hangar so that he could give a speech and try to make Max feel better. He started off by telling everyone that a major test would take place in the next day or so back in Texas. If the rocket passed, it would head to California. The crew at Vandenberg would be on the clock more than ever with the pad's readiness being the only thing gating the first launch. It was the people in the room versus Polyakov's bank account and patience.

> **MARKUSIC:** I just wanted to connect personally with you all on this whole thing. You know, I started this company about five years ago and have been through a lot of shit. Max Polyakov is here today. He's put in fucking over 150 million fucking dollars.

Applause.

> That right there is literally putting your money where your mouth is. If nothing else, we owe it to that guy to finish the job and to give him a big return on what he has done here and validate the risk that he took on us.

> Here's what's going to happen. We are going to go out this fall. This winter. And we are going to launch this rocket, and we are going to own the whole small-launch market. There's some smart people out there talking about it, but they're way freaking behind. You all know how hard this is. We have such an amazing chance to get a head start. The way we prove we are a legit company is to fly Alpha.
>
> How many of us here have been with a rocket company the first time it launched its rocket? None of us! Guess what? We have an unbelievable fucking opportunity here! This is not a normal job. You have something you will be able to reflect on for your entire life. I want you to embrace that and use that as part of your daily motivation. You are part of something incredibly special and God or the universe or whatever picked you out.
>
> It's August, and we are going to launch in November or December or something like that. We have time.

The flight to Austin-Bergstrom International Airport from Vandenberg took three hours, and it was a revealing trip. Markusic joined us and a cooler full of beers on the plane. Another bottle of Oban began making the rounds shortly after takeoff. Polyakov began the ride by expressing his deep disappointment in the lack of progress at Vandenberg. "My wife is tired of all the money that's going through Firefly, Tom," he said.

It turned out that Markusic and Polyakov were trying to raise money to bankroll Firefly over the next few years. An imminent deal was in place, according to Markusic, that would score hundreds of millions of dollars for Firefly.

You would think that Polyakov would have been happy about the prospect of such a deal, but he wasn't. On the one hand, he now hated cutting checks to Markusic each month. On the other, the deal would take place before Firefly's first launch. That meant that the company still had a ton of risk in front of it and that the investors could be stingier with their terms. If Markusic had done his job and moved faster, Firefly would have already launched, and Polyakov could have been sitting on billions. Instead, he would soon have to give up a huge chunk of the company at a bargain price. Just as bad, he'd be relinquishing his spot as Firefly's one clear supreme lord. The new investors might not be big fans of his. Or Markusic might work Max over and find a way to push him out.

Markusic, of course, wanted to keep the company alive and took pride in arranging the investment, which would lessen the financial burden on Polyakov. He felt awash in a sea of mixed messages from Max.

Thanks to the Oban and the altitude, everyone was getting pretty drunk by about halfway into the flight. Polyakov and Markusic were sparring, while Anisimov and I pretended we didn't notice and poured shots for each other. As heated as the conversation was, it took on a therapeutic twist. The two men were telling each other what they really thought. Polyakov complained about all the money he'd lost. Markusic complained about what a pain in the ass Polyakov was. Polyakov accused Markusic of stringing him along with an endless series of false promises. "I'm good at promising things," Markusic said as the Oban washed away all deference.

Once we landed in Texas, the party continued at a restaurant near our hotel, where Polyakov had the best suite. Everyone kept right on drinking, and Polyakov and Markusic kept right on fighting. Markusic's wife, Christa, turned up and didn't appear overly impressed with our general state. Her presence, though, changed the course of the meal. The tensions eased. People started laughing. Everyone was on the same team again and excited about Firefly's prospects. I drank my scotch and marveled at how rockets are made.

CHAPTER THIRTY-TWO

LIMITS

A new day. A new perspective. Or at least that was the plan.

Head pounding, I met up with Polyakov and Anisimov in the morning for our drive out to Firefly Farm. We'd already received word that the crucial engine test Markusic had mentioned back at Vandenberg had taken place and been a success. People wanted to celebrate, and Polyakov's physical presence in Austin provided the perfect excuse to go big. Firefly had hired a team of caterers to roast a whole pig outside at the test site in preparation for an evening barbecue. By the time we arrived at the complex, the pig was already there with tinfoil on its ears and a pole through its body. All the workers were talking about the pig and looking at the pig. The pig was a thing.

Polyakov did not love the pig as much as others did. I'd seen him have bad moods in the past and get hung up on a real or perceived transgression, but those bouts of anger had usually been fleeting. As it turned out, cranky Vandenberg Max was but the mild prelude for apoplectic and despondent Texas Max. He drove through the gates at Firefly Farm without a single rejoicing for his cows and aimed straight for the factory floors and the workers on them. Polyakov found a welder and did his thing, asking why so many deadlines had been missed. "Oh, man, don't start on me," the welder said.

The Firefly employees had not seen it coming. They'd been working for weeks without much rest to get the engine test done and had nailed it. All those involved were feeling pretty proud of themselves. One of the marketing people showed me a clip of a documentary being made about Firefly. It had a scene from years earlier in which Markusic was in tears, telling everyone that the company would have to be shut down. Now here they were on the verge of massive success.

Polyakov, however, mostly saw his money disappearing and his total control of the company possibly coming to an end. Things only worsened when Tom told Max that some of the delays he was so upset about had resulted from issues with the Ukrainian turbopump. Polyakov said that the US guys just didn't know how to build things right and were not skilled enough to grasp the nuances of the technology.

During a meeting, a couple of executives came in and asked Polyakov to sign a document that would allow auditors to pore over his financials. The potential investors had requested the step as part of the normal due diligence for such a deal. At the same time, he had to sign another document that he would continue to fund Firefly for the next year even if the new money did not come through. Polyakov felt taken advantage of, as Firefly needed him to ensure its survival but was also ready to shove him to the side. And people kept reminding him that his existence as a Ukrainian was making everything more difficult, as neither the US government nor other investors thought it was the best look for a US rocket company.

I left the room as Polyakov and Markusic really went at it. They were too consumed with their fight to notice my presence, but it seemed wrong to eavesdrop on what felt like a breakup between scorned lovers. Eventually, the two men came out of the room. "Fuck it," Polyakov said. "We're flying home." He would skip an upcoming test, the pig roast, and the party in his honor.

"This could have been a good day," Markusic said. "This could have been a great day."

On the flight home, Polyakov and Anisimov fell asleep. I fished around in the cooler for a beer, if for no other reason than that I was on a private jet and it seemed like the right thing to do.

I'd long had a soft spot for Max. He could come off as a bit mad. Okay, he *was* a bit mad. But he also often struck me as being the most realistic of the space moguls. He saw through much of the hype surrounding the market and talked about the business in ways that were more practical and less fantasy. He really could have been doing anything with all his money but had charged after rockets because it was a higher calling for him. It was in honor of his family, his country, and science. He'd taken a huge risk and convinced people such as Anisimov to dedicate their lives to the journey. All around him were loyal employees whom he obviously compensated well but who also believed in the man and his character. Unfortunately, the Firefly crew had never had a chance to get a better sense of all that. Understanding Polyakov required time and a measure of mental gymnastics to reshape his muddled English speeches into their real meaning. Those who had the time and opportunity to do so could see that he was thoughtful, authentic, and philosophical.

For all those reasons, I felt a deep empathy for Polyakov. The engineers at Firefly were doing their best and trying to build the rocket as fast as they could, but I could see things from Polyakov's perspective, too. He was the one who was risking everything. They were risking nothing. They were doing a job and took him for granted. Polyakov could have left this vast fortune to his kids and set his family up for centuries. Instead, he'd gotten into bed with Markusic. And ultimately, Markusic always seemed to be looking out for Markusic. Polyakov had known that going in, but facing it head-on and feeling it were two different things. It was clear that Polyakov had tired of playing games. He'd just wanted to build a rocket and not deal with all the problems.

We landed in Oakland and walked into the private airport lobby. Out of coincidence, Chris Kemp was standing right there. He'd been flying to keep up his hours for his private pilot license. I introduced the two men, and Kemp seemed to have no idea who Polyakov was.

It was an indignity on top of a day full of them. Polyakov had packed one of the Ukrainian turbopumps in a duffel bag and asked Kemp if he wanted to see it. Kemp peeked into the bag and expressed mild enthusiasm. The men exchanged awkward goodbyes, and we began driving back to Silicon Valley. "Fucking arrogant prick" was Polyakov's summary of the exchange.

IN NOVEMBER 2020, FIREFLY DID ship its big ole rocket from Texas to Vandenberg. There was no way it would be launched by year-end but perhaps by February or March. That milestone and the prospect of an end in sight might have improved Polyakov's mood under normal circumstances, but we were in the far-from-normal part of the show.

In February, the website Snopes had published a long story portraying Polyakov as the scum of the earth. A reporter had dug into a huge number of dating sites that appeared to be funded by Polyakov and determined that they were dating sites only in an abstract version of the term. We're talking about stuff such as plentyofhoes .com, iwantumilf.com, and shagaholic.com. The Snopes story argued that Polyakov and his business partners were bankrolling the sites through a bunch of shell companies and using them to bilk unsuspecting web surfers out of their money. People would sign up to the sites, the story said, and then often be spammed by messages from fake women. When the subscribers tried to cancel their monthly subscriptions, they would either not hear back from the sites or be misdirected all around the internet.

That was not a huge revelation. Firefly's original cofounders made similar claims earlier when they had sued Markusic and Polyakov. In the interim, however, Firefly had secured some large NASA and federal contracts, including a lunar mission. Those deals made things more complicated. Snopes argued that it might not make sense for someone supposedly operating in the internet underworld to be awarded government deals. In statements to the press, Polyakov categorically denied any fraudulent practice and said, "All of my businesses and investments operate within the law and, where applicable, disclose their terms of use. My focus is on space."

On top of the Snopes story, I'd written a magazine piece for *Bloomberg Businessweek* detailing my time with Polyakov in Ukraine and Firefly's aspirations to unite Soviet and US technology. Again, it was not a huge revelation to people who were in the know in the aerospace industry. The US government had blessed the one-way technology exchange between the two countries, and anyone who wanted to find the agreement could do so with little effort. Still, people were suspicious. During a meeting in Texas, for example, no less than a Firefly employee asked Polyakov in front of the whole company how he could be trusted not to send Texas-made engineering breakthroughs back to Ukraine.

According to Polyakov, those two stories helped derail all of Firefly's attempts to secure outside funding for the company. Lengthy negotiations would take place. A deal would almost be signed. And then, at the last minute, someone would balk because they'd read the Snopes story or come to believe that Polyakov was a double agent for Ukraine or Russia.

I didn't understand what all the fuss was about. It struck me as odd that Wall Street investors had suddenly developed such sympathy for horny guys on the internet. I mean, if you sign up for nastymams.com, you should fully expect to be scammed by someone. One could also easily make the case that Jeff Bezos had more to answer for about how Amazon workers were treated, and Musk was always doing questionable Musk things. The moral bar for space magnates was far from well established.

On top of that, Polyakov's work in Ukraine really could have been a major win for the United States. There was little doubt that Ukrainian technology had made its way to some of the United States' major enemies. No one else in Ukraine appeared to have any intention of trying to curb those leaks by investing in the country's aerospace industry. Meanwhile, you've got a guy living in Silicon

Valley with his nice, wholesome family who has spent $200 million on a rocket company and is offering to help fix this situation. To me that looked a hell of a lot better than the alternative.

The failed negotiations had left Polyakov enraged. Any other country would have lauded him as an economic hero and a technology celebrity. But in the United States, he was a subject of suspicion and unwelcome. He was ready to unload, and that was exactly what he did.

> **POLYAKOV:** This stuff is leaking into China and other bad countries, and no other company is going to do anything about it. I am the last hope of bringing the Soviet intellectual property to the US, where it can do good. Even Elon Musk is a pussy compared to me. This is some patriotic shit!
>
> This is not about Soviets against the USA anymore. It is about people who gave their lives to knowledge. It's about the noosphere. It's about knowledge. They shall respect the knowledge.
>
> I am coming to the end of my beliefs. People think I am a spy or whatever. It's very painful. I came to the US eight years ago with my family to show them the American view and the American way. I pay my taxes to the IRS.
>
> America has lost its shape. All of us are immigrants. I got drunk with my dad recently, and he said, "What did you expect? You jumped over two or three generations of the average American. You're involved in rockets. They don't understand you and hate you and will try and kill it all." I am an anomaly, unfortunately. At some stage, I will say, "Fuck it." Even my passion has its limits.

CHAPTER THIRTY-THREE

FLAMEOUT

On December 4, 2020, Polyakov called me and delivered a torrent of terrible news. The US government, he said, was blocking Firefly from securing a license to launch its rocket unless he stepped away from the company and sold off most of his shares to a more acceptable investor. The government had already demanded that he leave Firefly's board of directors and had several agencies applying pressure on him and insinuating that they would open investigations into his businesses or at least make life generally miserable for him.

Polyakov had just been to Vandenberg when all that happened. The feeling was that the presence of Firefly's rocket on the pad had triggered the sudden action against him. Few people had believed the company would get that far. They'd allowed the Ukrainian to pay for the experiment, but now that it looked real and the rocket launch appeared imminent, it was time to step in and assert control. Maybe someone high up in the government had pushed for this. Maybe Boeing or Lockheed Martin or Northrop Grumman had called in a couple favors because they didn't want to lose any more business to another start-up born on the model of SpaceX. Either way, Polyakov was fucked. He would cede control of the company and have to sell his stake before it launched.

At Vandenberg, Polyakov's badge had been downgraded to "Escort only" status. That meant he had to have a minder tracking his every move to see the rocket he had built. He could no longer talk to anyone who did technical work on the machine. "I cannot even go to the toilet without supervision," he said.

The impetus for those changes, according to Polyakov, had been a series of letters sent to him by the Department of Treasury, the air force, and NASA, among others, commanding him to relinquish control of the company. The letters said that he was no longer allowed to make any strategic or management decisions. Markusic would be the only board member for the time being until a couple of solid Americans could be found to join him. Firefly's Ukrainian factories were also to be shut down. Infuriated, Polyakov let rip again.

> **POLYAKOV:** It's all "Get Max the fuck out." There was only one decision: to give up everything, to sacrifice myself, basically, and get out of the company. And the soul, the Ukrainian passion or the Soviet Union passion, is brutally destroyed. America does not want to be in the position to make Max big money.
>
> It's normal for the USA. If you have strategic assets at this level and you start to achieve a certain level of success, they try to take it from you. It's a mix of the government, rival companies saying things about you, and intelligence. They want to know who is this Ukrainian who locates to the USA who put $200 million into this. In their minds, it's a Ukrainian, he comes from the internet, he's here to fly rockets. What the fuck is this guy who has never been killed by Russian spies? He's obviously part of some bad companies and organizations. Obviously. And he has this Belorussian guy who works next to him. And Belorussia is in love with Russia.
>
> All the offers I get to buy my stake right now are even less than what I invested. Maybe in the next two or three weeks, I will sell my stake for eighty cents on the dollar. Maybe I will do this just to have less troubles.

Tom doesn't care. He's done this same thing before.

They will for sure make me sell for a very simple reason: no one will put in more money if it gives me more reach. I have already been told that I will only be allowed to have twenty percent of Firefly, if I am lucky. Right now, I own eighty-five percent.

You want me to do like in *The Wolf of Wall Street*, right? He gets offered the deal by the government, and he comes and tells them, "Fuck you!" You know how he ended up, right? Do you want me to do it like that? But it will be more brutal this time.

So I'm leaving for Edinburgh on the twentieth of December with my family. It's done. The plane is ordered. We get a Christmas tree in Edinburgh on the twenty-first of December. My kids, my wife, my cat. We are getting a one-way ticket on the global express. Nonstop. So I got a big jet.

It was always going to turn out like this. It is our mistake. Even thinking that we would be allowed to do all of this because we are fucking patriots. But it's not allowed. When you are about to launch, you receive five letters in ten days. Bam! Bam! Bam!

You know what? I might give ninety percent of my stake to the University of Texas. Fuck it. I will give the other ten percent to Dnipro National University. UT will protect their position. No one will fuck with UT. Republicans. Ted Cruz. All that. I would rather do that than give it to some hedge fund or venture capitalist. I will give it for free.

You invest $200 million, spend four years of your life, bring this technology here, risk being killed by Russians. You never get any support from the US government. Never get any contracts from them. I've been told we will never get these contracts because of me. I've basically been thrown out of the USA. Fuck you, Max! Go on with your life.

Our rocket is ready to fly. It will fuck the market. It will probably end up being owned by something like Northrop Grumman or some private equity guys. They will just keep blocking our licenses until I go. And they just make you sit there and wait and bleed money.

> Once you get success, it's very hard to argue with you. Elon knows this. After the rocket is on the launchpad and launching, all of this other shit disappears. Before it launches, everyone thinks it's all fake and a scam to launder money or something. But if it's on the pad and it's real, then everybody shuts up. So they are trying to kill me before the launch. It's very aggressive.
>
> Don't worry. It's not the end of my money. I'm going to go to Scotland, and we're going to do internet stuff, financial tech, advertising tech, more games, gambling. I mean, look, I am a legal British citizen, for fuck's sake.

Just like that, Polyakov did, in fact, move his entire family to Scotland, where he bought a palatial estate. Actually, he bought a couple of them. He, however, did not donate most of Firefly to the University of Texas. That had been an overdramatic statement fueled by his rage. He also did not sell his stake in Firefly right away. He didn't plan on *Wolf of Wall Street*ing the situation because his spirit had been broken and he no longer had that level of fight. But he did want a decent deal, especially since tech companies were in the midst of raising mind-boggling amounts of money.

In May 2021, Firefly revealed that it had raised $75 million from a group of investors that included Jed McCaleb, a cryptocurrency billionaire. The company placed McCaleb on its board. It had added a couple of other directors, too, in the form of Deborah Lee James and Robert Cardillo. James had served as secretary of the air force for four years, while Cardillo used to run the National Geospatial-Intelligence Agency. The Ukrainian had been swapped for a military vet and a top spook. "It's a business decision," Markusic told reporters. "As Firefly continues to work more closely with the US government, Max and I decided it was best to have a leadership team made up of US citizens. Max is not a US citizen, but he is a very astute businessman."

Polyakov had been forced to part with about half his stake in Firefly, which he sold for $100 million to unnamed investors. He wasn't allowed to talk to anyone at the company that he'd saved and made possible. Firefly was now valued at more than $1 billion as a result of those deals.* On paper, Polyakov had been screwed out of hundreds of millions of dollars, but at least other people would be on the hook for Firefly's bills moving forward.

As thanks, NASA awarded Firefly a $93 million contract to build a lunar lander designed to carry scientific experiments to the moon in 2023. The only reason such a deal could exist was that Polyakov had earlier acquired the intellectual property rights to Israel's Beresheet lander, which had crashed into the moon in 2019. NASA hadn't seen value in the Israeli technology when Polyakov had been around, but it was happy to make use of his goodwill and smarts in acquiring the lander now. Markusic spun the situation and told the press that Firefly would build the lander from scratch, using the Beresheet as a vague guide. "This is one hundred percent American technology," he said.

Around the same time, another space start-up went through a similar situation. It had a Russian founder, who also had to relinquish control of his company and was replaced by red-blooded Americans. I wrote a story linking the two events and pointed out that the US government appeared to be cleaning house in the aerospace industry. My simple act of including both the Russian and Polyakov in the story effectively ended my relationship with Polyakov and Anisimov, who had become a friend, for quite some time. Polyakov considered the Russian a scoundrel and found it outrageous that I'd draw any parallels between their situations. He no longer returned my texts and calls.

Polyakov made a brief trip back to the United States in early June 2021. Firefly's rocket was still sitting unlaunched at Vandenberg. It had been there for months. Markusic blamed the delay on covid, saying that supply chain–caused mayhem had disrupted access to some key parts. Polyakov no longer really cared. He'd come back to Silicon Valley mostly to attend his daughter's high school graduation and sell his house. It'd been months since my story had run, and he reluctantly agreed to meet me.

* Markusic also told reporters that Firefly planned to raise hundreds of millions of dollars more by the end of the year.

When I stopped by his office on a Tuesday at around 2:00 p.m., Polyakov pointed me to a chair and began going over all the ways I'd wronged him in recent months. My stories had made it seem that he drank too much. The American public, he noted, did not trust people who drank too much. I'd also contributed to the narrative that he was a shady foreigner with shady dating websites and had given some people the general impression that he was goofy. Polyakov told me all that as he poured rounds of scotch for both of us into metal cups shaped like upside-down deer heads.

Though he spent plenty of time making my failings clear, the majority of his vitriol was directed at the rest of the world. His family had flown in before him on a private jet and had been questioned at length by customs officers. The officers had not relented until Polyakov's fourteen-year-old son had declared that he cherished his green card with all his heart because one day it would help him attend MIT. According to Polyakov, that brief burst of patriotism and ambition, coupled with the private jet, had convinced the officials that the family had not returned to the United States to freeload off the country. Polyakov, however, had not fared as well when he had flown in a couple of days later. Customs officials held him for about two hours, peppering him with questions about why he'd returned to the country.

Polyakov told me that he'd made some level of peace with what had become of Firefly. The US government had forced him to sell a large chunk of the company, but he still owned about 50 percent, and he expected that that share would be worth $1 billion or more if the company had a successful rocket launch. Eventually, he'd sell that stake, too, and add to his considerable wealth. Beyond all of that, his gaming businesses had thrived during the pandemic. He expected to sell one or two of the companies and clear another $1 billion to $2 billion. In Scotland, he had many acres of land with loads of animals and three lakes. He was also finally going to buy a distillery. It was good to be Max.

But the peace went only so far. Polyakov pivoted and started reeling off all the shady investments pulled in by his competitors.

Other US rocket and satellite makers had received money from Chinese investors and Russian investors. In some cases, they'd obfuscated the source of the money to make their backers seem more palatable. It seemed deeply unfair to Polyakov that those other companies should get a free pass, while his good name had been tarnished. On those points, he was not wrong. People in the space business will do just about anything to keep going. All sorts of money became available during the boom times of 2020 and 2021, and the US government seemed to turn a blind eye to all of it when it suited its interests.

During his most impassioned moment, Polyakov said he'd just canceled a phone call with a top official at the University of Texas. The plan had been for him to keep donating millions of dollars to an engineering program at the school because Polyakov liked to support engineers. But he'd had it. The school could go find money from someone else. Oh, and Markusic could go fuck himself. He'd called Max a Russian spy and a smut peddler to Max's face. According to Polyakov, Markusic was an ungrateful piece of shit who had milked money out of one set of investors and then out of Max and still couldn't launch a rocket after all those years. Markusic cared for no one other than Markusic, Polyakov said.

Near the end of our chat, Polyakov told me not to wrong him again. He joked that people on his team were deepfake experts and that videos of me doing all kinds of things could turn up on the internet. It was a typical Max moment when I was pretty sure he would not do something like that but not totally sure. The truth is that I loved it when Polyakov went on his tirades. Part of me also relished the idea that maybe I knew only a small fraction of what his life was really like. He was a true man of mystery.

As the fourth scotch went into the glass, what must have been the authentic Max appeared, at least for a moment. He'd gotten back onto the ungrateful-USA kick. At long last, the country had finally managed to kill his passion. "It's sad," he said. "It shouldn't end like this." His eyes filled with tears. Then he tried to lighten the mood by offering a toast to Uncle Sam.

Since he was getting ready to clean out his office, he insisted that I take something—a parting gift from what would be our last meeting in this country. He got up from behind his desk and began walking through the various rooms, past all of his sci-fi artwork, aerospace artifacts, and religious items. I waited near the lobby for a couple of minutes, and then he returned with a statue of a Viking. "You're Finnish, right," he said. "No," I said. "You're Finnish or Scandinavian," he said. "No, not really," I said. Polyakov looked at me with frustration. "Yes, I am," I said.

The statue showed a Viking holding a long animal horn in one hand with its opening hovering over the man's other hand. It was meant to show that the Viking had finished his drink and had not been tricking his companions by pretending to drain the vessel. Max explained that he was just like the Viking: a warrior, but one you could trust.

It took all the way until September 2021 for Firefly to get off its first launch—more than a year since we'd been to Texas and bailed on the pig roast. For a first shot, the rocket did remarkably well. It took off almost on time and flew for two and a half minutes. One of its four engines failed somewhere along the way, and the machine didn't have enough oomph to make it into orbit. No worries. For Rocket Land, it was a success.

Polyakov had flown out to watch the launch from Vandenberg. He had to stand in the viewing area with the rest of the plebs. While most people were busy celebrating Firefly, Polyakov had only grown more defiant. He threatened to take back his lunar lander technology, which would be a major blow to the United States' hope of returning to the moon. The passion was all gone. Space had turned into nothing but a moneymaking scheme for everyone else now. He was going to fuck Tom, fuck the United States, and take his turbopump elsewhere. "If they'd launched two or three years earlier, we would have killed the competition," he said. "I'm going to spend all my money now and die happy at fifty."

EPILOGUE

In the middle of 2022, I went to visit a company in Silicon Valley called LeoLabs. It had raised more than $100 million to build a network of radar stations all around the globe. The network was designed to look up into space and track every object in low Earth orbit, including satellites, old rocket bodies, and debris left over from collisions and explosions. The big stuff was easy enough to spot, but LeoLabs' technology was so good that it could pick out objects just a couple of centimeters in size.

Historically, governments and militaries had monitored the action taking place in low Earth orbit. Such groups wanted to know what their allies and enemies were up to in space, and, of course, they wanted to make sure that their rockets and satellites traveled to places where they would not collide with existing machines. The United States had developed its own radar systems for this type of work and shared much of its data openly. But by 2022, its systems could no longer keep up with everything being blasted into space.

Founded in 2015, LeoLabs had bet that low Earth orbit would need some management to avoid turning into a catastrophe. It had built its first four radar sites in Texas, Costa Rica, New Zealand, and Alaska. With just those, the company identified hundreds of thousands of objects already orbiting the earth. It could monitor the path of satellites and predict when they were likely to run into each other or into an old rocket body that was still stuck in space. In the years ahead, the company planned to build more and more radar stations so that it would have the ability to monitor all of the stuff all of the time.

The images produced by the LeoLabs tracking system and software were startling. You could see thousands of SpaceX's Starlink satellites arranged in a mathematical grid pattern around Earth. Hundreds more satellites from OneWeb and Planet Labs were tucked into the matrix. There were also debris fields stretching all the way around our planet. The most notable recent spray of debris had arrived in 2021, when Russia had chosen to shoot one of its own satellites with a missile to remind other countries that it could eliminate satellites on a whim. When the satellite was destroyed, it broke into more than 1,500 pieces.

To make sure that their satellites don't collide, companies such as SpaceX and Planet Labs pay LeoLabs to find their machines in space and keep track of their movements. If LeoLabs sees the possibility of a collision coming, it notifies the companies, and they take action by using the craft's propulsion system to adjust its orbit by a fraction. There are too many satellites and too many objects for this to be done manually. In 2022, LeoLabs was sending out an astonishing 400 million collision alerts per month. Computing systems at SpaceX, Planet, and other companies would receive the alerts and feed commands to their satellites so that they could move as needed. Back here on Earth, most of us went about our days blissfully unaware that any of that was happening.

"Low Earth orbit today is basically unmanaged," said Dan Ceperley, the CEO of LeoLabs. "You file a plan before a satellite is launched that shows you're going to avoid collisions, and you have to get a license to communicate down with the ground. But once you file the plan and it's approved, it's off to the races. You can go launch your satellite, and there's no follow-up. It's kind of crazy because these satellites can be up there for decades, and they're going into all kinds of orbits. It's just not organized. But with a bit of organization, I think we can put a lot more satellites into space."

The story of LeoLabs is the story of the new space race in a nutshell. We have reached a point where a fifty-person start-up has emerged as the air traffic control system for low Earth orbit. It's comforting that a company doing this type of work exists but also unsettling that such a task has been taken over by the commercial sector. I suspect that this will be the state commercial space will inhabit for quite some time—a position somewhere between exciting and harrowing.

It's clear that SpaceX has emerged as the dominant presence looming over the commercial space industry. It has the most impressive rocket fleet, and it builds and launches more satellites than any other company or country. Elon Musk may be fixated on Mars, but he's been busy proving out the economy of low Earth orbit as well. The Falcon 1 launch started SpaceX on that mission, and the company has never rested on its laurels.

For the rest of our characters, their future will depend on their ability to compete and on the evolution of commercial space itself. Billions upon billions of dollars have been pumped into hundreds of space start-ups. Planet has a dozen rivals. Rocket Lab, Astra, and Firefly have two dozen more. The orgiastic SPAC bubble increased investors' interest in commercial space. But as economies around the world slowed in early 2022 and a biting dose of reality returned to the financial markets, money became tight again. Highly speculative rocket and satellite companies faced more pressure to perform in an environment with less tolerance for risk.

Writing a book like this is dangerous. I've provided you with a deep look at what it's like to create an entirely new field of play for capitalism. And I've been following the story in near real time. There is a chance that one or more of the companies presented in these pages will no longer exist by the time this work reaches your hands or ears.

What's obvious to me, though, is that some form of this new

economy will be built, and it will play a major role in all of our lives. The space internet, the images, and the science emanating from low Earth orbit will be the basis of a new computing infrastructure. The big bet, as I've mentioned, is that other effects we cannot yet articulate or fathom will follow.

Plenty of people question the assumptions behind this gamble. They're quite sure that the commercial space bubble will eventually pop with little to show for all the excitement. But while there will be some painful moments along the way, I'm confident that the next chapter in our technological evolution will continue and result in profound changes to how our world functions. This is the nature of technology and of the human spirit when offered a new playground. As the epigraph of this book advises, "Look up: we have repealed the laws of gravity, torn off the ceiling of the world that was so very low."

And now back to our main characters.

As you may have guessed, things did not go great for Max Polyakov. Not long after Firefly's first launch, the US government continued to pepper him with letters that pretty much accused Polyakov of either being a Russian asset or one day becoming a Russian asset. The feds argued that Polyakov might choose to funnel American aerospace technology to Russia and that he represented a serious threat to US national security. "To the extent that national security can be summarized at an unclassified level, the concerns pertain to the influence of Polyakov on Firefly Aerospace, the potential transfer to Russia of proprietary non-public IP and technical information relating to sensitive United States Government customer information," the government wrote, according to a copy of the documents that I obtained.

The US government complaints were thin on specifics. In fact, there were *no* specific gripes against Polyakov. There was no mention of the dating sites or alleged nefarious business ties. The government simply spent page after page noting that Russia was an adversary to the US in space, that Polyakov hailed from Ukraine, and that Russia and Ukraine used to build spacecraft together. Despite presenting no hard evidence as to why Polyakov, who despised Russia, might help Russia, the government demanded that Polyakov dispose of *all* his Firefly shares as quickly as possible.

To prove that it meant business, the government blocked Firefly from trying to launch its next rocket. It cut the company off from the Vandenberg Space Force Base,* and it prevented Firefly from acquiring the licenses it needed to launch. It also thwarted the ability of some of Polyakov's other businesses to perform financial transactions by placing them on a federal blacklist.

One evening, in a fit of rage, Polyakov put up a post on social media saying he would sell all of his shares in Firefly to Tom Markusic for $1. In the post, he chided two dozen federal agencies for betraying him. "I hope you are happy now," he wrote. "History will judge all of you guys."

He did not actually follow through on the $1 sale, but he did get rid of his Firefly shares. On February 24, 2022, he revealed that a private equity firm had bought him out for an undisclosed sum. He had complained to me in the weeks leading up to the sale that the government had put him in an impossible position. Polyakov had to sell quickly and do a deal with the entire weight of the US government hanging over it. Very few people wanted to touch the transaction. Polyakov may have gotten his money back with a little extra on top, but not much more than that. He certainly did not earn back what Firefly was actually worth.

Was Polyakov a Russian asset? I would be shocked if that were the case, and the government certainly failed to back up its assertions with anything resembling proof.

My guess is that as it became clear that Firefly was very real competition, the company's rivals and detractors decided they'd had enough. Lobbyists were called. Favors were granted. Polyakov was an easy enough target to take down.

As word of Polyakov's sale reached the press, Vladimir Putin had just begun his assault on Ukraine. On that very first day of the

* In 2021, Vandenberg Air Force Base was renamed Vandenberg Space Force Base.

war, the Russians lobbed a bomb near the rocket factory in Dnipro. In the days after, my former tour guides gave up their public relations duties in favor of producing Molotov cocktails and learning how to fire machine guns. Ukrainian snipers took up posts on top of Polyakov's offices. Engineers who used to work for Firefly either joined the army or fled the country. Any hopes for a revival of the Ukrainian aerospace industry were in the process of being obliterated, along with much of the country. "Fucking fuck!!!!" Max wrote to me. "Russian motherfuckers!"

The only thing that brought Polyakov any solace during that period was that the very people who had acquired his shares in Firefly soon pushed Tom Markusic out as CEO.* The professional investors were less forgiving of delays and cost overruns than the Ukrainian space zealot had been. Firefly would continue on without either member of its odd couple.

In October 2022, Firefly performed its second attempt at a rocket launch and it was a huge success. It reached orbit and deposited a few satellites. The company became worth billions of dollars on the back of the launch, which Polyakov watched from afar via the internet. Firefly had reached orbit at incredible speed on the back of Polyakov's risk-taking, investment, and leadership, but the new CEO who had been freshly installed by the investors received all the glory.

By this point, Polyakov had other concerns.

From the moment Ukraine came under attack, Polyakov sprang into action. He led an effort to obtain a flood of commercial satellite imagery and funnel it to the Ukrainian military. Polyakov paid for much of the imagery himself, and his engineers analyzed the pictures for the army, telling them about Russian troop movements and other activities. Ukrainian military officials believed Polyakov's quick action played a major role in stopping the early siege on Kyiv and in aiding Ukraine's surprisingly strong resistance to the Russians. Polyakov began receiving a stream of awards and commenda-

* Markusic remained on the board of directors and continued to serve as Chief Technical Advisor.

tions from the highest echelons of the Ukrainian government. He had used commercial space technology to surprise and undermine the Russian military. A space superpower had been humbled by start-ups and quick thinking.

Even before Polyakov arrived on the scene, however, the massive influence of the satellite imagery had already been felt, as some of our other characters found themselves as significant players on the geopolitical scene. In the weeks leading up to the war, Russia had been denying that it planned to invade Ukraine. Its propaganda and politicking, though, stood no chance against the images being produced on a daily basis by Planet Labs. The world could see Russian forces gathering on the Ukrainian border. We all knew what was to come.

As the war progressed, Planet's images were on constant display on TVs, in newspapers, and on social media feeds. The world watched as forty-mile-long Russian convoys got stuck in the mud outside Kyiv. We saw the before and after images of hospitals and schools that had been destroyed. Russia tried to tell a different story about some of the bombings, claiming that the buildings had been military targets. But the pictures did not lie. And when other bombings or attacks occurred, open-source analysts began matching satellite images with photos and reports from the ground to try to establish more truths about what was going on in the war. No other conflict had ever been documented this way.

When the Russians attempted to destroy Ukraine's communications infrastructure, SpaceX sent Ukraine thousands of Starlink antennas. The space internet enabled Ukraine's military to keep operating in a fashion that would have been impossible a couple of years earlier. Military units could still talk to each other safely with the Russians unable to penetrate Starlink's encryption technology. The same Starlink systems allowed Ukrainian drone operators to orchestrate thousands of bombing missions from locations all around the country. Volodymyr Zelenskyy called Elon Musk to thank him, and Ukrainian generals posted similar thanks online.

While satellite technology had been used in past conflicts, this was the first true Space War. The tools built by commercial space companies gave Ukraine advantages that humbled the Russian military and altered the course of the conflict.

Planet Labs went public in December 2021, during the height of the SPAC frenzy. It raised hundreds of millions of dollars and became valued at close to $3 billion. The company revealed that it had eight hundred customers and around $130 million in annual revenue. Will Marshall and Robbie and Jessy Kate Schingler became multimillionaires and still live in their commune together. Most of the key figures from Planet's early days mentioned in these pages remain at the company.

Chris Boshuizen, however, left Planet in 2015 after feuding with Marshall and Robbie. "I often felt like the role I played between those two was finding the truth among their idealism," he said. "It was finding the pragmatic approach to what they wanted to do. We were all tired, and Will and Robbie had decided they wanted to go at it alone. We chatted, and I told them that I supported their decision. I could tell it was hard for them to even ask me. We had a big hug after.

"It was probably the cost of me living by myself versus living in the house with them. I've pretty much come to terms with things. I still think starting Planet is one of the best things I ever did, and I'm proud of it. But I always tell people, 'Don't have three founders.'"

Boshuizen went on to become a venture capitalist. One of his first investments was in a rocket start-up from New Zealand called Rocket Lab. In late 2021, he flew to space as a tourist aboard a Blue Origin rocket.

Outside of Planet Labs, Marshall, Robbie, and Jessy Kate have spent much of the last few years trying to start a human colony on the moon, as you do. They led an organization called the Open Lunar Foundation that sought to build the first privately funded lunar settlement. The group, which also included Chris Kemp, Pete Worden, Creon Levit, Ben Howard, and Steve Jurvetson, thought that rockets had come down in price far enough that individuals, rather than nations, could consider conducting their own lunar missions.

They hoped to start a new civilization on the moon that would be open to people from all nations and would have new rules of governance outside the traditions here on Earth.

I attended the Open Lunar meetings for a couple of years. The initial idea was to send a couple of Doves to orbit the moon and discover the best spot for a settlement. Then there would be a couple of missions in which robot probes would be placed on the surface. After that, a habitat would be built. Sergey Brin and the Russian tech investor and early Planet Labs backer Yuri Milner were proposed as the financiers of the program, and the meetings often took place at Milner's $100 million mansion in Silicon Valley.

Open Lunar eventually scaled back its hopes and dreams and is more of a policy project now. Jessy Kate remains one of the heads of the group and is trying to influence the politics and strategies of the various lunar settlements being proposed.

The space friends still gather every New Year as part of their 4D—Dream, Drive, Develop, and Deliver—club. Will Marshall, Jessy Kate, Robbie, Chris Kemp, and the rest of the chums meet to discuss their hopes and ambitions aloud in a group setting and take stock of their lives. The event is an annual reminder of the special bond that the friends have formed. It's rare to find people who keep their friendship so tight over so many years and take the effort to put ritual around it. "In my mind, Jessy is the spiritual leader of the group," said one of the Rainbow Mansion friends who has attended 4D. "She doesn't say a lot, but I think she pushes things in certain directions. You can see that they've all been in this, been in life together."

Pete Worden left Ames in 2015, and the place has not been the same since. In fact, NASA reorganized itself after Worden's departure so that executives at the various centers must report to NASA headquarters rather than to the center directors. It was a strategy designed to stop a NASA center director from going rogue and keep a Pete Worden from ever arising again. As you might have guessed, Ames is now quite boring.

Worden and Ames/Rainbow Mansion alum Kevin Parkin have

been working for Yuri Milner on projects related to deep-space exploration. "Expansion into the solar system and the beyond. That is where I am," Worden said. "Eventually we should be settling the planets around the nearby star systems."

Al Weston still works at Ames but not for NASA. He's moved into one of the hangars that Google took over and is busy building a fleet of airships for Sergey Brin.

If I were writing a spy novel or a fictionalized version of this tale, Pete Worden would be the mastermind driving the plot. Worden spent decades dreaming about chucking a satellite on a cheap rocket at a moment's notice so that the military could spy on whatever it wanted to. Without Worden, there might be no Planet Labs or Astra. It takes some squinting, but there might not even be a SpaceX or Rocket Lab. After all, it was Worden who talked the government into backing Musk in the Falcon 1 days and Al Weston, Worden's emissary in New Zealand, who talked people into investing in Beck. Worden more or less conjured this reality into existence. "Black ops general befriends a bunch of bright, young kids and convinces them to do his bidding without their even realizing it" would make for a pretty good story.

Peter Beck has kept right on being Peter Beck and has remained a step or several ahead of the competition.

Rocket Lab went public in mid-2021, raising hundreds of millions of dollars on its way to being valued at billions. The IPO established Beck as one of New Zealand's premier entrepreneurs and made him one of the country's wealthiest citizens. It also, however, revealed that the rocket business remains tough. Despite being the second coming of SpaceX, as of 2022, Rocket Lab was still not turning a profit on its launches.

The company will use much of its newfound funds to build a large, reusable rocket called Neutron. It will be a direct competitor to SpaceX's Falcon 9. Rocket Lab has also been perfecting techniques to make Electron reusable so that it can keep increasing the pace of its launches, while keeping costs down. On top of those moves, it built another launchpad in New Zealand and one in the United States. It has flown to space successfully dozens of times and put up hundreds of satellites.

Rocket Lab is no longer just about rockets, either. The company has started making most of the common parts of satellites at its own factories. Customers can simply add the tech and science bits that make their payloads special and plop them into Rocket Lab's satellite platform. This has increased the company's revenue and profits by moving it into the more lucrative satellite business and helping set it up as a one-stop space shop.

Rocket Lab has not managed to launch rockets every three days as Beck desired. Covid slowed the company down with New Zealand having one of the strictest lockdowns, and, well, it's just really hard to launch tons of rockets. Nonetheless, Rocket Lab is the only company that has neared SpaceX's pace and has a similar track record of successful launches.

In July 2022, Rocket Lab carried a payload to the moon on behalf of NASA. It was the most ambitious mission in history for a small-rocket maker, and the company's rocket performed perfectly. Just as Beck had predicted, New Zealand needed to add lunar considerations to its space legislation. Rocket Lab has more contracts for lunar missions and some for Mars and Venus, too.

In a move that gave me great pleasure, New Zealand also put legislation into place to ensure that satellites launched from its shores are tended to throughout their life spans. That is to say, the country demanded that the machines be monitored in space and disposed of well. It is the only country I'm aware of that has such rules.

Astra went on a wild ride. In March 2022, it succeeded in flying satellites into orbit on behalf of paying customers. Despite all of the previous explosions and considerable drama, it had raced to space at record speed. It also joined SpaceX and Rocket Lab in the most elite of clubs.

Astra held an event at its factory for investors in May 2022 to celebrate the good news. I was astounded as Alameda's mayor turned up and welcomed the guests by singing the company's praises. "I am thinking that Chris Kemp and Astra could do for Alameda what

Elon Musk and Tesla did for Fremont," she said, referring to Tesla's car factory in a nearby town. "But without the controversies, right, Chris?"

Alameda had, in fact, tried to kick Astra out of its factory a few months earlier. But it was now impossible to deny what the company had built. The factory had expanded yet again, and there were several rockets ready to go into space. Astra had also filed paperwork to begin building its own space internet constellation made up of 13,600 satellites. It would design and build the machines in a new nearby factory.

When Kemp took the stage, he pointed to the accomplishments of commercial space during the war in Ukraine. He noted all the space companies that had gone public. He predicted a $1 trillion space economy by 2040. He also took a shot at Rocket Lab for still not launching rockets as often as Peter Beck had predicted.

He closed the event by championing the work of his employees. Astra had been the smallest team ever assembled to fly a satellite into orbit "four years faster than any company in history." Its thesis about making small, cheap rockets might have been wrong, but it was going to see things through no matter what.

Just a few weeks later, in June 2022, Astra's plans changed considerably. The company had another launch in which it tried to ferry some weather-tracking satellites into space on behalf of NASA, but the satellites were lost when the rocket didn't quite make it into orbit. Everyone at Astra had been hoping for one successful commercial launch after another, but that was not to be the case.

At first, Kemp put a positive spin on things, saying that the company would hunker down and fix its rocket. But by August, it decided to scrap its small rocket altogether and concentrate instead on building a bigger machine. Astra's new rocket would be able to carry 1,300 pounds to orbit and begin test launches "sometime in 2023," Kemp said.

Kemp did not acknowledge out loud the key technology that would let Astra make a larger rocket so fast. The engines powering the bigger machine would be designed not by Astra but by Firefly, and they would rely on a Ukrainian-made turbopump for their power.*

Kemp revealed Astra's new plans as the company reported its second-quarter results. It posted a net loss for the quarter of $82 million and still had $200 million cash on hand. The money would be enough, he said, to build a big rocket and then stamp them out by the dozens in the factory in Alameda. "We learned what we needed to learn from Rocket 3," he texted me. "Rocket 4 is why we took the company public."

Chris Kemp still wears all black.

---

* In another twist, the military contractor Northrop Grumman announced in 2022 that its old-line Antares rocket would shift from using Russian-made engines to Firefly's engines. If Polyakov had been out to help the Russians, he'd done a terrible job by paying to develop the technology now embraced by two US companies.

THANKS AND ACKNOWLEDGMENTS

Working on a book for five years requires a lot—a lot of patience and goodwill from your subjects, a lot of support from your friends, and a lot of tolerance from your family.

My wife, Melinda, and sons, Bowie and Tucker, saw all the best and worst moments of this project. I made the boys laugh by reading about multidicked cows in my Max voice, taught them about aerospace, and showed them a rocket or two. But I was locked away in my writer's cave for many, many months—too many—and missed a lot of time with them that I'll never get back. Nonetheless, their smiles and pep talks kept me going. I'm a lucky dad.

On good days, Melinda saw me run into the house filled with excitement because I'd witnessed something amazing or had a fantastic interview. On bad days, she saw me buckling under the stress and on the verge of a breakdown. All the while, she did nothing but encourage me and keep our family functioning. This might sound sappy, but I am 100 percent sure that I would never have written much of consequence had it not been for Melinda. She saved me from my worst tendencies and gave me everything in return. She's part muse, part better angel right fair. I love her.

There were people along this journey whom I badgered again and again for information and who still talked to me anyway: Chris Kemp. Max Polyakov. Will Marshall. Robbie and Jessy Kate Schingler. Adam London. Artiom Anisimov. Pete Worden. Peter Beck. Morgan Bailey. Trevor Hammond. I deeply appreciate the time they gave me and can never adequately repay them. The same goes for so many of the employees at the various companies and particularly to Team Astra, who let me turn up week after week and interrupt their days.

I put my literary agent, David Patterson, into some very awkward positions during this endeavor. He handled everything like the pro he is. In addition to being an ace navigator of the publishing world, he's a terrific psychologist. Thank you, David, for always being there and for helping me breathe.

Howie Sanders, my man in Hollywood, is relentless with his optimism and encouragement. Every time he pops onto the phone, I just feel good and like everything is going to be okay. He believed in this sucker right from the start and helped me see all kinds of new possibilities with the material. Thank you, Howie, for thinking big and always having my best interests at heart.

I managed to torture my editor, Sarah Murphy, and she responded with kindness and support. During the editing process, it often felt as though we had some kind of mind meld. Bless you, Sarah, for putting up with me and giving this tome so much love and affection.

One of the luckiest things to happen in my life was running into Brad Stone. We've worked together and been close friends for a very long time now. He's as decent, supportive, and wise as humans come. Even though each of us does his own thing, I think of us as a team navigating journalism and writing together.

I was also fortunate to end up working at Bloomberg. I'm not aware of any company that better supports reporters. The first tip of the cap should go to Mike Bloomberg, who has allowed me to scour the four corners of the world for stories and given me every opportunity to maximize my talents. That might sound like ass kissing, but I doubt that Mike will even see this, and, well, it's true. Working at Bloomberg has also meant having unmatched colleagues, editors, and friends. Jim Aley, Kristin Powers, Jeff Muskus, Max Chafkin, Alan Jeffries, and Victoria Daniell have all been kind enough to tolerate my eccentricities and help make the things that appear under my name so, so much better for their efforts. I adore them all.

Also in the adoration club are Meghan Schale, Francesca Kustra, and Shirel Kozak. These tremendous filmmakers stood by my side as I fumbled my way through learning how to make a documentary on some of the subjects in this book. To say that the learning process was steep would be a heroic understatement. I am not a skilled enough writer to express the amount of gratitude I have for these three women. I'm in awe of their talent and will never forget their generosity.

David Nicholson and Diana Suryakusuma had the misfortune of being stuck with me for many hours in many far-off lands, as the third scotch was poured at a bar and I inevitably retold them stories about the oddities of the new space race. Their bad luck was my gain. They're more kin than friends at this point. My life would not be the same had I not met them, and I'd do anything for them. If you're ever being poisoned by a shaman in the Chilean desert, these are the people you want by your side.

The first person that I send a draft of any new book to is Keith Lee. We met by accident on a tennis court, and our families have been intertwined ever since. It pleases me to no end that my two boys have gotten to know Keith and see such a fine example of a caring father and husband. Ever generous in spirit, Keith is smart and thoughtful with his comments on my work. He helps give me the courage to keep going.

I'd also like to thank the good people of Idaho and Pete and Marianne, in particular, for giving me a place to write parts of this book amid beautiful surroundings. The same goes for New Zealand and its fine denizens. I've been to many countries, and you won't find one better than Aotearoa. It's magical.

Last, I want to thank my mum, Margot, and dad, John, for nurturing every creative impulse that I ever had and forever being full of surprises. Thank you also to Blase and Judy for letting me luck into such a warm and loving family.

My parents moved to Mexico not too long ago on what seemed like a whim and settled in among a collection of wonderful human beings. Two of those humans—mis amigos Julián y Andrés—are now tennis companions, neighbors, and lifelong friends. The majority of this book was written in Mexico, where great food and great people helped pull me away from the page and reset both my body and mind. Covid sure as shit sucked, but without it, I would not have had all of the unexpected extra time with my parents (and my gatito writing companions, Uno, Dos, y Tres), nor would I have had the chance to fall in love with Mexico.

If I forgot anyone, I guess you weren't that memorable. Just kidding. I adore you, too.

INDEX